COMPUTER SCIENCE, TECHNOLOGY AND APPLICATIONS

FOCUS ON INFORMATION, INTELLIGENCE AND KNOWLEDGE

Computer Science, Technology and Applications

Additional books in this series can be found on Nova's website under the Series tab.

Additional E-books in this series can be found on Nova's website under the E-books tab.

COMPUTER SCIENCE, TECHNOLOGY AND APPLICATIONS

FOCUS ON INFORMATION, INTELLIGENCE AND KNOWLEDGE

RAFIQ ISLAM
EDITOR

Nova Science Publishers, Inc.
New York

For permission to use material from this book please contact us:
Telephone 631-231-7269; Fax 631-231-8175
Web Site: http://www.novapublishers.com

Library of Congress Cataloging-in-Publication Data

Focus on information, intelligence, and knowledge / editor, Rafiq Islam.
p. cm.
Includes bibliographical references and index.
ISBN 978-1-61209-979-8 (hardcover)
1. Adaptive control systems. 2. Biotechnology. 3. Artificial intelligence. 4. Systems engineering. 5. Manufacturing processes. 6. Traffic engineering. 7. Telecommunication systems. I. Islam, Rafiqul, 1959-
TJ217.F63 2011
658.4'038011--dc22
2011005295

Published by Nova Science Publishers, Inc. † New York

CONTENTS

PREFACE

This new book presents and discusses current research in the study of computer science, with a particular focus on information systems, intelligence and knowledge. Topics discussed include intelligent systems for threatening e-mail detection; infrastructure of a context adaptive and pervasive multimodel multimedia computing system; adaptive business intelligence for information and communication technology management; efficiency of machine learning techniques in biomedical data and computer vision and artificial intelligence in flight control of unmanned aerial vehicles

Chapter 1 - This paper reviews the development of a scientific approach that makes possible, when describing and analyzing natural phenomena, some incorporation of intangible elements with tangible ones. Over the millennia that have passed since the emergence of human civilizations in Africa and Asia, this problem has posed a most daunting challenge. Apart from the difficulty of the task itself, there are many hidden assumptions to be tackled in the Eurocentric (*i.e.*, European-centered) traditions of conventional science and mathematics. These assumptions have tended, over time, to sandbag the effort. The discussion of this paper aims to elaborate some of the most critical steps of the task itself. Along with indicating where conventional science took a path that actually contradicts natural processes, this paper points out unexpected or hitherto unconsidered strengths of certain alternative approaches, long buried or dismissed either as religious hocus-pocus or "uncivilized", *i.e.*, non-European in origin. These latter may rectify and overcome the obstacles to tangibles intangibles integration. This paper examines critically whether the human race has progressed in terms of knowledge in the last few centuries. It looks into the reasons why we still don't know how a dead body can be preserved without resorting to *industrially-engineered* chemicals — something actually accomplished with mummies from the days of ancient Egypt. The paper discusses how the notion that, somehow, after Newton, everything had changed and we evolved into our modern-day civilized status doesn't withstand the first serious scrutiny of what *kind* of science should be used to model Nature. This paper examines the first premise of all major theories and 'laws' that were introduced in the so-called 'New Science' of post-Renaissance Europe and thenceforth to the rest of the world. The paper deconstructs the aphenomenal assumptions behind numerous 'laws' and numerical solution schemes that are touted as the only solutions to engineering problems. Finally, the paper presents a framework for the science of intangibles, showing that only with the science of intangibles does a pro-Nature technology development and knowledge-based analysis become possible.

Chapter 2 - For environment impacts assessment, several methods have been proposed in literature, such as Life cycle analysis, Multi criteria analysis and Cost benefit analysis. Recently, other methods combining Multi Criteria Decision Analysis (MCDA) and Artificial Intelligence (AI) have been explored to develop enhanced methodologies for knowledge based decision support. In this paper, a hybrid approach is based on multi criteria analysis and evidence theory is presented for impacts assessment of urban mobility. Various criteria are evaluated frequently using complex data obtained from several information sources. A global evaluation on criteria set is done by the Choquet integral aggregation taking into account the interaction between criteria. Finally, each evaluation process is illustrated by using cases studies.

Chapter 3 - In this chapter we present on line tuning of PID controller through bio-mimicry of bacteria foraging. Genetic Algorithm is currently a standard for optimization. Bio-mimicry is a new technique proposed for optimization by Kevin M. Passino of Ohio State University, Ohio. A comparison of famous Genetic Algorithm with bio-mimicry of bacteria foraging has been made. This comparison is done by comparing their application to the problem of tuning a PID controller. In this chapter we observe that bio-mimicry is equally applicable to the problem of control system design (here tuning PID controller). The results of bio-mimicry algorithm are better than to that of genetic algorithm.

Chapter 4 - Data mining is the quest for knowledge in databases to uncover previously unimagined relationships in the amount of data. . This paper proposes to detect malicious emails (i.e. related to terrorism, fraud, etc.). Deception theory suggests that deceptive writing is characterized by reduced frequency of first person pronouns and exclusive words and elevated frequency of negative emotion words and action verbs. We apply this model of deception to the e-mail dataset, then applied decision tree algorithm to generate the decision tree .The decision tree that is generated is used to test the e-mail as threatening or not. In particular we are interested in detecting fraudulent and possibly criminal activities from e-mail corpora.

Chapter 5 - The aim of pervasive multimodal multimedia computing is to realize anytime, anywhere computing using various modes of human-computer interaction. The current state-of-the-art pervasive systems and solutions, however, do not include applications that are related to pervasive multimodality. Also, the current multimodal interfaces were designed with pre-defined modes of human-machine interaction that were not chosen based on the given context of the user, of his environment and of his computing system. This paper addresses these weaknesses by proposing a pervasive multimodal multimedia computing system in which its modalities are chosen based on their suitability to the given interaction context. This same system chooses a multimodal (or unimodal) interface based on the given context, available media devices and user preferences. This paper discusses the challenges in designing the infrastructure of such computing system and illustrates how we addressed those challenges. This work is our contribution to the ongoing research that aims at realizing pervasive multimodality.

Chapter 6 - Value Stream Mapping (VSM) is a lean principle, which is used for analyzing and controlling the flow of material and information through production systems. This particular lean principle allows documenting current working methods, recording inventory levels and computing throughput time. These data are used to design a future vision of production and information flow that can be utilized as bases for lean thinking. Ultimately, this allows prioritizing the opportunities for waste elimination and time saving. The aim of this paper is to recognize the bottlenecks when producing a specific product in turn to achieve improvements through average number of jobs (WIP), throughput time, machine delay time, and product queuing time. A product was specified, production line for producing this product was monitored, and data were collected and analyzed, consequently, current state map for producing the primal products was developed and constructed, then it was discussed and brainstormed with senior engineers of the plant that produces the product, the weakness points were found in the flow of production within the system, therefore the future state map for the product was developed and presented. Finally, discussion, results, recommendations and future work were suggested and presented.

Chapter 7 - With the rapid advancement of both business techniques and technologies in recent years, knowledge has become an important and strategic asset that determines the success or failure of an organisation. Analyses showed that a competitive advantage in business environment depends on the accessibility to adequate and reliable information in shortest time possible and the high selectivity in the creation and utilisation of information. An effective instrument to create, aggregate and share knowledge in an organisation has therefore become a key target of management. As the organisations today can no longer spend money on excessive infrastructure and technology that does not provide a quick impact on the business' bottom-line, Business Intelligence (BI) is becoming indispensable to an organistion's success in the emerging global economy. BI is currently one of the fastest developing directions in information and communication technology. In this paper, we introduce the concept of BI and address the importance of BI in revolutionising knowledge towards economics and business advancement. Ensuing sections then explore some specific BI applications, including discussion surrounding the technologies and frameworks which may facilitate BI in an e-business initiative. In the concluding remarks, the future of BI is drawn to wrap up the studies.

Chapter 8 - Truss and tower systems are normally designed for specified loads by using respective codes. In certain cases, they may be subjected to loads more than the design values. Thus a need arises to protect these systems when the specified loads are exceeded by certain margins. This paper focuses on providing in-built smartness in a three-dimensional, two-panel tower system to handle both force and deformation when unanticipated loads up to 55 percent increase over a short duration act on these systems. Analysis is made for loads at the rate of 1.25, 1.35, 1.45 and 1.55 times the design load of the tower. The paper explains suitable control forces that are to be generated and how the system under combined action of unanticipated and control forces are balanced to keep the structural system integral during short duration of unanticipated loads. This has been highlighted with an example.

Chapter 9 - The objective of this chapter is to study the quantitative forecasting method of induced traffic volume in Chinese situation. With sufficient consideration of the mechanism and evolution of induced traffic, the relation between induced traffic volume and induced economic growth is discovered. Considering their close relation, adaptive forecasting of induced traffic volume in keeping with induced economic growth is feasible. Accordingly, the modified spatial economic approachable model is adopted to bring forward the forecasting method. The technical route is expounded step by step. Empirical study is provided by preliminary test of the method on the Ha-Shuang freeway. The test result shows the applicability of the proposed method.

Chapter 10 - Machine learning and statistical techniques applied to gene expression data have been used to address the questions of distinguishing tumor morphology, predicting post treatment outcome, and finding molecular markers for disease. Today the classification of different morphologies, lineages and cell histologies can be performed successfully in many instances. The performance in predicting treatment outcome or drug response has been more limited but some of the results are quite promising. The scope of the research reported here is to investigate the efficiency of machine learning techniques in a number of different biomedical problems. To this end, a number of experiments have been conducted using representative learning algorithms. The decision of which particular method to choose is a complicated problem. A good alternative to choosing only one method is to create a hybrid forecasting system incorporating a number of possible solution methods as components (an ensemble of classifiers). For this purpose, we have implemented a hybrid decision support system that combines the representative algorithms using a stacking variant methodology and achieves better performance than any examined simple and ensemble method.

Chapter 11 - Many reference models have been suggested for the adaptive hypermedia systems but they lack the capability to share their information about the user with other AH systems. In the present paper, we propose a reference model where AH systems have two user models – one integral to the storage layer and having domain dependent characteristics and another 'shared user model' having domain independent attributes and characteristics. Similarly, AH systems would have two adaptation models – one integral having adaptation rules for domain dependent data and another 'shared adaptation model' having domain independent adaptation rules. The shared user model and shared adaptation models are part of the 'shared module' which is shared among different AH systems. 'Shared Module' provides the benefit of initializing the AH systems with more mature and accurate user model, saving of time and effort in creating individual user models for different AH systems and enforcing some sort of standardization and synergistic behaviour among them. Apart from the 'Shared Module', the paper also suggest the division of runtime layer into input acquisition component and presentation component, also the division of the presentation specification interface into inference engine and adaptation engine in individual AH systems.

Chapter 12 - In this paper, we study the multi – carrier direct – sequence code division multiple – access system when the received signal is divided into predetermined spreading sequence, these sub - bands modulates M carriers. The receiver produces the correlation for each carrier. Because of the no – knowledge of the phase of the received signal in each sub - band, the receiver cannot coherently combine outputs of the correlators, we consider in this paper two non - coherent combining schemes; the equal gain combining and the selecting combining. The calculation of the probabilities of detection and false alarm for both multi - carrier and single carrier systems in a Fading Rayleigh channel demonstrate that the performance of the multi - carrier system is better than the performance of the single carrier in equal gain combining, while the performance of the multi - carrier system in the selecting combining has the same performance as the single carrier.

Chapter 13 - The Single Electron Transistor is a nano electronic device; it has inherent multiple valued logic and re-configurable properties. It's a symmetrical device, when provide a back gate it acts as an n-SET, and if the back gate terminal is grounded the same device acts as a p-SET. This setup greatly reduces the number of component required to implement the same functionality. In this paper a Multiple Valued Logic Multiplier using reconfigurable-SET has been proposed. It has been found that p-SET and n-SET can be effectively combined into a single SET having dynamically reconfigured as n-SET or p-SET using an external voltage being applied to the back gate. This approach does not require any variance in the method of fabrication for different units or on any other parameters of the device. A thorough analysis on Eight Valued logic reconfigurable SET based Multiplier circuit has been performed and its performance in terms of area, delay and power consumption has been estimated.

Chapter 14 - In this paper the thermal analysis has been performed considering the human head exposed to cellular phones and focusing the attention on the temperature increase and its distribution in the skin, gray matter, ear and skull. The temperature increase and its distribution are calculated at different power levels 0 mW, 600 mW and 1000 mW. It is well known that some significant thermal damage can occur in sensitive tissue under conditions of partial-body exposure to intense EM waves.

Chapter 15 - It is anticipated that recent advances in nanotechnology, distributed micro-electromechanical system (MEMS) applications, fiber optic sensors, and recent developments in smart structures and designer materials will play a key role in the near future in enhancing the functionality, serviceability, and increased lifespan of civil infrastructure systems. A multidisciplinary research front, represented by material scientists, physicists, chemists, biologists, and engineers from diverse disciplines, especially civil, electrical, computer and mechanical engineering, is required to integrate the multiple interactive component systems related to the design, construction and maintenance of highway infrastructure into a meta smart road system.

This paper reviews some of the latest developments related to the applications of smart devices for distributed monitoring of highway infrastructure systems. These include embedded wireless sensors to investigate the concrete mixing process leading to optimization of mix designs and mixers, measurement of subsequent compaction and shrinkage after placement of concrete, and detecting and identifying several deteriorations in concrete structures, including the presence of chloride ions, carbonation, and leaching of calcium hydroxide.

Potential futuristic applications which involve a combination of wireless sensors, advanced concrete pavement materials, fiber optic sensors, sophisticated data acquisition and processing systems, etc. are discussed in light of a recently released long-term plan for concrete pavement research and technology based on the concepts of pavement performance and systems integration.

Chapter 16 - Applications of computer vision systems (CVS) in the flight control of unmanned aerial vehicles (UAV) are considered. In many projects, CVS are used for precision navigation, angular and linear UAV motion measurement, landing (in particular shipboard landing), homing guidance and others. All these tasks have been successfully solved separately in various projects. The development of perspective CVS can be divided into two stages. The first stage of perspective CVS development is the realization of all the above tasks in single full-scale universal CVS with acceptable size, weight and power consumption. Therefore, all UAV flight control tasks can be performed in automatic mode on the base of information that is delivered by CVS. All necessary technologies exist and the degree of its maturity is high. The second stage of CVS development is integration of CVS and control systems with artificial intelligence (AI). This integration will bring two great benefits. Firstly it will allow considerable improvement of CVS performance and reliability due to accumulation of additional information about the environment. Secondly, AI control system will obtain a high degree of awareness about the state of the environment. This allows the realization of a high degree of control effectiveness of the autonomous AI system in a fast changing and hostile environment.

Chapter 17 - Diverse worldviews directly shape ontological and epistemological understandings related to what counts as valid theory and who is able to theorize. If valid knowledge development is believed to rest within the purview of professionals in the academy then the everyday theorizing capabilities of individuals within the community remain unrecognized or if recognized, discounted. However, if theory generation is viewed as a product of ordinary people's discovery of personal meaning within the experiential world then theory and knowledge generation become different in content but not in kind to that of the academic. Everyday theorizing results in practical knowledge generation that is no longer the singular domain of those who work in the academy but is instead a common and natural activity of daily life. Community Based Research (CBR) is one means by which those who engage in academic inquiry may partner with individuals and groups in the community to work together in a participatory manner. Participatory research values each person's knowing of the world and how this knowing may contribute to knowledge generation. Researching with people rather than on them is integral to the process of CBR. Inclusive grassroots research increases knowledge generation that can then be constructed into everyday theory. In this way the experiential knowledge of community members constructively critiques social inequities and may address social issues most effectively at the local community level.

In: Focus on Information, Intelligence and Knowledge
Editor: Rafiq Islam
ISBN 978-1-61209-979-8
2011 Nova Science Publishers, Inc.

Chapter 1

DELINEARIZED HISTORY OF TIME, SCIENCE AND KNOWLEDGE

G. M. Zatzman**, M. M. Khan, A. B. Chhetri and M. R. Islam

Department of Civil and Resource Engineering,
Dalhousie University, Canada

ABSTRACT

This paper reviews the development of a scientific approach that makes possible, when describing and analyzing natural phenomena, some incorporation of intangible elements with tangible ones. Over the millennia that have passed since the emergence of human civilizations in Africa and Asia, this problem has posed a most daunting challenge. Apart from the difficulty of the task itself, there are many hidden assumptions to be tackled in the Eurocentric (*i.e.*, European-centered) traditions of conventional science and mathematics. These assumptions have tended, over time, to sandbag the effort. The discussion of this paper aims to elaborate some of the most critical steps of the task itself. Along with indicating where conventional science took a path that actually contradicts natural processes, this paper points out unexpected or hitherto unconsidered strengths of certain alternative approaches, long buried or dismissed either as religious hocus-pocus or "uncivilized", *i.e.*, non-European in origin. These latter may rectify and overcome the obstacles to tangibles-intangibles integration. This paper examines critically whether the human race has progressed in terms of knowledge in the last few centuries. It looks into the reasons why we still don't know how a dead body can be preserved without resorting to *industrially-engineered* chemicals — something actually accomplished with mummies from the days of ancient Egypt. The paper discusses how the notion that, somehow, after Newton, everything had changed and we evolved into our modern-day civilized status doesn't withstand the first serious scrutiny of what *kind* of science should be used to model Nature. This paper examines the first premise of all major theories and 'laws' that were introduced in the so-called 'New Science' of post-Renaissance Europe and thenceforth to the rest of the world. The paper deconstructs the aphenomenal assumptions behind numerous 'laws' and numerical solution schemes that

* EEC Research Organization, Corresponding author, Email: eecro@eastlink.ca

are touted as the only solutions to engineering problems. Finally, the paper presents a framework for the science of intangibles, showing that only with the science of intangibles does a pro-Nature technology development and knowledge-based analysis become possible.

Keywords: Scientific research, Newtonian mechanics, aphenomenal model, Nature science, history of science, natural computation.

1. Introduction

It is becoming increasingly clear that the current mode of technology development is not sustainable (Mittelstaedt, 2007). Most agree with Nobel Chemistry laureate Dr Robert Curl that the current technological marvels are 'technological disasters'. Few, however, understand the root cause of this failure. This failure now endangers the future of entire species, including our own. Many theories put forward— including those branded as 'conspiracy theories', 'pseudoscience', 'creationism', etc. —but none provide answers to the questions that face the current perilous state of the world.

This lack of sound, fundamentally correct theories is felt in every sector. The outcome is seen in the prevailing popular mood throughout Western countries that "*t*here *i*s *n*o *a*lternative". This so-called TINA syndrome was recently identified once again by Zatzman and Islam (2007a), who credited former British Prime Minister (now Baroness) Margaret Thatcher with its enunciation during the 1980s as the secular equivalent of Holy Write among Anglo-American policymakers. The essence of this approach upholds and promotes the status-quo as the only way to move forward.

The situation has meanwhile become so desperate that some Nobel laureates (e.g., Robert Curl) or environmental activists (e.g., David Suzuki) openly despair Humanity's acquisition of technology in the first place as a form of societal 'Original Sin'. Others have pinpointed the severe pollution burdens mounting especially in the energy resource exploitation sector to single out petroleum operations and the petroleum industry for special blame. In the context of increasing oil prices, the petroleum sector has become an almost trivially easy target, and this situation has been compounded by more recent pronouncements from current US President George W. Bush about "oil addiction". Even his most ardent detractors have welcomed this particular comment as a signal insight (Khan and Islam, 2007a). Numerous alternate fuel projects have been launched — but they propose the same inefficient and contaminated processes that got us into trouble with fossil fuel.

Albert Einstein famously stated, "The thinking that got you into the problem, is not going to get you out." Today, there is no evidence that modern Eurocentric civilization is ready to propose a way out of this technological conundrum. The symptoms are ubiquitous, from toxic addiction (Mittelstaedt, 2006) to global warming (Chhetri and Zatzman, 2008). How is the crying need to treat these symptoms being addressed, however? Soon after it was revealed that the farmers of India have been committing suicide in record numbers, and that the much-vaunted 'green revolution' was actually a fraud keeping humanity in a chokehold (Saunders, 2007), the Congressional Medal of Honour was bestowed on Dr Norman Borlaug. This is the individual who orchestrated the key experimental research undertaken with Mexican maize in the 1940s that underpinned the eventual introduction of costly chemical pesticides, herbicides and other 'aids' for enhancing the productivity of individual tillers throughout the Third World, in one crop after another (Editorial, 2007).

In the chemical sector (food, drug, and lifestyle), similar absurdities continue to abound. Recently, chemical companies 'celebrated' 100 years of PVC, as though PVC has done Humanity good. Yet, China is being accused of using PVC and other toxic chemicals in children's toys, leading to the recall (only in USA and Europe) of millions of toys (CNN, 2007). Weeks before, the head of China's Food and Drug Administration was executed for allowing unwanted chemicals in drugs. That followed weeks of other major scandals (or accusations) regarding Chinese chemical treatment of food and health products (*e.g.*, toothpaste) earmarked for export to United States. In this same year, debates have been raging around the possible connections to chemical fertilizer, chemical pesticides, genetically modified crops, and you-name-it to honey bee Colony Collapse Disorder (CCD) among European honey bees.

At one level, the official discourse is rigidly maintained that hormones are no more just hormones and stem cell research is absolutely necessary because there is no substitute to real (natural) cells. Yet, even Prince Charles "gets it". Recently, he joined the choir of pro-nature scientists by asking the Sheikhs of Abu Dhabi (site of the second-highest rate of child diabetes in the world), "Have you considered banning McDonald's?" The Prince might not be willing to admit the connection of British-invented sugar culture to diabetes, but he is delighted to report he owns an organic farm.

Somehow, when it comes to scientists, trained with the science of tangibles, this organic product is 'not necessarily' better than other products, such as chemically-grown using toxic pesticide. One food scientist wrote to one of the authors that "I have yet to be convinced that the history of a molecule affects its function and that natural is necessarily better." Apart from the all-pervasive culture of fear that prevents one's taking a stand in case a lawsuit looms on the horizon (in geology, the identical syndrome hides behind the frequently-encountered claim that 'the possibility of finding oil cannot be ruled out entirely'), this comment from a food scientist contains a wry reminder of what has corrupted modern science in the most insidious manner. This has been the dilemma of modern age. Scientists today do not, or find that they cannot, simply uphold the truth. Those that speak, when they speak, uphold self-interest and the status quo. If natural is not necessarily better than artificial, and if chemicals are just chemicals, why should we have a different branch called food science? For that matter, why differentiate chemical engineering, nuclear engineering, pharmacy, military science, or anything else any longer? With such an attitude, why not declare defeat and place them all hereafter under a single umbrella: the Science of Disinformation. Only recently, by the way, comparing opinion polling methodologies in general with the results of actually surveying in detail a defined population from and in a developing country, Shapiro *et al.* (2007) were able to elaborate in some detail the *modus operandi* of this science of disinformation. Has the overall state of the scientific world become so grim that the general public has to resort to trusting social-political activists — *e.g.*, Al Gore's *An Inconvenient Truth*, Michael Moore's *Sicko!* and even a relative novice like Morgan Spurlock and his documentary, *Supersize Me!* — ahead of professionally knowledgeable scientists? This paper attempts to find the cause of such helplessness in a science whose modernity seems to consist entirely of promoting only the culture of tangibles under the slogan "More is better — because there is more of it."

2. Delinearized History of Time, Science, and Truth

Thousands of years ago, Indian philosophers commented about the role of time, as a space (or dimension), in unraveling the truth, the essential component of knowledge (Zatzman and Islam, 2007a). The phrase used by Ancient Indian philosophers was that *the world reveals itself*. Scientifically, it would mean that time is the dimension in which all the other dimensions completely unfold, so that truth becomes continuously known to humans, who use science (as in critical thinking). Another very well-known principle from Ancient India is the connection among *Chetna* (inspiration), *dharma* (inherent property), *karma* (deeds arising from *Chetna*), and *chakra* (wheel, symbolizing closed loop of a sustainable life style). Each of these concepts scientifically bears the intangible meanings, which cannot be expressed with conventional European mathematical approach (Joseph, 2000). Only recently, Ketata et al. (2006a; 2006b; 2006c; 2006d) recognized this fact and introduced a series of mathematical tools that can utilize the concept of meaningful zero and infinity in computational methods.

These ancient principles contain some of the most useful hints, extending far back into the oldest known human civilizations, of true sustainability as a state of affairs requiring the involvement of infinite time as a condition of maintaining a correct analysis as well as ensuring positive pro-social conclusions (Khan and Islam, 2007b). Moving from Ancient India to Ancient China, the Chinese philosophers provide one with some very useful insight into very similar principles of sustainability and knowledge. The well-known statement, although rarely connected to science, of Confucius (551-479 B.C.) relates unraveling of the truth to creating *balance* (the statement is: *Strive for balance that remains quiet within*). For Confucius, balance had the essential condition of 'quiet within'. This idea is of the essence of intangibles in the "knowledge" sense (Zatzman and Islam, 2007b).

In The *Qu'ran* (first and only version compiled in mid-7th century), humans' time on earth and time in nature are all part of one vast expanse of time. This position is entirely consistent with the notion that the world reveals itself. In terms of the role of intention, the most famous saying of The Prophet — the very first cited in the Bukhari's collection of the *hadiths* — is that any deed is based on the intention (Hadiths of The Prophet, 2007). A review of human history reveals that what is commonly cast or understood as "the perpetual conflict between Good and Evil" has always been in fact about opposing intentions. The Good has always been characterized by an intention to serve a larger community, while Evil has been characterized as the intention to serve a self-interest. What was known in Ancient India as the purpose of life (serving humanity) is promoted in the Qur'an as serving self interest in the long-term (as in infinity, see discussion later). Because nature itself is such that any act of serving others leads to serving the self in the long term, it is conceivable that all acts of serving others in fact amount to self-interest in the long-term (Islam, 2005b). In terms of *balance*, The *Qu'ran* promoted the notion of *qadar* (as in *Faqaddarahu*, فقدّرهُ, meaning 'thereby proportioned him', 80:19, The Qur'an), meaning proportionate or balanced in space as well as time. The *Qu'ran* is also specific about the beginning and end of human life. Personal communication with a PhD from MIT (as reported in Abou-Kassem et al., 2007) resulted in the following discussion. The word 'aakherat'(اخرة) is repeatedly used in The *Qu'ran* whenever the need for taking a long-term approach is highlighted. The word "*aakherat*" (as in *Al-aakherat*[1], الآخِرَة , 2:4 The Qur'an) is the feminine form of the adjective "*aakher*" (آخِر), meaning "last", as in last one to arrive, or generally last one to take place. This is the simple or immediate meaning. A wider connotation: the word *aakher*, meaning the last when it describes a period of time, also connotes the infinite in that period. Thus, *Aakherat* referring to the life after, means it is an infinite life. This calls for the notion that while man is considered mortal, in fact man is eternal. Man's body is mortal but his *self* is eternal. This is the only kind of creation with this quality, being eternal. Hence, man approaches divine nature with the attribute of being eternal, that is the existence in $+\infty$ of time while differs with the attribute of having a beginning, meaning man lacks the existence in time = 0. The same after-life is sometimes referred to as *ukhraa* (أخرى), the feminine form of *aakhar*. Note the slight difference from the previous *aakher*, meaning last. The word *aakhar* means 'second', but used only when there is no 'third' (Arabic and Sanskrit distinguish grammatically between one, two and more than two, unlike European languages that distinguish between one and more-than-one, as reported by Islam and Zatzman, 2006). This *ukhraa* is not used in The *Qu'ran*. In contrast to *aakherat*, two words are used in The *Qu'ran*. One is *ulaa* (as in Al-Ulaa, الْأولى, meaning The First, 92:13, The Qur'an). So, the life in this world is the 'first life' in contrast to the life after. The other word is *dunya* (as in Al-Dunya,الدُّنْيَا, 79:38, The Qur'an), which means the 'nearer' or 'lower'. So this life is nearer to us time-wise and also of lower quality. This Qur'anic description of human life, his role in this world and his eternal life is quite consistent with other oriental philosophies that considerably predate the period in which The *Qur'an* was first compiled in its tangible form.

There is a notion widespread in the Western world that the monotheistic premises of each of the three Abrahamic religions — Judaism, Christianity and Islam — point to broad but unstated other cultural common ground. The historical record suggests such has not been the case, however, when it comes to certain fundamental premises of the outlook on and approaches taken to science and scientific method.

The position of mainstream Greek, *i.e.*, Eurocentric, philosophy on the key question of the nature of the existence of the world external to any human observer is: *everything is either A or not-A*. That is Aristotle's law of the excluded middle which assumes time t = "right now" (Zatzman and Islam, 2007b). Scientifically, this assumption is the beginning of what would be termed as steady-state models, for which Δt approaches 0. This model is devoid of the time component, a spurious state even if a time-dependent term is added in order to render the model 'dynamic' (Abou-Kassem et al., 2007). Aristotle's model finds its own root in Ancient Greek philosophy (or mythology) that assumes that 'time begins when Chaos of the Void ended' (Islam, 2005b). Quite similar to what Aristotle's law of excluded middle, the original philosophy also disconnected both time function and human intention by invoking the assumption, "the gods can interrupt human intention at any time or place". This assertion essentially eliminates any relationship between individual human acts with a sense responsibility. This particular aspect was discussed in details by Zatzman and Islam (2007a) who identified the time function and the intention as the most important factors in conducting scientific research. Their argument will be presented later in this section.

The minority position of Greek philosophy, put forward by Heraclitus, was that matter is essentially atomic and that, at such a level, everything is in endless flux. Mainstream Greek philosophy of Heraclitus' own time buried his views because of their subversive implication

[1]"Al" stands for "The"

that nature is essentially chaotic. Such an inference threatened the Greek mainstream view that Chaos was the Void that had preceded the coming into existence of the world, and that a natural order came into existence putting an end to chaos.

What Heraclitus had produced was in fact a most precise description of what the human observer actually perceives of the world. However, he did not account for time at all, so changes in nature at this atomic level incorporated no particular direction or intention. In the last half of the 18th century, John Dalton reasserted the atomic view of matter, albeit now stripped of Heraclitus' metaphysical discussion and explanations. Newton's laws of motion dominated the scientific discourse of his day, so Dalton rationalized this modernized atomic view with Newton's object masses and, and we end up with matter composed of atoms rendered as spherical balls in three-dimensional space, continuously in motion throughout three-dimensional space ... within time considered as an independent variable. This line of research seals any hope for incorporating time as a continuous function, which would effectively make the process infinite-dimensional.

Zatzman and Islam (2007b) have offered extensive review of Aristotle's philosophy and provided one with scientific explanation of why that philosophy is equivalent to launching the science of tangibles. In economic life, tangible goods and services and their circulation provide the vehicles whereby intentions become, and define, actions. Locked inside those tangible goods and services, inaccessible to direct observation or measurement, are intangible relations – among the producers of the goods and services and between the producer and Nature – whose extent, cooperativeness, antagonism and other characteristic features are also framed and bounded by intentions at another level, in which the differing interests of producers and their employers are mutually engaged. In economic terms, Zatzman and Islam (2007b) identified two sources of distortion in this process. They are: 1) linearization of complex societal non-linear dependencies (functions and relationships) through the introduction of the theories of marginal utility (MU); and 2) lines in the plane intersect as long as they are not parallel, *i.e.*, as long as the equation-relationships they are supposed to represent are not redundant. The first source removes very important information pertaining to social interactions and the second source enables the use of the "=" sign, where everything to its left is equated to everything to its right. Equated quantities cannot only be manipulated, but – especially – interchanged, according to the impeccable logic, as sound as Aristotle (who first propounded it), which says: two quantities each equal to a third quantity must themselves be equal to one another or, symbolically, that 'A = C' and 'B = C' implies that 'A = B'. Scientific implications of this logic will be discussed in the latter part of this section. Here, in philosophical sense, the introduction of this logic led to the development of a 'solution'. As further arguments will be built on this 'a solution', soon this 'a solution' will become 'the solution' as all relevant information are removed during the introduction of the aphenomenal process. This would lead to the emergence of 'equilibrium', 'steady state', and various other phenomena in all branches of New Science. It won't be noticeable to common people that these are not natural systems. If anyone questions the non-existence of such a process, s/he will be marginalized as 'conspiracy theorists', 'pseudo-scientists', and numerous other derogatory designations. This line of thinking would explain why practically all scientists up until Newton had tremendous difficulty with the Establishment in Europe. In the post-Renaissance world, the collision between scientists and the Establishment was erased — *not* because the Establishment became pro-science, but more likely because the New Scientists became equally obsessed with tangibles, devoid of time function as well as intention

(Zatzman and Islam, 2007a). Theoretically, both of these groups subscribed to the same set of misconceptions or aphenomenal bases that launched the technology development in the post-renaissance era. Khan and Islam (2007a) identified these misconceptions as: 1) Chemicals are chemicals or energy is energy (meaning they are not function of the pathway and are measured as a function of a single dimension); 2) If you cannot see, it doesn't exist (only tangible expressions, as 'measurable by certain standard' are counted); 3) Simulation equals emulation (if there is agreement between reality and prediction at a given point in time, the entire process is being emulated). Zatzman and Islam (2007a) and Khan and Islam (2007a) attributed these misconceptions to the pragmatic approach (whatever works must be true), which can be traced back original Greek philosophy.

The immediate consequence of the science of tangibles is that every decision that emerges is scientifically false. The removal of a dimension and ignoring the consequences is inherent to this consequence (Mustafiz, 2007). Because time is the most important dimension, the omission of this dimension has the severest consequences. Time also forms the pathway of the science of intangibles. Intention, on the other hand, forms the root or foundation of the science of intangibles. Ignoring any of these would render the process aphenomenal. Zatzman and Islam (2007b) cited a number of examples to establish this assertion. One of them is as follows:

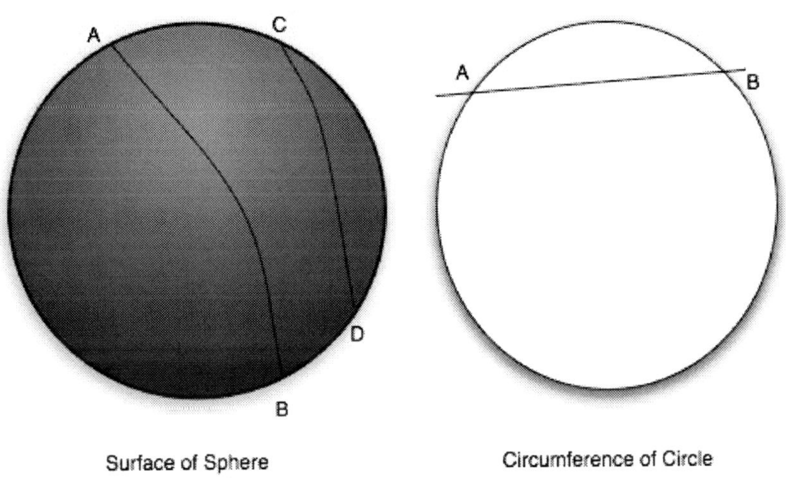

Figure 1. Sphere and Circle Compared (redrawn from Zatzman and Islam, 2007b).

Compare the circle to the sphere, in Figure 1 above. The shortest distance between two points A and B on the circle's circumference is a straight (secant) line joining them, whereas the curved arc of the circumference between the two points A and B is always longer than the secant. The shortest distance between two points on the surface of a sphere, on the other hand, is always a curve and can never be a straight line. Furthermore, between two points A and B lying on any great-circle of the sphere (*i.e.*, lying along any of an infinite number of circles that may be drawn around the surface of the sphere whose diameter is identical to that of the sphere itself) and between any other two points, say C and D, a curved line of the same length joining each pair of points will subtend a different amount of arc. For the points not on the great-circle, the same length-distance will subtend a greater arc than for points on a great-circle; curvature k in each case is uniform but different. On the other hand, what makes a circle circular rather than, say, elliptical, is precisely the condition whereby equal distances along its circumference subtend equal arcs, because curvature k anywhere along the circumference of a circle is the same.

Is anyone whose comprehension remains confined to cases of the circle in flat planar space likely to infer or extrapolate from such knowledge whatever they would need in order to grasp and use such differences found on the curved surface of the sphere? Of course not: indeed, any solution or solutions obtained for a problem formulated in two-dimensional space can often appear utterly aphenomenal when transferred or translated to three-dimensional space. In terms of the originating example of the fundamental metrics of the two kinds of surfaces: to those working in the environment of spherical surface-space, it becomes quickly obvious the shortest distance between two points that stand really close to one another on the surface of a sphere will approximate a straight line. In fact, of course, it is actually a curve and the idea that "the shortest distance between two points is a straight line" remains something that only approximates the situation in an extremely restricted subspace on the surface of a sphere. In a pragmatic sense, it is quite acceptable to propose that the shortest distance between two points is a straight line. However, in a real world which is anything but linear, this statement cannot form the basis for subsequent logical development as it would falsify all subsequent logical trains. Zatzman and Islam (2007b) called this attempt to simplify a model used to account for complex phenomena by chopping dimensions away at the outset (with the idea of adding them back in at the end) was "about as practicable as trying to make water run uphill."

The above point is further expounded by giving a simple, yet thought-provoking example (Zatzman and Islam, 2007). This example involves an experiment carried out without the knowledge of the findings of Galileo and Newton. Someone could observe and write up everything about an experiment involving dropping a weighty lead ball from the top of the Tower of Pisa. On another occasion the same could be done for an experiment entailing dropping a feather from the top of the Tower. Comparing the two, the key observation might well be that the lead ball fell to the ground faster and, from that fact, it might even be speculated that the difference in the time taken to reach the ground was a function of the difference in the weight of the feather and the lead ball. They argued that no matter how much more precise the measuring of the two experiments became, in the absence of any other knowledge or discovery it would be difficult to overthrow or reject this line of reasoning. Now, let's say the measuring devices available for subsequent repetition of this pair of experiments over the next 10, 20 or 30 years develop digital readouts, to three or four decimal places of accuracy. Quite apart from what engineering marvels these 'precision clocks' have led to, the fact would remain that this advancement in engineering does not help correcting the conclusion. In fact, the engineering advancement without regard to the basis of fundamental science would actually increase confidence toward the wrong conclusion. This is the situation for which engineering (making of precision clocks) only serves to strengthen the prejudice of the science of tangibles.

Another example was cited by Abou-Kassem et al. (2007). This one involves the discussion of the speed of light. It is 'well recognized' in the New Science world that the speed of light is the maximum achievable speed. The argument is based on the famous Einstein's equation of $E=mc^2$. However, scientific scrutiny of this 'well recognized' statement in Physics shows that this statement is aphenomenal. When it is stated what is the speed of light, it is asserted that this is the speed of light (or radiation) within a vacuum. Then, when one asks what is a vacuum, it emerges that the vacuum in true sense[2] does not exist and one must resort to a definition, rather than a phenomenal state.

Historically, there were always a debate about the limit of speed of light, but only in the post-renaissance culture that speed of light became a 'constant'. For example note the following quote from Wikipedia (Website 1, 2007):

> "Many early Muslim philosophers initially agreed with Aristotle's view that light has an infinite speed. In the 1000s, however, the Iraqi Muslim scientist, Ibn al-Haytham (Alhacen), the "father of optics", using an early experimental scientific method in his *Book of Optics*, discovered that light has a finite speed. Some of his contemporaries, notably the Persian Muslim philosopher and physicist Avicenna, also agreed with Alhacen that light has a finite speed. Avicenna "observed that if the perception of light is due to the emission of some sort of particles by a luminous source, the speed of light must be finite."
>
> The 14th century scholar Sayana wrote in a comment on verse Rigveda 1.50.4 (1700–1100 BCE—the early Vedic period): "Thus it is remembered: [O Sun] you who traverse 2202 yojanas [ca. 14,000 to 30,000 km] in half a nimesa [ca. 0.1 to 0.2 s]", corresponding to between 65,000 and 300,000 km/s, for high values of *yojana* and low values of *nimesa* consistent with the actual speed of light."

In modern science, the justification of a constant *c* is given through yet another definition – that of vacuum itself. The official NIST site (on Fundamental Physics constant) lists this constant as having a value of 299 792 458 m/s, with a standard uncertainty of zero (stated as 'exact'). Now, all of a sudden, a definition, rather than a phenomenon (measurable quantity) becomes the basis for future measurements. It is so because the fundamental SI unit of length, the meter, has been defined since October 21, 1983[3], one meter being the distance light travels in a vacuum in 1/299,792,458 of a second. This implies that any further increase in the precision of the measurement of the speed of light will actually change the length of the meter, the speed of light being maintained at 299,792,458 m/s. This is equivalent to changing the base of a logical train from reality to aphenomenality and effectively creating a process that penalizes any improvement in discovering the truth (the true speed of light in this case). Abou-Kassem et al. (2007) argued that this definition of meter is not scientifically any more precise than the original definition that was instituted by the French in 1770's in the following form: one meter is defined as 1/10,000,000 the distance from the North Pole to the Equator (going through Paris). They also discussed the spurious arrangement of introducing the unit of time as a second. It wasn't until 1832 that the concept of second was attached to the SI arrangement. The original definition was 1 second = 1 mean solar day/864,000. As late as

---

[2] Merriam-Webster online dictionary lists the four meanings of vacuum, none phenomenal:

1. emptiness of space
2. a space absolutely devoid of matter b : a space partially exhausted (as to the highest degree possible) by artificial means (as an air pump) c : a degree of rarefaction below atmospheric pressure
3. a : a state or condition resembling a vacuum : VOID <the power *vacuum* in Indochina after the departure of the French -- Norman Cousins> b : a state of isolation from outside influences <people who live in a *vacuum*...so that the world outside them is of no moment -- W. S. Maugham>
4. a device creating or utilizing a partial vacuum

[3] Decided at the Seventh Conférence générale des poids et mesures (CGPM; the same acronym is used in English, standing for "General Conference on Weights and Measures". It is one of the three organizations established to maintain the International System of Units (SI) under the terms of the Convention du Mètre (Metre Convention) of 1875. It meets in Paris every four to six years.

1960 the ephemeris second, defined as a fraction of the tropical year, officially became part of the new SI system. It was soon recognized that both mean solar day and mean tropical year both vary, albeit slightly, more 'precise' (apparent assertion being more precise means closer to the truth) unit was introduced in 1967. It was defined as 9,192,631,770 cycles of the vibration of the cesium 133 atom. The assumption here is that vibration of cesium 133 atom is exact, this assumption being the basis of Atomic clock. Only recently, it has been revealed that this assumption is not correct, creating an added source of error in the entire evaluation of the speed of light. On the other hand, if purely scientific approach is taken, one would realize that the true speed of light is neither constant nor the highest achievable speed. Clayton and Moffat (1999) discussed the phenomenon of variable light speed. Also, Schewe and Stein (1999) discussed the possibility of very low speed of light. In 1998, the research group of Lene Hau showed that the speed of light can be brought down to as low as 61 km/hour (17 m/s) by manipulating the energy level of the medium (Hau et al., 1999). Two years later, the same research group reported near halting of light (Liu et al., 2001). The work of Bajcsy et al. (2003) falls under the same category except that they identified the tiny mirror-like behavior of the media, rather than simply low energy level. More recent work on the subject deals with controlling light rather than observing its natural behavior (Ginsberg et al., 2007). Abou-Kassem et al. (2007) used the arguments provided by previous physicists and constructed the following graph. It is clear from the graph (Figure 2) that the assumption that 'speed of light', 'vacuum', and 'unit of time', 'unit of distance' are some arbitrarily set constants do not change the true nature of nature, which remains continuously dynamic. Note that media density can be converted into media energy, only if continuous transition between energy and mass is considered. Such transition, as will be clearer in the following section, is rarely talked about in the context of engineering (Khan et al., 2007). This graph also reveals that that once definitions and assertions have been accepted in face values and are not subject to further scrutiny, the possibility of increasing knowledge (as in being closer to discover the truth about nature) is diminished.

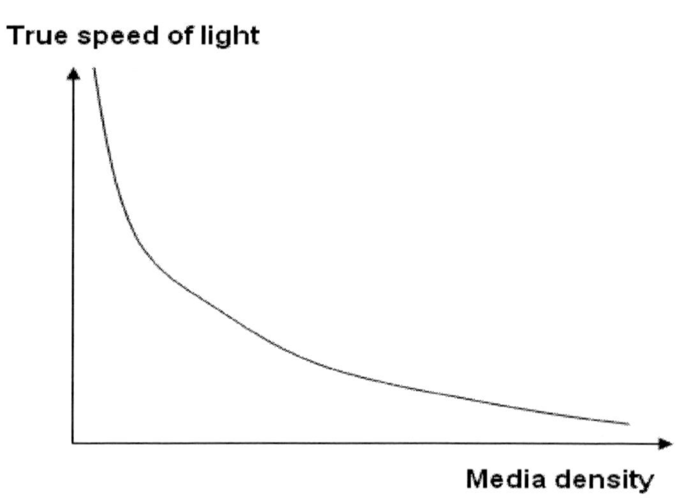

Figure 2. Speed of light as a function of media density (redrawn from Abou-Kassem et al., 2007).

Recently, a chemical engineer argued that the above graph is only a trivial manifestation of what Bose-Einstein theory would have predicted some 100 years ago. According to him, Bose-Einstein theory predicts the speed of light within an infinitely dense medium must be zero. Similar to the attitude expressed by the food scientist (in the Introduction of the present paper), this line of thinking unfortunately characterizes modern-day education in which scientists become so focused on obvious tangible expressions that they are incapable to thinking beyond the most tangible aspect of the research. Recently (2001), Eric Cornell, Wolfgang Ketterle, and Carl Wieman were awarded Nobel Prize for "for the achievement of Bose-Einstein condensation in dilute gases of alkali atoms, and for early fundamental studies of the properties of the condensates". Their work, however, far from vindicating science of tangibles, has in fact made it more necessary to take an alternate approach. Eric Cornell's most popular invited lecture is titled: Stone Cold Science: Things Get Weird Around Absolute Zero. This 'weird'-ness cannot be predicted with the science of tangibles. One further citation of Eric Cornell is: "What was God thinking? Science can't tell" (Cornell, 2005). Far from being content that all discoveries have been made and, therefore, there is nothing more to discover, a true researcher in pursuit of knowledge readily picks up on these shortcomings of conventional thinking, and as history frequently shows, it is within the inquisitive mind that answers reside. Ironically, this would also mean, discoveries are made because one is not afraid to make mistakes and, more importantly, theorize without fear of the Establishment. Satyendranath Bose, who did not have a PhD, himself did his part of the mistake. In fact, as the following quote from Wikipedia shows (Website 2, 2007), his findings were initially discarded by mainstream journals and he had to resort to sending Einstein his original work. It was only then that Einstein introduced the idea from light to mass (photon to atom) and the paper was finally published. If Bose did not have the tenacity and if Einstein did not have the decency to consider thinking "outside the box", one could not begin to think what would be the state of laser and all the gadgets that we take for granted today.

> "Physics journals refused to publish Bose's paper. It was their contention that he had presented to them a simple mistake, and Bose's findings were ignored. Discouraged, he wrote to Albert Einstein, who immediately agreed with him. His theory finally achieved respect when Einstein sent his own paper in support of Bose's to Zeitschrift für Physik, asking that they be published together. This was done in 1924." (from Website 2, 2007)

In the core of any scientific research, the attitude must be such that there is no blind faith or automatic acceptance of an existing principle. In this particular case of Nobel Prize winning work of Eric Cornell, he would not be able to observe anomalies if he took Bose-Einstein theory as absolute true (Tung *et al.*, 2006). In fact, all experimental observations indicate, there has to be major adjustments, if not re-formulation, made to the Bose-Einstein theory, as evidenced by follow up research that won Nobel Prize in 2005 (discussion below).

As recently as 2005, Nobel Prize in Physics was awarded jointly to Roy Glauber 'for his contribution to the quantum theory of optical coherence' and to John Hall and Theodor Hänsch 'for their contributions to the development of laser-based precision spectroscopy, including the optical frequency comb technique'. Indeed these discoveries have something to with Bose-Einstein theory, but far from being a regurgitation or redemption of the theory, they open up opportunities for new discoveries.

Even after the above series of Nobel Prize quality research, following questions remain unanswered. They are:

1. Bose envisioned dissimilar characteristics of any two photons and Einstein extended that work to study with what degree of freedom two dissimilar atoms would behave within a balanced system. The work of the Nobel Physics laureate discovered the separation of certain atoms under very cold conditions, yet they concluded "all atoms are absolutely identical". The reason behind this conclusion? The difference between atoms could not be measured. Fleischhauer (2007) began to question this assumption by comparing milimetre distance in a lab in quantum scale with the distance between the moon and the earth in mega scale and calls 'indistinguishability' as applicable to 'far apart'. However, soon (in the same article) he yields to the pragmatic approach and concludes, "But it also shows that we are entering a state of unprecedented experimental control of coherent light and matter waves. That could bring very real technological benefits: applications that spring to mind include quantum information interfaces that allow the transfer of a quantum bit encoded in a photon to a single atom, as well as ultra-sensitive rotation sensors and gravity detectors". One must wonder What would happen if this pragmatic reasoning was eliminated?
2. With the above conclusion, how can we observe asymmetry in experimental observations (as reported by Tung et al., 2006)? It must be noted that perfect symmetry is aphenomenal: Nature does not have any example of perfect symmetry. Such asymmetry is rarely taken in account in conventional theory (Dorogovtsev et al., 2000).
3. Later work of 2005, comes up with similar conclusion about photons, with tremendous implications on light source development and monitoring (e.g. GPS). How would this explain the difference in various light sources, particularly between natural light and artificial light?
4. For both sets of discoveries, the focus is on 'engineering' phenomena rather than observing. What are the possible implications of this 'engineering' on quality of energy source or matter, *e.g.*, nano materials that are being engineered?
5. What do these theories say about how transition between mass and energy takes place? Satyendra Nath Bose wrote (Website 2, 2007) the real gas equation of state some 100 years ago, what can be said today about real fluids that are still modeled with aphenomenal (unreal) first premises?
6. What does variable speed of light mean in terms of the definition of the unit of time. What would a paradigm shift in standards and technology development entail if one were to consider truly natural unit of time (Zatzman, 2007). Should the shift be to more 'engineering' or more natural?

Website 3 shows more recent observation of solar orbit being 'dented' rather than being 'round'. The conventional computations of Tycho Brahe and Isaac Newton about planetary orbits subject to inverse-square laws are just-so-much the Hard Science equivalent of the first premise "all Americans speak French". The question here to be asked is, why doesn't the distortion of distance created by effect(s) of these magnetic field affect the applicability of an inverse-square law derived in terms of gravitational field effects due to attraction-repulsion of the very different masses involved. Obviously, the magnetic-field effects are going to affect forms of matter in the Energy-mass spectrum. If the aphenomenal assumption that energy and mass are separable is revoked, light will be affected. Consequently, whatever we think we see will be affected. In addition, questions, such as, why does space "curve" mainly or only or at all under the effect of gravitational field(s), or does space curve mainly or only or at all under the effect of (electro-)magnetic field(s), or neither or both? Do these fields interact at all with one another (on earth, and viewed from earth, any such interaction has remained undetected)? Curiously, even the latest probes do not ask these questions as the first premise has never been questioned in the modern age (Islam, 2008).

Nearly a millennium ago, long before the Renaissance reached Europe, Averröes (1126-1198 AD, known as Ibn Rushid outside of the western world) pointed out, that Aristotelian logic of the excluded middle cannot lead to increasing knowledge unless the first premise is true. In another word, the logic can be used only to differentiate between true and false, as long as there is a criterion that discerns the truth from falsehood. The difficulty, all the way to the present modern age, has been the inability to propose a criterion that is time-honored (Zatzman and Islam, 2007a). For Averröes, the first premise was the existence of the (only) creator. It was not a theological sermon or a philosophical discourse, it was purely scientific. Inspired by The *Qur'an* that cites the root word *ilm* (علم, meaning "science", the verb, *yalamu*, يلم, standing for 'to understand', something that is totally subjective and is the first step toward acquiring knowledge), second most frequently (only second to the word, *Allah*), he considered The *Qur'an* as the only available, untainted communication with the creator and linked the first premise to the existence of such a communication. Averröes based his logic on post-Islamic Arabia (630 AD onward). As stated earlier, similar Ancient India and China, Islamic philosophy outlined in The *Qur'an* did include intangibles, such as intention and the time function. The modern-day view holds that knowledge and solutions developed from and within nature might be either good, or neutral [zero net impact] in their effects, or bad – all depending on how developed and correct our initial information and assumptions are. The view of science in the period of Islam's rise was rather different. It was that, since nature is an integrated whole in which humanity also has its roles, any knowledge and solutions developed according to how nature actually works will be *ipso facto* positive for humanity. Nature possesses an inbuilt positive intention of which people have to become conscious in order to develop knowledge and solutions that enhance nature. On the other hand, any knowledge or solutions developed by taking away from nature or going away from nature would be unsustainable. This unsustainability would mark such knowledge and solutions as inherently anti-nature.

Based on those intangibles, great strides in science, medicine, and all aspects of engineering was made in the Islamic era. Much of this was preserved, but by methods that precluded or did not include general or widespread publication. Thus, there could well have been almost as much total reliable knowledge 1400 years ago as today, but creative people's access and availability to that mass of reliable knowledge would have been far narrower. Only recently it is discovered that Islamic scholars were doing mathematics some 1000 years ago of the same order that are thought to be discovered in the 1970s (Lu and Steinhardt, 2007) – with the difference being that our mathematics can only track symmetry, something that does not exist in nature (Zatzman and Islam, 2007a). Knowledge definitely is not within the modern age. Recently, a three-dimensional PET-scan of a relic known as the 'Antikythera Mechanism' has demonstrated that it was actually a universal navigational computing device – with the difference being that our current-day versions rely on GPS, tracked and maintained by satellite (Freeth et al., 2006). Only recently, Ketata et al. (2007a) recognized that computational techniques that are based on ancient, but nonlinear counting techniques, such as Abacus, are far more superior to the linear computing. Even in the field of medicine, one would be shocked to find out what Ibn Sina ('Avicenna') said regarding nature being the source of all cure still holds true (Crugg and Newman, 2001) – with the proviso that not a single quality given by nature in the originating source material of, for example, some of the most advanced pharmaceuticals used to "treat" cancer remains intact after being subject to mass production and accordingly stripped of its powers actually to cure and not merely "treat", *i.e.*, delay, the onset or progress of symptoms. Therefore, there are examples from the history that show that knowledge is directly linked with intangibles and in fact, only when intangibles are included that science leads to knowledge (Vaziri et al., 2007).

Thomas Aquinas (1225-1274 AD) took the logic of Averröes and introduced it to Europe with a simple yet highly consequential modification: he would color the (only) creator as God and define the collection of Catholic church documentation on what eventuated in the neighborhood of Jerusalem some millennium ago as the only communication of God to mankind (hence the title, *bible* – the (only) Book). If Aristotle was the one who introduced the notion of removing intention and time function from all philosophical discourse, Thomas Aquinas is the one who legitimized the concept and introduced this as the only Science (as in process to gaining knowledge). Even though, Thomas Aquinas is known to have adapted the logic of Averröes, his pathway as well as prescribed origin of acquiring knowledge was diametrically opposite to the science introduced by Averröes. This is because the intrinsic features of both God and *bible* were dissimilar to the (only) creator and *Qu'ran*, respectively (Armstrong, 1994). For old Europe and the rest of the world that it would eventually dominate, this act of Thomas Aquinas indeed became the bifurcation point between two pathways, with origin, consequent logic, and the end being starkly opposite. With Aristotle's logic, something either is or is not: if one is 'true', the other must be false. Because, Averröes' 'the creator' and Thomas Aquinas's 'God' both are used to denominate monotheist faith, the concept of science and religion became a matter of conflicting paradox (Pickover, 2004). Averröes called the (only) creator as 'The Truth' (In Qu'ranic Arabic, the word 'the Truth' and 'the Creator' refer to the same entity). His first premise pertained to the book (*Qu'ran*) that said, "Verily unto Us is the first and the last (of everything)"(89.13). Contrast this to a "modern" view of a creator. In Carl Sagan's words (Hawking, 1988), "This is also a book about God…or perhaps about the absence of God. The word God fills these pages. Hawking embarks on a quest to answer Einstein's famous question about whether God had any choice in creating the universe. Hawking is attempting, as he explicitly states, to understand the mind of God. And this makes all the more unexpected the conclusion of the effort, at least so far: a universe with no edge in space, no beginning or end in time, and nothing for a Creator to do."

This divergence in pathways was noted by Zatzman and Islam (2007a). Historically, challenging the first premise, where the divergence is set, has become such a taboo that there is no documented case of anyone challenging it and surviving the wrath of the Establishment (Church alone in the past, Church and Imperialism after the Renaissance). Even challenging some of the cursory premises have been hazardous, as demonstrated by Galileo. Today, we continue to avoid challenging the first premise and even in the information age it continues to be hazardous, if not fatal, to challenge the first premise or secondary premises. It has been possible to keep this modus operandi because new "laws" have been passed to protect 'freedom of religion' and, of late, 'freedom of speech'. For special-interest groups, this opens a Pandora's box for creating 'us *vs* them', 'clash of civilizations' and every aphenomenal model now in evidence (Zatzman and Islam, 2007b).

Avoiding discussion of any theological nature, Zatzman and Islam (2007a) nevertheless managed to challenge the first premise. Rather than basing the first premise on the Truth *à la* Averröes, they mentioned the importance of individual acts. Each action would have three components: 1) origin (intention); 2) pathway; 3) consequence (end). Averröes talked about origin being the truth; they talked about intention that is real. How can an intention be real or false? They equate real with natural. Their work outlines fundamental features of nature and shows there can be only two options: natural (true) or artificial (false). The paper shows Aristotle's logic of anything being 'either A or not-A' is useful only to discern between true (real) and false (artificial). In order to ensure the end being real, the paper introduces the recently developed criterion of Khan (2006) and Khan and Islam (2007b). If something is convergent when time is extended to infinity, the end is assured to be real. In fact, if this criterion is used, one can be spared of questioning the 'intention' of an action. If any doubt, one should simply investigate where the activity will end up if time, t goes to infinity.

This absence of discussion of whatever happened to the tangible-intangible nexus involved at each stage of any of these developments is no merely accidental or random fact in the world. It flows directly from a Eurocentric bias that pervades, well beyond Europe and North America, the gathering and summation of scientific knowledge everywhere. Certainly, it is by no means a property inherent - either in technology as such, or in the norms and demands of the scientific method *per se*, or even within historical development - that time is considered so intangible as to merit being either ignored as a fourth dimension, or conflated with tangible space as something varying independently of any process underway within any or all dimensions of three-dimensional space. Recently, Mustafiz et al. (2007) identified the need of including a continuous time function as starting point of acquiring knowledge. According to them, the knowledge dimension does not get launched unless time as a continuous function is introduced. They further show that the knowledge dimension is not only possible, it is necessary. The knowledge is conditioned not only by the quantity of information gathered in the process of conducting research, but also by the depth of that research, *i.e.*, the intensity of one's participation in finding things out. In and of themselves, the facts of nature's existence and of our existence within it neither guarantee nor demonstrate our consciousness of either, or the extent of that consciousness. Our perceptual apparatus enables us to record a large number of discrete items of data about the surrounding environment. Much of this information we organize naturally and indeed unconsciously. The rest we organize according to the level to which we have trained, and-or come to use, our own brains. Hence, neither can it be affirmed that we arrive at knowledge directly or merely through perception, nor can we affirm being in possession at any point in time of a reliable proof or guarantee that our knowledge of anything in nature is complete.

Historically, what Thomas Aquinas model did to European philosophy is the same as what Newton's model did to the New Science. The next section examines Newton's models. Here it would suffice to say that Newton's approach was not any different from the approach of Thomas Aquinas or even Aristotle. One exception among scientists in Europe was Albert Einstein, who introduced the notion of time as the fourth dimension. However, no one followed up on this aspect of Einstein's work and it was considered that the addition of a time term in the Newton's so-called steady state models will suffice. Mustafiz (2007) recognized the need of including the time dimension as a continuous function and set the stage for modeling science of intangibles (Abou-Kassem et al., 2007).

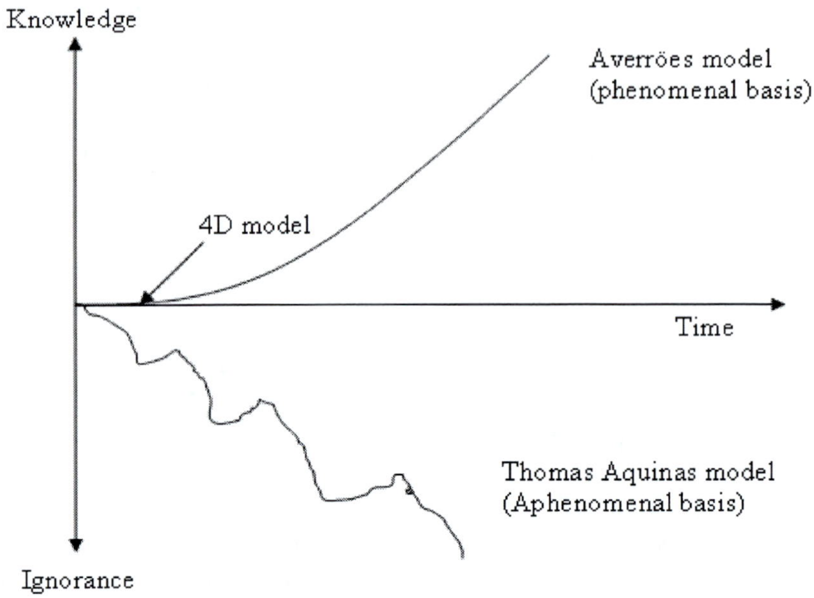

Figure 3. Logically, a phenomenal basis is required as the first condition to sustainable technology development. This foundation can be the Truth as the original of any inspiration or it can be 'true intention', which is the essence of intangibles (modified from Zatzman and Islam, 2007a and Mustafiz et al., 2007).

Table 1 summarizes the historical development in terms of scientific criterion, origin, pathway and consequences of the principal cultural approaches to reckoning, and reconciling, the tangible-intangible nexus.

Table 1. Criterion, origin, pathway and end of scientific methods in some of the leading civilizations of world history

People	Criterion	Origin	Pathway	End
Zatzman and Islam (2007)	$\Delta t \rightarrow \infty$	Intention	*f(t)*	Consequences
Khan (2006)	$\Delta t \rightarrow \infty$	Intention	Natural	Sustainability
(Zatzman and Islam, 2007a)	$\Delta t \rightarrow \infty$	Intention	Natural	Natural (used $\Delta t \rightarrow \infty$ to validate intention)
Einstein	t as 4th-D	"God does not play dice…"	Natural	N/A
Newton	$\Delta t \rightarrow 0$	"external force" (1st Law)	No difference between natural and artificial	Universe will run down like a clock
Aquinas	Bible	Acceptance of Divine Order	All *knowledge* and *truth* reside in God; *choice* resides with Man	Heaven and Hell

People	Criterion	Origin	Pathway	End
Averröes	Al- Furqan الفرقان (meaning The Criterion, title of Chapter 25 of *The Qur'an*) stands for *Qur'an*	Intention (first *hadith*)	*Amal saliha* (good deed, de-pending on good intention)	Accomplished (as in *Muflehoon*, اَلْمُفْلِحُونَ , 2:5), Good (+∞) Losers (as in *Khasheroon*, اَلْخَاسِرُونَ, 58:19), Evil (-∞)
Aristotle	*A* or *not-A* (Δt=0)	Natural law	Natural or arti-ficial agency	(*Eudaimonia*, tr. "happiness", actually more like "Man in harmony with universe")
Ancient India	Serving others; "world reveals itself"	Inspiration (*Chetna*)	*Karma* (deed with inspiration, *chetna*)	Karma, salvation through merger with Creator
Ancient Greek (pre-Socratics)	*t* begins when Chaos of the void ended	the Gods can interrupt human intention at any time or place	N/A	N/A
Ancient China (Confucius)	N/A	Kindness	Quiet (intangible?)	Balance

With this above table, we present a table (Table 2) that shows the characteristic features of Nature. These are true features and are not based on perception. They are true because there are no example of their opposites.

Table 2. Typical features of natural processes as compared to the claims of artificial processes (Adapted from Khan and Islam, 2007a)

Features of Nature and natural materials	
Feature no.	Feature
1	Complex
2	Chaotic
3	Unpredictable
4	Unique (every component is different), *i.e.*, forms may appear similar or even "self-similar", but their contents alter with passage of time
5	Productive
6	Non-symmetric, *i.e.*, forms may appear similar or even "self-similar", but their contents alter with passage of time
7	Non-uniform, *i.e.*, forms may appear similar or even "self-similar", but their contents alter with passage of time
8	Heterogeneous, diverse, *i.e.*, forms may appear similar or even "self-similar", but their contents alter with passage of time
9	Internal
10	Anisotropic
11	Bottom-up

Table 2. (Continued)

Features of Nature and natural materials	
Feature no.	Feature
12	Multifunctional
13	Dynamic
14	Irreversible
15	Open system
16	True
17	Self healing
18	Nonlinear
19	Multi-dimensional
20	Infinite degree of freedom
21	Non-trainable
22	Infinite
23	Intangible
24	Open
25	Flexible

Now, in Table 3, we summarize many currently used 'laws' and theories, all of which emerged from the New Science after renaissance. Note how the first premises of practically all of these theories violate fundamental features of Nature. Only conservation of mass that in fact has root in ancient times and theory of relativity do not have an aphenomenal first premise. It is important to note that only recently Kvitco (2007) discredited Einstein's relativity altogether. However, he did not elaborate on the first premise of the theory. Our contention is, Einstein's relativity theory appears to be spurious if processed through the science of tangibles. So far, there is no evidence that the first premise of the theory of relativity, as Einstein envisioned, is aphenomenal.

Table 3. How the Natural features are violated in the first premise of various 'laws' and theories of the science of tangibles

Law or theory	First premise	Features violated (see Table 2)
Conservation of mass	Nothing can be created or destroyed	None
Lavoisier's deduction	Perfect seal	15
Phlogiston theory	Phlogiston exists	16
Theory of relativity	Everything (including time) is a function of time	None
$E = m c^2$	Mass of an object is constant	13
	Speed of light is constant	13
	Nothing else contributes to E	14, 19, 20, 24
Planck's theory	Nature continuously degrading to heat dead	5, 17, 22
Charles	Fixed mass (closed system), ideal gas, Constant pressure,	24, 3, 7
Boyles	A fixed mass (closed system) of ideal gas at fixed temperature	24, 3, 7

Law or theory	First premise	Features violated (see Table 2)
Kelvin's	Kelvin temperature scale is derived from Carnot cycle and based on the properties of ideal gas	3, 8, 14, 15
Thermodynamics 1st law	Energy conservation (The first law of the thermodynamics is no more valid when a relationship of mass and energy exists)	None
Thermodynamics 2nd law	Based on Carnot cycle which is operable under the assumptions of ideal gas (imaginary volume), reversible process, adiabatic process (closed system)	3, 8, 14, 15
Thermodynamics 0th law	Thermal equilibrium	10, 15
Poiseuille	Incompressible uniform viscous liquid (Newtonian fluid) in a rigid, non-capillary, straight pipe	25, 7
Bernouilli	No energy loss to the sounding, no transition between mass and energy	15
Newton's 1st law	A body can be at rest and can have a constant velocity	Non-steady state, 13
Newton's 2nd law	Mass of an object is constant Force is proportional to acceleration External force exists	13 18
Newton's 3rd law	The action and reaction are equal	3
Newton's viscosity law	Uniform flow, constant viscosity	7, 13
Newton's calculus	Limit $\Delta t \rightarrow 0$	22

Now, if New Science has given us only theories and 'laws' that have spurious first premise, Averröes criterion would make New Science aphenomenal. This is indeed found in modern technologies that have resulted in 'technological disasters', reversing originally declared 'intention' for every technology. This is shown in Table 4.

Table 4. Analysis of "breakthrough" technologies

Product	Promise (knowledge at t= 'right now'	Current knowledge (closer to reality)
Microwave oven	Instant cooking (bursting with nutrition)	97% of the nutrients destroyed, produces dioxin from baby bottles
Fluorescent light (white light)	Simulates the sunlight and can eliminate 'cabin fever'	Used for torturing people, causes severe depression
Prozac (the wonder drug)	80% effective in reducing depression	Increases suicidal behavior
Anti-oxidants	Reduces aging symptoms	Gives lung cancer
Vioxx	Best drug for arthritis pain, no side effect	Increases the chance of heart attack
Coke	Refreshing, revitalizing	Dehydrates; used as a pesticide in India
Transfat	Should replace saturated fats, incl. high-fiber diets	Primary source of obesity and asthma
Simulated wood, plastic gloss	Improve the appearance of wood	Contains formaldehyde that causes Alzheimer

Table 4. (Continued)

Product	Promise (knowledge at t= 'right now'	Current knowledge (closer to reality)
Cell phone	Empowers, keep connected	Gives brain cancer, decreases sperm count among men.
Chemical hair colors	Keeps young, gives appeal	Gives skin cancer
Chemical fertilizer	Increases crop yield, makes soil fertile	Harmful crop; soil damaged
Chocolate and 'refined' sweets	Increases human body volume, increasing appeal	Increases obesity epidemic and related diseases
Pesticides, MTBE	Improves performance	Damages the ecosystem
Desalination	Purifies water	Necessary minerals removed
Wood paint/varnish	Improves durability	Numerous toxic chemicals released
Leather technology	Won't wrinkle, more durable	Toxic chemicals
Freon, aerosol, etc.	Replaced ammonia that was 'corrosive'	Global harms immeasurable and should be discarded

Source: Zatzman and Islam, 2007b.

3. Other Considerations in Mathematics and Science

In both Sanskrit and Arabic, grammatical 'number' is: singular, two-fold, or 'plural' (meaning: more than two). This is important because adjectives and verbs have to agree with the number of the noun serving as the subject of the sentence. In *all* European languages, there is simply: singular, meaning one only or fraction-of-one, and plural, meaning more-than-one.

In the tribal languages that we know about or that still exist and that go back to the period from 40 000 to 12 000 years ago (before last Ice Age), there is no way to enumerate quantities past about 20 or 30. This might have something to do with how many lived together or in close contact in communities BEFORE versus AFTER the emergence of settled locales based on agriculture and animal husbandry.

The ancient Greeks have only very general ideas of large numbers, thus there is the myriad, meaning 10,000 --- but it is for counting much larger numbers in "units" of 10, 000.

Ancient India houses several mathematical traditions that tackle very large numbers, most interestingly the Jaina mathematicians. It seems instead of fractions/rational numbers, the Jaina mathematicians redefine a whole so it has a very very large number of "units" which exist in nature in some meaningful form. So for example the circuit of the sum around the earth can be measured in minutes of arc; that becomes the unit. Then the sun's circuit during the year is defined by the hundreds of millions of minutes in the total arc. The next threshold number is the number of such years that "the world" has existed according to the Jain priests' calculations, which is then crores of lakhs of minutes-of-arc.

Although schemas of this kind neatly skirt the apparent contradictions of fractions (what logic does a fraction have within a system that enumerates units by a decimal positioning system?), decimal representation was entirely normal in Indian maths of the late classical period, because this was or could be made consistent with the positioning system of representing number-values. However, these were numbers that existed as the result of subdividing by 10s, 100s, 1000s etc.

What the Indian mathematicians did not come up with was anything that would "reduce" 25 over 100, which could be represented as "0.25", to "1/4". And going the other way, Egyptian and Babylonian mathematics had fractions, but not positional representation of numerical value --- so they too could not make the link.

Well into the 19th century, a value such as "66-2/3" was meaningful mainly only as the aliquot third part of 200. As "66.666..." on a number line, just ahead of 66.5 and just less than 66.75, it would have been meaningful mainly to people like the research mathematician Riemann, trying in the 1860s to define the real line as an infinitely divisible and ultimately non-denumerable continuum of numbers. There was really little if any notion of physically measuring such a quantity from any natural phenomenon.

3.1. Numbers in the Qur'an

As mentioned earlier, the Qu'ran is the only document that has survived in its original form for over 1400 years. This is also the only document that has used the word *ilm* (Science, as in the process) a great many times. With this science, great strides were made for many centuries toward increasing knowledge and making tangible progress in human society. However, focus on tangible was not the reason for this tangible progress. It was instead focus on intangibles. Just after the introductory chapter, Chapter 2 (verses 3-4) outlines the conditions for true guidance. The first condition is set out to be 'believing in the intangibles' (the Arabic word being, "Al-Ghaib", الغيب).

ٱلَّذِينَ يُؤْمِنُونَ بِٱلْغَيْبِ وَيُقِيمُونَ ٱلصَّلَوٰةَ وَمِمَّا رَزَقْنَٰهُمْ يُنفِقُونَ (٣)

(approximate translation of the meaning: those who believe in the intangible, establishes communication, and spends in charity from their lawful belongings)

It is no surprise that this document has the mention of very few numbers, all of which, nevertheless have very significant meaning that is the subject to future research. In this paper, we will simply state the numbers that appear in the Qur'an. To begin with there is no word in Arabic that would bear similar meaning as in European mathematics. The word most commonly used as an equivalent to number is, 'rukm', رقم, as in Chapter 83, verse 9, كِتَٰبٌ مَّرْقُومٌ, *kitabum markum*, which means 'written record'. The word rukm has a number of other meanings, such as ornament, engrave, mark or sign, all indicating for 'something to be known'. It seems, this word in Arabic stands for tangible expressions of an intangible and not something that is an image or aphenomenal, valid only for a time, t= 'right now'. Note that there is no mention of zero in the Qur'an, even though the Arabic equivalent word, *cipher*, existed even before the Qur'an came into tangible existence. There are *hadiths* of the prophet mentioning the word, *cipher*, which stood for 'nothing' or 'empty'. This word, in fact, has been adapted into the English language (as used in 'decipher'), once again standing for 'meaningless' or empty zeros. (At the same time, it is a place-holder: it can serve a function during computation without itself being a number. This raises an interesting point. According to the science of tangibles, in the matter of chemical reactions that can only take place in the presence of certain catalysts, it is repeated everywhere with almost the force of religious fervour: a catalyst does not take part in the reaction! Why don't the proponents of this "logic" not insist just as fervently that 0 as a placeholder takes no part in the computation?) The introduction of 0 in computation was introduced by the Arab who adapted the concept and the symbol from Ancient India. The influence of India in post-prophet Muhammad Arab mathematics is so dominant that even the Indian numerals replaced original Arab numerals (now adopted in European mathematics) that were actually more scientific than than the Indian counterpart (e.g. every numeral represents the numeric value by the number of angle that it forms).

The Quran did not use any numeral, but used the following numbers in various concepts (some examples are given).

Number One (in the context of 'there is only one *ilah*[4]):
(قل إنما هو اله **واحد** وإنني بريء مما تشركون) (سورة الأنعام, الآية 19) Chapter 6, verse 19

Number Two (in the context of 'do not choose two *ilah*'s)
(قال تعالى:(وقال الله لا تتخذوا إلهين **اثنين** إنما هو إله واحد
(سورة النحل, الآية 51) Chapter 17, verse 51

Number Three (in the context of 'not uttering Three (Trinity), stop because Allah is the only One *ilah*)
(ولا تقولوا **ثلاثة** انتهوا خيرا لكم)
(سورة النساء, الآية 171) Chapter 4, verse 171

Number Four (in the context of allowing aphenomenal model worshippers for months for free movement)
(فسيحوا فالأرض **أربعة** أشهر) Chapter 9, verse 2
(سورة التوبة, الآية 2)

Number Five (in the context of the actual number of men in the story of the people of the Cave)
(ويقولون **خمسة** سادسهم كلبهم رجما بالغيب)
(سورة الكهف، الآية22) Chapter 18, verse 22

Number Six (in the context of creating the Universe in six *yaum*'s[5])
(إن ربكم الله الذي خلق السماوات و الأرض في **ستة** أيام)
(سورة الأعراف,الآية 54) Chapter 7, verse 54

Number Seven (in the context of seven gates in hell)
(ها**سبعة** أبواب لكل باب منهم جزء مقسوم) Chapter
(سورة الحجر, الآية 44) Chapter 15, verse 44

Number Eight (in the context of eight angels upholding the Throne on the day of judgment, when there would be no secret)
(ويحمل عرش ربك فوقهم يومئذ **ثمانية**)
(سورة الحاقة, الآية 17) Chapter 69, verse 17

---

[4] The word, *ilah*, is erroneously translated as 'god', the actual meaning is closer to shelter-giver, as in husbandry. *Al-Ilah* stands for 'the only shelter giver', which might explain why it is commonly translated as the 'God'.

[5] The word, *yaum* is errorneously translated as 'day'. It is closer to phase. This is supported by other statements that are indicative of phase and not 'day', for instance, 'the day of judgment' is called '*yaum addin*'. For 'day', as in daylight, there is a different word, *nahar* in Arabic. Also, in Arabic, Earth day begins when the sun sets.

Number Nine (in the context of nine persons making mischief and not reforming, during the period of *Thamud*, the ones who built the crystal valley in Jordan)
(وكان في المدينة***تسعة*** رهط يفسدون في الأرض)
(سورة النمل, الآية48) Chapter 27, verse 48

Number 10 (in the context of fasting for 10 days if a pilgrim cannot afford to sacrifice an animal)
(تلك ***عشرة*** كامل)
(سورة البقرة, لآية196) Chapter 2, verse 196

Number 11(in the context of Prophet Joseph's description of his dream of 11 planets[6], the sun, and the moon prostrating toward him):
أحد عشر كوكبا والشمس و القمر رأيتهم لي ساجدين)
(سورة يوسف, لآية 4)Chapter 12, verse 4

Number 12 (in the context of the number of months in a year as per Divine order)
(إن عدة الشهور عند الله ***اثنا عشر*** شهرا في كتاب الله
(سورة التوبة, الآية 36)Chapter 9, verse 36

Number 19 (in the context of
(عليها***تسعة عشر***
(سورة المدثر, الآية 30) Chapter 74, verse 30

Number 20 (in the context of 20 steadfast righteous people overcoming attack from 200 unrighteous ones)

(إن يكن منكم ***عشرون*** صابرون يغلبوا مائتين
(سورة الأنفال,الآية 65) Chaper 8, verse 65

Number 30 (in the context of ordained bearing and weaning of a child for 30 months and maturity for man by 40 years)
(وحمله وفصاله***ثلاثون*** شهرا
(سورة الأحقاف, الآية15) Chapter 46, verse 15

Number 40 (in the context of assigning 40 nights of solitude)
(وإذ واعدنا موسى ***أربعين*** ليلة ثم اتخذتم العجل من بعده وأنتم ظالمون
(سورة البقرة, الآية 15) Chapter 2, verse 15

Number 50 (in the context of Noah spending 950 (1000 minus 50) years before the flood struck)
(ولقد أرسلنا نوحا إلى قومه فلبث فيهم ألف سنة إلا ***خمسين*** عاما
(سورة العنكبوت, الآية 14) Chapter 29, verse 14

Number 60 (in the context of feeding 60 needy people in case someone is unable to fast two consecutive months as penalty to violating family code of conduct)
(فمن لم يستطع فإطعام***ستين*** مسكينا

[6] The Arabic word is: كَوْكَبًا, *kawakeb*. There are other words in Arabic to describe planets as well. The distinction of these terminologies is beyond the scope of this paper.

(سورة المجادلة, الآية 4) Chapter 58, verse 4

Number 70 (in the context of the length of the chain holding captives in hell)
(ثم في سلسة ذرعها ***سبعون*** ذراعا فسلكوه
(سورة الحاقة, الآية 32) Chapter 69, verse 32

Number 80 (in the context of punishment in the form of number of lashes for unfounded allegation against a chaste woman)
فاجلدوهم ***ثمانون*** جلدة ولا تقبلوا منهم شهادة أبدا
(سورة النور, الآية 4) Chapter 24, verse 4

Number 90 (in the context of two disputants (brothers) coming to David and asking to settle the dispute, one of them having nine and 90 ewes still wanted the other one's only one)
وهذا أخي له تسع ***وتسعون*** نعجة ولي نعجة واحدة
(سورة ص, الآية 22) Chapter 38, verse 22

Number 100 (in the context of someone being dead for 100 years yet answering "a day or part of a day" when asked "How long did you tarry (thus)?"
قل بل لبثت ***مائة*** عام
(سورة البقرة, الآية 259) Chapter 2, verse 259

Number 200 (In the context of 20 righteous with steadfastness vanquishing 200 of the attackers)
إن يكن منكم عشرون صابرون يغلبوا ***مائتين***
(سورة الأنفال,الآية 65) Chapter 8, verse 65

Number 300 (In the context of habitants of the cave sleeping for 300 years and nine days)
ولبثوا في كهفهم ***ثلاث مائة*** سنين و ازدادوا تسعا
(سورة الكهف, الآية 25) Chapter 18, verse 25

Number 1000 and 2000 (in the context of 1000 righteous people that are steadfast vanquishing 2000 attackers)[7]
وإن يكن منكم ***ألف*** يغلبوا ***ألفين*** بإذن الله
(سورة الأنفال, الآية 66) Chapter 8, verse 66

Number 3000 (in the context of 3000 angels sent down to help the righteous who were under attack by wrongdoers)
إذ تقول للمؤمنين ألن يكفيكم أن يمدكم ربكم ***بثلاثة آلاف*** من الملائكة منزلين
(سورة آل عمران, الآية 124) Chapter 3, verse 124

Number 5000 (in the context of increasing the number of angels from 3000 to 5000 in order to help the righteous (see above))
هذا يمددكم ربكم ***بخمسة آلاف*** من الملائكة مسومين
(سورة آل عمران, 125) Chapter 3, verse 125

[7] Note how the same verse stated earlier 20 righteous people vanquishing 200 attackers, as compared to 1000 righteous vanquishing 2000 attackers – the relationship is definitely not linear

Number 100,000 (in the context of Jonah being dispatched to a locality of 100,000 or more who were initially misguided)
وأرسلناه إلى ***مائة ألف*** أو يزيدون
(سورة الصافات, الآية 147), Chapter 37, verse 147

3.1.1. Fractions

Fractions in the Qur'an are principally used to denote natural division, as applied in sharing or just division of properties. Consequently, most divisions deal with family heritage laws. Following are some examples.

Number One Over Two (in the context of family heritage law)
ولكم ***نصف*** ما ترك أزواجكم إن لم يكن لهن ولد
(سورة النساء, الآية 12) Chapter 4, verse 12

Number One over Three (in the context of family heritage law)
إن لم يكن له ولد وورثه أبواه فلأمه ***الثلث***
(سورة النساء, الآية 11) Chapter 4, verse 11

Number One over Four (in the context of family heritage law)
(قال تعالى:(فإن كان لهن ولد فلكم ***الربع*** مما تركن
(سورة النساء, الآية 12) Chapter 4, verse 12

Number One over Five (in the context of distribution of war booty[8])
واعلموا إنما غنمتم من شيء فإن لله ***خمسه***
(سورة الأنفال, الآية 41) Chapter 8, verse 41

Number One over Six (in the context of family heritage law)
فإن كان له إخوة فلأمه ***السدس***
(سورة النساء,الآية 11) Chapter 4, verse 11

Number One over Eight (in the context of family heritage law)
(قال تعالى:(فإن كان لكم ولد فلهن ***الثمن*** مما تركتم
(سورة النساء, الآية 12) Chapter 4, verse 12

3.1.2. Natural Ranking

Various numeral rankings are used in the Qur'an. Following are some of the examples:

First (in the context of first one to be obedient to Allah)
قل إني أمرت أن أكون أول من أسلم
(سورة الأنعام, الآية 14) Chapter 6, verse 14

Second (in the context of the second one as a companion of the prophet when he was migrating from Mecca to Medina, being driven out by Arab pagans)[9]

---

8 It is important to note the war booty in Quranic context applies strictly to defensive war. There was only one offensive war during the entire period of the revelation of the Qur'an. This war (the last one) had no casualty, generated no war booty, and was followed by general amnesty for the very people who attacked the Prophet's group in the form of four wars in less than 10 years.

إلا تنصروه فقد نصره الله إذ أخرجه الذين كفروا ثاني ثنين
(سورة التوبة, الآية 40) Chapter 9, verse 40

Third (in the context of adding a third messenger because first two were not heeded)
إذ أرسلنا إليهم اثنين فكذبوهما فعززنا بثالث فقالوا إنا إليكم مرسلون
(سورة يس, الآية14) Chapter 36, verse 14

Fourth (in the context of no secret can be kept from Allah as he makes the fourth when three human beings consult in secret)
ما يكون من نجوى ثلاثة إلا هو رابعهم
(سورة المجادلة, الآية 7) Chapter 58, verse 7

Fifth (in the context of taking a fifth oath invoking one's innocence in absence of material witness)
والخامسة أن لعنة الله عليه إن كان من الكاذبين
(سورة النور, الآية 7) Chapter 24, verse 7

Sixth (in the context of the sixth companion (a dog) of the inhabitants of the cave)
ويقولون خمسة سادسهم كلبهم رجما بالغيب
(سورة الكهف, الآية22) Chapter 18, verse 22

Eighth (in the context of the eighth companion (a dog) of the inhabitants of the cave, as hearsay)
ويقولون سبعة وثامنهم كلبهم
(سورة الكهف, الآية 22) Chapter 18, verse 22

3.1.3. Mathematical Operations

Mathematical operations in the Qur'an are all natural, i.e., addition to increase, subtract to decrease, multiply to strike or counter, and divide to breakdown in among natural recipients or compartments. No operation that is spurious would yield any result. These operations were mainly used by 7th century onward Muslims to calculate compulsory charity as well as just share of heritage properties. Examples are given below:

Addition (in the context of 300 and 9 years spent by the inhabitants of the cave)
(قال تعالى:(ولبثوا في كهفهم ***ثلاث مائة سنين و ازدادوا تسعا***
(سورة الكهف, الآية 25) Chapter 8, verse 25
The operation is: 300 + 9

Subtraction (in the context of Noah spending 1000 minus 50 years with his tribe)
ولقد أرسلنا نوحا إلى قومه فلبث فيهم ***ألف سنة إلا خمسين عاما***
(سورة العنكبوت, الآية 14) Chapter 29, verse 14
The operation: 1000 – 50

Multiplication (in the context of righteous people fighting attackers and vanquishing them even when outnumbered)[10]

[9] Note that second, ثاني (*thani*), in Arabic only applies to ranking. The unit of time, 'second' is incorrectly translated in Arabic as *thani*. Such artificial time unit does not exist in Quranic Arabic (Zatzman, 2007)

[10] The root word used for 'multiplication' in Arabic is ضرب, which stands for numerous meanings. For example (Chapter and verse numbers from the Quran are indicated):

إن يكن منكم **عشرون** صابرون يغلبوا **مائتين**
وإن يكن منكم **ألف** يغلبوا **ألفين** بإذن الله
(سورة الأنفال, الآية 65, 66) Chapter 8, verses 65, 66
The operation: 20 striking against 200; 1000 striking against 2000

Division (in the context of fractions allocated to various recipients of a heritage)
إن لم يكن له ولد وورثه أبواه فلأمه **الثلث** فإن كان له إخوة فلأمه **السدس**
(سورة النساء, الآية 11) Chapter 4, verse 11

3.1.4. The Number Seven and its Multiple in the Qur'an

In the last year, a book called (إشراقات الرقم سبعة في القران الكريم), which won The Dubai International Holy Quran Award, talked about the number seven in the Qur'an. The book mentioned many things about number seven and its multiple and here a few examples:

1.The fact that there are seven heavens[11] were said seven times in the Quran:
(سورة البقرة, الآية 29) Chapter 2, verse 29
(سورة الإسراء, الآية 44) Chapter 17, verse 44
(سورة المؤمنون, الآية 86) Chapter 23, verse 86
(سورة فصلت, الآية 12) Chapter 41, verse 12
(سورة الطلاق, الآية 12) Chapter 65, verse 12
(سورة الملك, الآية 3) Chapter 67, verse 3
(سورة نوح, الآية 15) Chapter 71, verse 15

2. The fact of creation the earth and heavens in six *yaums* also mention seven times:
(سورة الأعراف, الآية 54) Chapter 7, verse 54
(سورة يونس, الآية 3) Chapter 10, verse 3
(سورة هود,الآية 7) Chapter 11, verse 7
(سورة الفرقان, الآية 59) Chapter 25, verse 59
(سورة السجدة,الآية 32) Chapter 32, verse 32
(سورة ق, الآية 38) Chapter 50, verse 38
(سورة الحديد, الآية 4) Chapter 57, verse 4

3. The number of السور and the number of الايات between the first time number seven was mention and the last time are a multiple of number seven:
First time (سورة البقرة (2), الآية 29) Chapter 2, verse 29
Last time (سورة النبأ (78), الآية 12)Chapter 78, verse 12
So, that:

To travel, to get out: 3:156; 4:101; 38:44; 73:20; 2:273
To strike: 2:60,73; 7:160; 8:12; 20:77; 24:31; 26:63; 37:93; 47:4
To beat: 8:50; 47:27
To set up: 43:58; 57:13
To give (examples): 14:24,45; 16:75,76,112; 18:32,45; 24:35; 30:28,58; 36:78; 39:27,29; 43:17; 59:21; 66:10,11
To take away, to ignore: 43:5
To condemn: 2:61
To seal, to draw over: 18:11
To cover: 24:31
To explain: 13:17

[11] In Qur'an, paradise (*jannah* سَمَوَاتِ :The word used is .(celestial)is different from heaven (or garden which also stands for sky or a layer of stars.

78 – 2 = 77 = 7 * 11
The numbers of الآيات between them are 5649 = 7 * 807

4. The number of السور (Chapters) that begin with the praise of Allah are seven :
(سورة الإسراء, الآية 1) Chapter 17, verse 1
(سورة الحديد, الآية 1) Chapter 57, verse 1
(سورة الحشر, الآية 1) Chapter 59, verse 1
(سورة الصف, الآية 1) Chapter 61, verse 1
(سورة الجمعة, الآية 1) Chapter 62, verse 1
(سورة التغابن, الآية 1) Chapter 64, verse 1
(سورة الأعلى, الآية 1) Chapter 87, verse 1

4. Policy Making and Governance

Please introduce this section. The idea is to show that there is such thing as natural policy making and natural governance. There is an example from the Islamic era.

It is highly unconventional for papers on scientific topics to veer into matters normally considered part of "politics". Yet, in the context of the matters to be elaborated in the present paper, the authors would be derelict in the responsibility they have undertaken is they did not take up the matter of "policy making and governance".

One of the achievements claimed by Eurocentric "civilization" and its champions, and taken for granted by many, has been the apparent separation of "science" — shorthand for truthful knowledge derived from methods of data gathering that are objective and reliable, from "politics" — understood as a shorthand for gathering knowledge and power based on extensive resort to deceptions or other ruses of various kinds. If, however, the world of phenomena in nature is objective and knowable only in the course of a principled and methodical exposure, separation and rejection of errors and falsehoods, why shouldn't a similar discipline not be applicable and possible in human affairs, such as governance?

This natural approach to policy-making was seriously attempted under the early Muslim caliphates, and always upheld in Muslim societies at least notionally as a supreme objective. However, such an approach never resonated with Christian rulers or communities of Christian believers. Among Christian rulers and their subjects, Law came from God, the soundness of a ruler's judgment was proof of the ruler's devotion to God, and the progress of a people was similarly evidence of their goodness in God's eye. As the tyrannical essence of the feudal system established on these foundations became insupportable and was overthrown in one European country after another, any and every connection of religious beliefs or personages with political rule came to be rejected. With the English and French revolutions of the 17th and 18th century respectively, the notion of social class interest thereafter took the place formally occupied by "God's grace", and politics and policy-making increasingly became objects of contestation between opposing social class interests.

During the middle and late European Middle Ages, matters of governance and rule in Islamic societies degenerated into various kinds of private arrangements among the leading, richest families which bore many similarities in appearance to contemporary monarchies in Europe. However, the European feudal system did not exist in these societies as an instrument enslaving the vast majority of people on the land. Individuals and groups seeking power for their family or group either made deals among the elites or tried to mobilize military formations based on clan and tribal loyalties, but with rather less of the redivision of territory and none of the certification of such redivision by a papal-type figure. Under the Ottoman system, privileges assigned by the Sultan could trump either existing arrangements governing land use put in place by administrative arms of the Sultan or private ownership. Rules governing all other aspects of administration and policy were adapted from the *Qu'ran*. Private property enjoyed neither special nor foundational rights. Not surprisingly, eighteenth-century advocates of the Enlightenment, caught up with doctrines elaborated from the 17th century revolution in England, looked at the Ottoman system's of rule with its lack of prioritization of the rights of property as an example of extreme arbitrariness. Not only was there nothing there to learn from, but it was deemed a kind of cancer that should never be allowed into Europe.

In our own time, two simultaneous developments are bringing notions of natural systems of governance back to the fore. On the one hand, systems of governance in Western countries have degenerated into a continual terrorist assault on their own populations in the name of anti-terrorism. At the same time, imperial aggression against Arab and Muslim countries has sparked the ouster of Western imperial power from key countries such as Iran, as well as the reassertion of Islamic-principled systems, such as sharia, as forms of natural governance that can protect the *ummah* from the enslaving consequences of "democracy" dictated by local puppets of Western overlords.

In Muslim societies, there exists a *Qu'ran*-ic tradition that intimately connects the notion of governance with certain ideas about the natural powers of human judgment and how best to preserve the pro-social content of that power in all circumstances. The practical content of the position of "ruler" could just as well be that of a principal investigator on a scientific research project or that of the chief of a commercial corporate entity — but the general principles applicable to matters involving judgment would be the same.

Recently, a United Nations Development Program report (UNDP, 2002) made explicit reference to the role of politics in the lifestyle ordained by the *Qu'ran*. The following quote clearly makes the distinction between natural justice and policy making and Europe-centric sense of ruling.

"Governance" Terminology in the Arabic Language

The derivation of words relating to governance in the Arabic language is fascinating and instructive in that it embodies essentially all the elements of modern good governance. Nearly all the names of modern governance institutions are derived from one three-letter root (حكم), pronounced as *hokm*, which corresponds to "govern". "Govern", in turn, means "to judge" between people. The same root, with minor phonetic modification, leads to our preferred Arabization of "governance" (Al-hokm). It also signifies "deep knowledge of the law and its interpretation", which determines criteria for judgment. Another phonetic modification results in "justice" as well as "wisdom", which pave the way for ethical considerations in judgment (mercy and public welfare, above justice?). Another simple derivation from the root yields "court"--not so in English or French, for example. Yet another simple derivation produces "government", which in the original Arabic usage literally means "lifting injustice"--a meaning that governments, especially in repressive regimes, would be well advised to contemplate. Finally, a "ruler" is a "person appointed to judge among people". The other side of appointment is, naturally, the possibility of dismissal, which implies accountability of the ruler. "Tyranny" is another simple derivation from the root, and since justice is a supreme value in this (linguistic) governance system, tyranny must be grounds for dismissal."

The report (UNDP, 2002) further quotes Imam Ali Bin Abu Talib, the fourth Caliph of Islam and is the first Imam of the Shia sect. The quotation is as below:

"Imam Ali Bin Abi Taleb: On Governance

1. He who has appointed himself an Imam of the people must begin by teaching himself before teaching others, his teaching of others must be first by setting an example rather than with words, for he who begins by teaching and educating himself is more worthy of respect than he who teaches and educates others.
2. Your concern with developing the land should be greater than your concern with collecting taxes, for the latter can only be obtained by developing; whereas he who seeks revenue without development destroys the country and the people.
3. Seek the company of the learned and the wise in search of solving the problems of your country and the righteousness of your people.
4. No good can come in keeping silent as to government or in speaking out of ignorance.
5. The righteous are men of virtue, whose logic is straightforward, whose dress is unostentatious, whose path is modest, whose actions are many and who are undeterred by difficulties.
6. Choose the best among your people to administer justice among them. Choose someone who does not easily give up, who is unruffled by enmities, someone who will not persist in wrongdoing, who will not hesitate to pursue right once he knows it, someone whose heart knows no greed, who will not be satisfied with a minimum of explanation without seeking the maximum of understanding, who will be the most steadfast when doubt is cast, who will be the least impatient in correcting the opponent, the most patient in pursuing the truth, the most stern in meting out judgment; someone who is unaffected by flattery and not swayed by temptation and these are but few.

5. Modeling Natural Phenomena in Multiple Dimensions

In relatively recent history, Einstein was the first scientist to recognize time as the fourth dimension. This recognition explained the transition from mass to energy and made it possible for scientists to consider combining mass- and energy-balance equations. The first use of the fourth dimension, however, was not by a physicist. Hermann Minkowski (1864-1909), a mathematician, used the fourth dimension to 'solve' the space-time continuum. In 1914, Gunnar Nordström (1881-1923), a theoretical physicist, included four dimensions in his gravitational theory, a theory that can be called the first general theory of relativity (Ravndal, 2003). His work on splitting of five-dimensional space into Einstein's and Maxwell's equations in four dimension was the first recognition of more than 4 dimensions in modeling reality. Years later (1921), Kaluza (1885-1954) combined Maxwell's theory of electromagnetism and Einstein's theory of general relativity to develop a unified theory (Kaku and O'Keefe, 1994). This theory was useful for modeling galaxies, solar systems, and spacecraft for outer space travel. In 1926, Oskar Klein (1894-1977) hypothesized that the fourth spatial dimension is curled up in a circle of very small radius, so that a particle moving a short distance along that axis would return to where it began. This is perhaps the first non-linear, albeit circular, approximation of infinitely small distance in multi-dimensional analysis (as opposed to Newton's Δx approaching linearity). The distance a particle can travel before reaching its initial position is said to be the size of the dimension. This extra dimension is a compact set, and the phenomenon of having a space-time with compact dimensions is referred to as compactification. Following this initial development, many researchers worked on the development string theory and its variations (e.g. superstring theory), a detailed description of which is available in a recent textbook (Becker et al., 2007). It is generally recognized that the string theory or its variations do not explain physical phenomena. However, only Abou-Kassem et al. (2007) attributed this short-coming to spurious assumptions of various models and presented them in line with aphenomenal models that are incorrectly promoted as the correct models. One of these assumptions involves the cylindrical shape assumption. The research group of Wessen (e.g. Wessen et al., 2000) eliminated this assumption and was 'successful' in solving space-time-continuum problems (Wessen, 2002). With time, many more dimensions have been recognized (as many as 11, as discussed by Caroll, 2004), but few have ventured into the possibility of modeling infinite dimension. Yet, this is precisely what is needed in solving natural phenomena (Mustafiz et al., 2007).

In groundwater and hydrology applications, Bear (1972) is credited to highlighting the need for including greater dimensions in order for a natural value to be representative. He introduced the concept of Representative Elemental Volume (REV) that would essentially mean that unless a certain dimension is used, the value can fluctuate to such an extent that it is not meaningful. Islam (2002) used the same concept to introduce through numerical modeling that the volume of the sample is of extreme importance, both in physical and numerical modeling. Figure 4 shows his finding.

In the previous section, the shortest distance between two points has been discussed in the context of dimensionality. Such discussion is also of importance in determining the relationship between truth and dimensionality. Decades ago, in a discussion on fractals, a professor asked the allegorical question: How long is the Mississippi River? (Prof. M.R. Islam, personal communication). It was in line with the ground breaking work of Benoît Mandelbrot (1967) who would later be known as the father of fractals. The intention behind this question was to demonstrate that the distance between two points depends on the scale used. It is easily understood by discussing the distance measured by an airplane, by a surveyor that uses a scale of 10s of meters, a surveyor that uses scale of 1 meter, a surveyor that uses his feet, then all the way down to an ant that measure its feet to even measure distances over pebbles. Figure 5 shows the effect of distance reported between two points for various dimensions used. To-date, 12 dimensions have been identified. It is conceivable that this trend will continue and in the future many more dimensions will be revealed. The question one should ask is what is the true distance between the two points are being monitored in the following figure? As the dimension is increased, the distance is increased. Is there a limit to this trend? This was not the question that was answered by Benoît Mandelbrot.

Instead, he introduced the concept of self-similarity and revolutionized 20th century mathematics. However, a close scrutiny of nature would reveal such self similarity is non-existent, making the fractal mathematics aphenomenal.

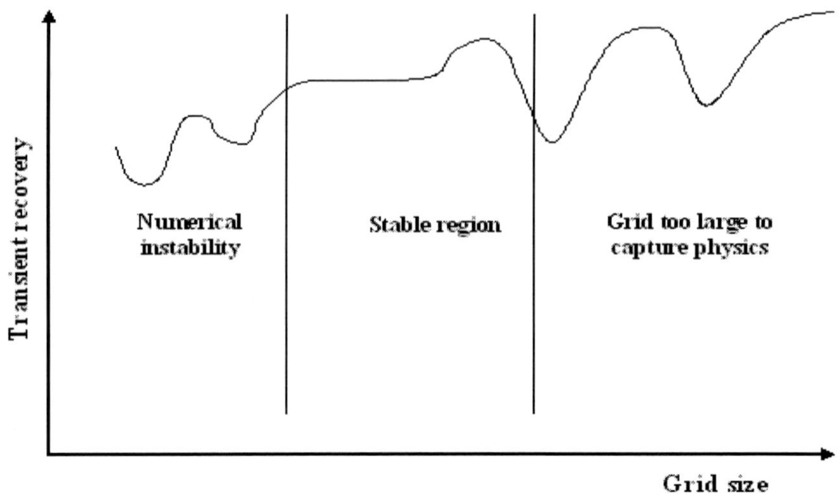

Figure 4. The effect of numerical grid sizes on the prediction of fluid flow behavior (From Islam, 2002).

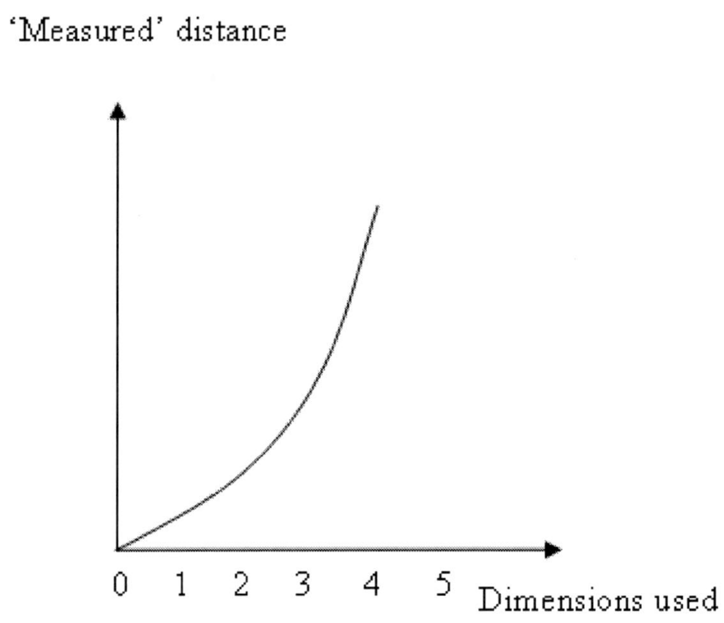

Figure 5. Relationship between dimensions used and measured distance (modified from Abou-Kassem et al., 2007).

Mustafiz et al. (2007) discussed the effect of linearization (which means the dimensions are reduced to 1) on the predicted value. This work has been further extended to multiphase flow, for which more dimensions (philosophically) are added (Islam et al., 2007). As expected, the effect of linearization of a higher order phenomenon gives even greater discrepancy between 1D and 3D results, quite similar to what was observed by Islam and Chilingar (1995) over a decade ago. The important point to be made in this graph is that the inclusion of time as a continuous function, there is no need to include a finite number of dimensions, as has been the practice thus far in Physics and Cosmology. The use of time as a continuous function introduces infinite dimension. This dimension was characterized as 'Knowledge dimension' by Mustafiz (2007).

In engineering applications, the need for modeling with multiple dimensional approach comes from practical need of explaining the existence of chaotic behavior (Ketata et al., 2007b; 2007c). Even though it has been long been recognized that nature is chaotic, very little modeling work has been extended in order to simulate chaos deterministically (Ketata et al., 2006c; 2006d). In the past, Islam and Nandakumar (1986) attempted to model multiple solutions as well as the bifurcation point for porous media problems with mixed convection. This work was later expanded by Islam and Nandakumar (1990) and others. More recently, similar attempt has been made in other disciplines (e.g. Coriell et al., 1998).

One of the first attempts to relate multiple solutions to dimensional analysis was reported by Islam and Chilingar (1995). In the context of microbial movement within a porous medium, a prime candidate for nonlinear dynamics, they showed that by simply changing the dimension of the model, the outcome will change drastically (Figure 6) Such recognition was found to be useful in other analyses involving porous media characterization, which is prone to multiple solutions (Katz et al., 1995).

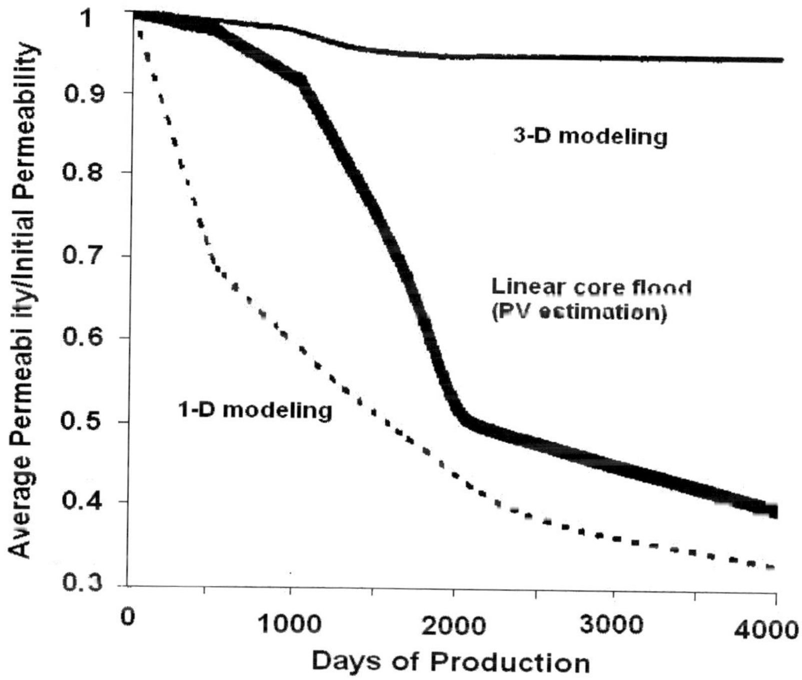

Figure 6. The effect of dimensionality on outcome (from Islam and Chilingar, 1995).

5.1. Transition from Mathematics of Tangibles to Mathematics of Intangibles

The publication of the book, *Principia Mathematica* by Sir Isaac Newton at the end of 17th century has been the most significant development in European-centered civilization. It is also evident that some of the most important assumptions of Newton were just as aphenomenal as the assertion of Thomas Aquinas, except Newton did not talk about Theology (Zatzman and Islam, 2007a). By examining the first assumptions involved, Zatzman and Islam (2007b) were able to characterize Newton's laws as aphenomenal, for three reasons that they 1) remove time-consciousness (nature is truly dynamic); 2) recognize the role of 'external force' (equivalent to 'gods playing with human intention in pre-Aristotle Greek philosophy); and 3) do not include the role of intention. In brief, Newton's law ignore, albeit implicitly, all intangibles from nature science.

Zatzman and Islam (2007b) identified the most significant contribution of Newton in mathematics as the famous definition of the derivative as the limit of a difference quotient involving changes in space or in time as small as anyone might like, but not zero, *viz.*

$$\frac{d}{dt}f(t) = \lim_{\Delta t \to 0} \frac{f(t+\Delta t) - f(t)}{\Delta t} \qquad [1]$$

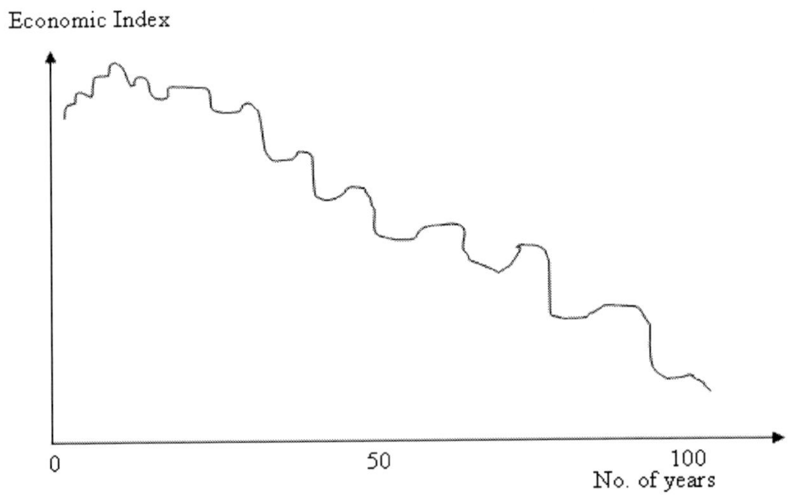

Figure 7. Economic wellbeing is known to fluctuate with time (adapted from Abou-Kassem et al., 2007).

Without regards to further conditions being defined as to when and where differentiation would produce a meaningful result, it was entirely possible to arrive at "derivatives" that would generate values in the range of a function at points of the domain where the function was not defined or did not exist. Indeed: it took another century following Newton's death before mathematicians would work out the conditions – especially the requirements for continuity of the function to be differentiated within the domain of values – in which its derivative (the name given to the ratio-quotient generated by the limit formula) could be applied and yield reliable results. Kline (1972) detailed the problems involving this breakthrough formulation of Newton. However, no one in the past did propose an alternative to this differential formulation, at least not explicitly. The following figure (Figure 7) illustrates this difficulty.

In this figure, economic index (it may be one of many indicators) is plotted as a function of time. In nature, all functions are very similar. They do have local trends as well as global trend (in time). One can imagine how the slope of this graph on a very small time frame would quite arbitrary and how devastating it would be to take that slope to a long-term. One can easily show the trend, emerging from Newton's differential quotient would be diametrically opposite to the real trend. In order to offer a substitute taking derivative at an infinitely small space, Zatzman and Islam (2007b) provided the following procedure. They took a simplified example to the implication of Newton's formula (Figure 8).

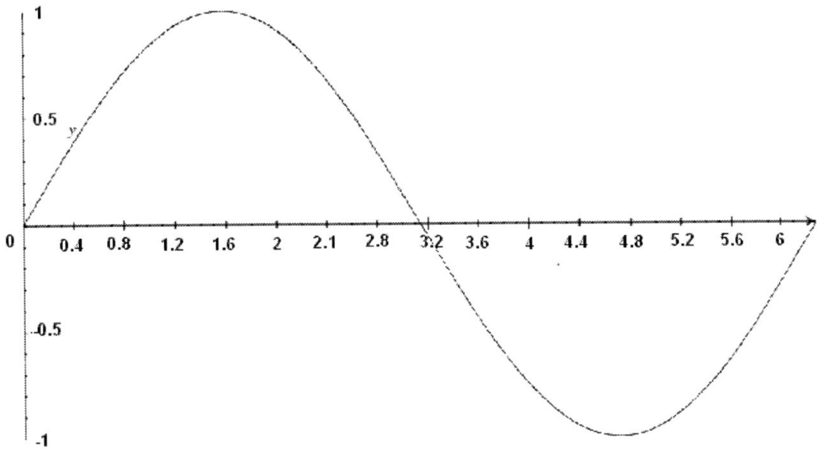

Figure 8. Graphic representation, in Cartesian coordinates, of the classic simple function f(t)=sin t.

It is assumed that the sine function illustrated above represents some physical phenomenon. Using Newton's difference-quotient formula (see Eq. 1), the instantaneous rate of change anywhere along the graph-line of this function, which will be continuous anywhere within the interval $(-\infty, +\infty)$, *i.e.*, $-\infty \leq t \leq +\infty$, can be computed stepwise as follows (Figure 9):

$$\frac{d}{dt}f(t) = \lim_{\Delta t \to 0} \frac{\sin(t+\Delta t) - \sin t}{\Delta t} = \lim_{\Delta t \to 0} \frac{\sin t \cos \Delta t + \sin \Delta t \cos t - \sin t}{\Delta t}$$

This means that, as one moves continuously along the domain t, the *instantaneous rate of change* along the curve represented by the graph for f(t) can be computed by evaluating the cosine of t at that value on the horizontal axis. What is being described is change *within* the function; *the function itself, of course, has not changed.* As this particular function happens to be periodic, it will cycle through the same values as the operational output described by this graph proceeds through subsequent cycles. This makes it quite easy to see that the function itself describes a steady-state condition, which may not be phenomenal, but by taking the derivative one is not imposing any additional feature to the function. In fact, however, even if the function were some polynomial, anything lying on the path of its graph (for instance,

Figure 7) would represent the steady-state operation of that function: steadiness of state is not reducible to some trait peculiar to periodic functions. Contrast this to Newton's method, which implicitly invokes the notion of approximating instantaneous moments of curvature, or infinitely small segments, by means of straight lines.

As Δt approaches 0, cos Δt approaches cos 0, which is 1. Meanwhile, because sin x approaches x for decreasingly small values of x, the term $\frac{\sin\Delta t}{\Delta t}$, also becomes unity. So:

$$\frac{d}{dt}f(t) = \lim_{\Delta t\to 0}\frac{\sin(t+\Delta t)-\sin t}{\Delta t} = \lim_{\Delta t\to 0}\frac{\sin t\cos\Delta t+\sin\Delta t\cos t-\sin t}{\Delta t} =$$

$$\lim_{\Delta t\to 0}\frac{\sin t+\sin\Delta t\cos t-\sin t}{\Delta t} = \lim_{\Delta t\to 0}\frac{\sin\Delta t\cos t}{\Delta t} = \lim_{\Delta t\to 0}\frac{\sin\Delta t}{\Delta t}\cos t = \lim_{\Delta t\to 0}\cos t = \cos t$$

$$\lim_{\Delta t\to 0}\frac{\sin\Delta t\cos t}{\Delta t} = \lim_{\Delta t\to 0}\frac{\sin\Delta t}{\Delta t}\cos t = \lim_{\Delta t\to 0}\cos t = \cos t$$

Figure 9. Generating the first derivative $f'(t)$ for the function $f(t)$ = sin t using Newton's difference-quotient formula.

This linearization is done at an infinitely small scale, so that common observers cannot argue about the validity of the linearization. However, when the same function is integrated, people are surprised that the original function is not recovered except of special cases (see further development below). Zatzman and Islam (2007b) argued that this linearization within a bold and utterly unprecedented Newton's approach, contains a trap for the unwary: going backward or forward in space or in time is a matter of indifference. This 'time travel' has the potential of invoking aphenomenal state (contrary to natural or phenomenal state), later to legitimize it as the 'only solution', falsifying the entire scientific process. If natural reality is to be modeled as it actually unfolds, however, the requisite mathematics has to close the door on, and not permit the possibility of, treating time as reversible. In lieu of invoking linear slope, they suggested

the construction of a difference quotient based on evaluating the limit at some arbitrary common value like 0, a finite constant real value c were used instead. A new derivative may be defined thus:

$$\underset{\Delta t\to c}{f'(t)} = \frac{d_c}{dt}f(t) = \lim_{\Delta t\to c}\frac{\sin(t+\Delta t)-\sin t}{\Delta t} = \lim_{\Delta t\to c}\frac{\sin t\cos\Delta t+\sin\Delta t\cos t-\sin t}{\Delta t}$$

Now, as Δt approaches c, cos Δt approaches cos c, which is anywhere in the interval [-1,+1]. Meanwhile, the term $\frac{\sin\Delta t}{\Delta t}$ may fall anywhere in the interval [$-\frac{\sqrt{3}}{2}$,$+\frac{1}{c}$]. Applying these maxima and minima generates the open interval -(2 sin t + $\frac{1}{c}$ cos t) ≤ $\frac{d_c}{dt}f(t) \leq \frac{1}{c}$ cos t, in which:

- at $t = 0$ ($+2k\pi$), $\frac{d_c}{dt}f(t)$ converges to a single value, *viz.*, $\frac{1}{c}$, which is positive (> 0);
- at $t = \frac{\pi}{6}$: $-(2 + \frac{\sqrt{3}}{2c}) \leq \frac{d_c}{dt}f(t) \leq \frac{\sqrt{3}}{2c}$, which straddles 0;
- at $t = \frac{\pi}{4}$: $-\sqrt{2}(1+\frac{1}{2c}) \leq \frac{d_c}{dt}f(t) \leq \frac{\sqrt{2}}{2c}$, which straddles 0;
- at $t = \frac{\pi}{3}$: $-(\sqrt{3}+\frac{1}{2c}) \leq \frac{d_c}{dt}f(t) \leq \frac{1}{2c}$, which straddles 0;
- at $t = \frac{\pi}{2}$: $-2) \leq \frac{d_c}{dt}f(t) \leq 0$, which is mainly negative (≤ 0);
- at $t = \frac{2\pi}{3}$: $-(\sqrt{3} - \frac{1}{c}/) \leq \frac{d_c}{dt}f(t) \leq -\frac{1}{2c}$, which is entirely negative (<0);
- at $t = \frac{3\pi}{4}$: $-\sqrt{2}(1-\frac{1}{2c}) \leq \frac{d_c}{dt}f(t) \leq -\frac{\sqrt{2}}{2c}$, which is entirely negative (<0) and reduces,
- at $c=1$, to $-\frac{\sqrt{2}}{2}$.

From here, heading towards $t = \pi$, other features emerge:

- At $t = \frac{5\pi}{6}$, $\frac{d_c}{dt}f(t)$ lies somewhere between $-(1 - \frac{\sqrt{3}}{2c})$ and $-\frac{\sqrt{3}}{2c}$, in which:
- for $c = \frac{\sqrt{3}}{2}$, $-1 \leq \underset{\Delta t \to c}{f'(t)} \leq 0$;
- for $\frac{\sqrt{3}}{2} < c < \sqrt{3}$, while for $c > \sqrt{3}$, $\underset{\Delta t \to c}{f'(t)} < 0$; and
- for $c = \sqrt{3}$, $\underset{\Delta t \to c}{f'(t)} = -0.5$;
- At $t = \pi$: $-\frac{1}{c} \leq \frac{d_c}{dt}f(t) \leq \frac{1}{c}$

Figure 10. Generating family of first derivatives, $\{\underset{\Delta t \to c}{f'(t)} = \frac{d_c}{dt}f(t)\}$, for $f(t) = \sin t$ using modified difference-quotient

Two immediate notes must accompany this development. First, even though c is stated to be a positive number greater than zero, it is not arbitrary. The actual value of c should be at least greater than characteristic value of the process. The notion of Representative Elemental Volume (REV), as first promoted by Bear (1972) is useful in determining a reasonable value for c. The other note is, using c that approaches 0 (Newton's approximation) cannot be introduced, even when the characteristic value is very small (e.g. phenomena at nano scale). This is because, zero in European mathematics refers to 'nothing' or 'void', which is aphenomenal (Ketata et al., 2006b). On the other hand, when it comes to 'meaningful' (Ketata et al., 2006b; 2006c) the value of *c* can not only approach zero or even infinity, it can actually be set to those values. Scientifically, 'meaningful' zero and infinity are both intangibles and do qualify to be used in the science of intangibles.

From Figure 10, it becomes clear that we are dealing with multiple, in fact: infinite, solutions. Mustafiz et al. (2007) argued that such should be the case when one is modeling Nature or intangibles. Abou-Kassem et al. (2007) argue that Newton's derivative is also prone to infinite solutions, with the exception that this possibility is not apparent and is not discussed by even the most prolific researchers. This argument is discussed later in this section.

In a different discourse, Abou-Kassem et al. (2006) introduced the so-called Engineering approach in order to solve reservoir engineering problems. Abou-Kassem et al. (2006) pointed out that there is no need to go through this process of expressing in differential equation form (avoiding Newton's ratio-quotient), followed by discretization. In fact, by setting up the algebraic equations directly, one can make the process simple and yet maintain accuracy (Mustafiz et al., 2007).

Figure 11 shows how formulation with the engineering approach ends up with the same linear algebraic equations if the inside steps are avoided. Even though the engineering approach was known for decades (known as the control volume approach), no one identified in the past the advantage of removing in-between steps.

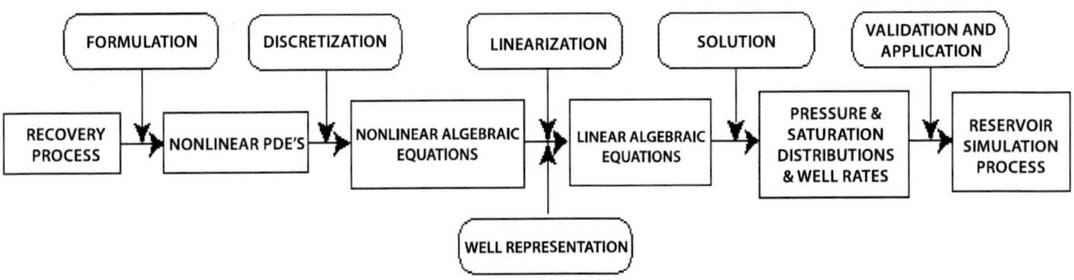

Figure 11. Major steps used to develop reservoir simulators (redrawn from Abou-Kassem et al., 2006).

In analyzing further the role of mathematical manipulation in solving a natural problem, Abou-Kassem et al. (2007) cited the example of the manipulation of the simple function, $y = 5$. Following steps show how this function simple function can take route of knowledge or ignorance, based on the information that is exposed or hidden, respectively.

Step 1: y = 5. This is an algebraic equation that means, y is a constant with a value of 5. This statement is an expression of tangibles, which becomes clear if the assumptions are pointed out. The assumptions are (Islam and Zatzman, 2006): a) *y* has the same dimension as 5 (meaning dimensionless); b) nothing else matters (this one actually is a clarification of the condition a). Therefore, the above function implies that *y* cannot be a function of anything (including space and time). The mere fact that there is nothing in nature that is constant makes the function aphenomenal. However, subsequent manipulations (as in Step 2) make the process even more convoluted.

Step 2: dy/dx = 0. This simple derivation is legitimate in calculus that originates from Newton's ratio of quotient theory. In this, even partial derivative would be allowed with equal legitimacy as nothing states in conventional calculus that such operation is illegitimate. By adding this derivative in x (as in x direction in Cartesian coordinate), a spurious operation is performed. In fact, if Step 1 is true, one can add any dimension to this the differential would still be 0 – statement that is 'technically' true but hides background information that becomes evident in Step 3.

Step 3: If one integrates dy/dx, one obtains, y = C, where C is a constant that can have infinite number of values, depending on which special case it is being solved for. All of a sudden, it is clear that the original function. All of a sudden, it is clear that the original function (y=5) has disappeared.

Step 4: One special case of y=C is, y=5. To get back the original and unique function (as in Step 1), one now is required to have boundary conditions that are no longer attached to the mathematical formulation. If a special case of C=5 is created, similarly one can have y=1, 2, 3, 4, 5, 6,... How does one know which solution will give the 'true' solution. Based on pragmatism (whatever works is true, see Zatzman and Islam (2007a) for further details), one resorts to eliminating solutions that do not meet the immediate need. In this particular case, one but one solution are called 'spurious' solutions, because they failed to match the solution of interest, i.e., y=5.

This simple example shows how imposing Newton's differential and integrating procedure convolutes the entire process, while losing information that would allow anyone to trace back the original algebraic function. One the other hand, if one is looking at an actual phenomenon, then $dy/dx = 0$ could mean that we are the very start of something, or at the very end of something. However, we actually don't know, because, in physical nature, spatial transformations that do not incorporate a time element are like the two-dimensional person who could be sleeping or dancing. If we look at $\partial y/\partial x$, on the other hand, then we have to look also at $\Delta y/\Delta t$, and then we also have to consider the situation where $\Delta y/\Delta x=0$ but $\Delta y/\Delta t$ is non-0. This might very well be a branch-point, a point of bifurcation or, generally speaking, something marking a change from an old state to a new state. Branch-points in physical natural reality clearly imply infinite solutions, since the process could go, literally, anywhere from that branch-point. This approach of locating bifurcation phenomena has eluded previous researchers engaged in modeling chaos (Gleick, 1987).

The engineering implication of the Newtonian approach was highlighted recently by Abou-Kassem et al. (2007). The following steps were highlighted. Note the similarity of these steps with the one shown above regarding a simpler function.

Step 1) Mass balance + Darcy's law → It is an algebraic equation.

Step 2a) Time variable is added to it through Newton's differential quotient → time increment is allowed to approach to 0.

Step 2b) Space variable is added through Newton's differential quotient → space increment is allowed to approach 0.

Step 2c) differential equations emerge, burying assumptions added in 2a and 2b.

Step 3) Differential equation integrated analytically, original solutions will not be recovered. The integrated values will have the possibility of having infinite solutions, depending on the boundary conditions, initial conditions, etc.

Step 4) Three special cases (out of infinite number of solutions possible) are identified as part of the integrated solution of the Differential equation in Step 3. They are:

Step 4a) Steady state, compressible. This is Mass balance and Darcy's law as in Step 1.

Step 4b) Unsteady state, slightly compressible fluid. This is the typical equation that gives rise to the Diffusivity equation. This one would have liquid density as the only equation of state (density as a function of pressure).

Step 4c) Unsteady state, compressible fluid case. This is the typical gas flow equation. This would have gas compressibility (z factor that is a function of pressure and temperature). Here the term 'unsteady' means there is a time derivative. As time extends to infinity, this term drops off, making the model steady state. This alone shows that these 'unsteady state' models are not dynamic models and do not emulate nature that is dynamic at all time.

This development is seen as a great accomplishment of modern mathematics. We started off with one equation, but ended up with three equations that describe other fluids. The pragmatic approach says, if the results are okay, the process must be okay.

Step 5) Because analytical methods are very limited in solving a PDE and require their additional assumptions regarding boundary conditions, numerical techniques were introduced. This one essentially involves discretization through Taylor series approximation and subsequent elimination of higher order terms, arguing that at higher order they are too small (e.g. if Δx is less than 1, Δx^2 is << 1; Δx^3 <<< 1).

Step 6) The removal of higher order terms, as in Step 5, is equivalent to eliminating the additions of space and time variables as in Steps 2a and 2b. We, therefore, recover original mass balance and Darcy's law (Momentum balance equation) as in Step 1. The Engineering approach works with these algebraic equations (Step 1) rather than working with PDE and subsequent discretized forms. This explains why Engineering approach coefficients are the same as the 'mathematical approach' coefficients.

When we go from y=5 to dy/dx = 0, we add the possibility of adding an infinite series. When we then go from dy/dx to y = constant, we are left with infinite solutions, all because of the previous addition. On the other hand, if we do this integration numerically (by invoking Taylor series approximation), we end up having the original solution only if we ignore the left over terms of the infinite series that may or may not be convergent. It is important to see the actual derivation of Taylor series expansion. This really *is* magical --- as in *aphenomenal!*

SUMMARY AND CONCLUSIONS

The approach to knowledge in this paper has challenged many previous assumptions, not only in their contents but especially their "linearity" and implicit assumption of a more or less smooth continuous path of knowledge from the past into the present and the future. That is what this paper has set out to "de-linearize".

From the evidences presented in this paper, does it seem reasonable to maintain that new knowledge accumulates on the basis of using earlier established findings, with the entire body of knowledge then being passed on to later generations? To begin to answer that question, we must first know: on what basis does an individual investigator cognize the existing state of knowledge? If the individual investigator cognizes the existing state of knowledge on the basis of his/her own re-investigation of the bigger picture surrounding his/her field of interest, that is a conscious approach that shows the investigator operating according to conscience.

If, on the other hand one accepts as given the so-called conclusions reached up to now by others, such a consideration could introduce a problem: what were the *pathways* by which those earlier conclusions were reached? An investigator who declines to investigate those pathways is negating conscience.

Enormous blackmail was exercised against scientists aimed at compelling them to negate their conscience. This accounts for Galileo's resorting to defensive maneuvers (claiming he was not out to disprove Scripture) — a tactic of conceding a small Lie in order to be able to continue nailing down a larger more important Truth. Why mix such hypocrisy into such matters? Because *it had worked for other investigators in the past*. What was new in Galileo's case was the decision of the Church of that time not to permit him that private space in which to maneuver, in order to make of him an example with which to threaten less-talented researchers coming after him. The worst we can say against Galileo after that point is that, once an investigator (in order to get along in life) goes along with this, s/he destroys some part of her/his usefulness as an investigator.

Prior to the conflict with Church authority occasioned by the matter of Galileo's teachings, conflicts had already broken out about the relative weight of current knowledge discovered experimentally and what appeared to be the meaning and direction of earlier and most ancient knowledge. The Church's authority was vested in the unchanging character of the key conclusions of earlier investigators. This authority was never vested in the integrity and depth of probing by earlier investigators and investigations into all the various pathways and possibilities.

In medieval Europe, especially during in the two centuries preceding Galileo, the resort to experimental methods did not arise on the basis of rejecting or breaking with Church authority. Rather it was justified instead by a Christian-theological argument, along the following lines:

- knowledge of God is what makes humans right-thinking and good and capable of having their souls saved in Eternity;
- this knowledge should be accessible wherever humans live and work;
- the means should be at hand for any right-thinking individual to verify the Truth or eliminate the Error in their knowledge.

These "means" were then formulated as the starting point of what would become "scientific method".

Combining the matter of the absence of any sovereign authority for the scientific investigator's conscience, and the Christian-theological justification for certain methods of investigation that might not appear to have been provided by any previously-existing authority: what was the result? Even with scientific methods, such as experiments, the conscience of an investigator who separated his/her responsibility for the Truth from the claims of Church authority — but without opposing or rebelling against that authority — could not ensure that his/her investigation could or would increase knowledge of the truth.

As long as knowledge was being published and disseminated by monks copying scrolls of works compiled hundreds of years and even millennia earlier, this did not surface as a threat to Church authority. However, with the establishment in Europe of printing by means of movable type — a technology brought from China — all this would change. The basis on which publication of new findings and research would now take place became a burning issue. While every case has its own features, certain elements run commonly throughout. Publication served to advance knowledge in rapid and great strides *IF and ONLY IF* authority was vested in the integrity and depth of probing by earlier investigators and investigations into all the various pathways and possibilities. Otherwise, the societal necessity and usefulness for publication would be readily and easily subverted. The best known form of that subversion is what is sometimes called the "the Culture of Patents".

Today, under the conditions of what is known as the "Information Age", this takes increasingly the form of what is known as "monopoly right"". Under present conditions and for the foreseeable future, *IF and ONLY IF* we put first the matter of the *actual conduct* of scientific investigations, as well as the 'politics' attaching to that conduct (meaning: the *ways and means* by which new results are enabled to build humanity's store of knowledge), *THEN and ONLY THEN* can we hope to reconstruct the actual line of development. With the knowledge of this actual line of development, for any given case, we can then proceed to critique, isolate and eliminate the thinking and underlying ideological outlook that keep scientific work and its contents travelling down the wrong path on some given problem or question.

ACKNOWLEDGMENTS

Funding for provided by the Atlantic Innovation Fund (AIF) of ACOA as well as NSERC is gratefully acknowledged. Mehedi Al-Mahrouqi of Sultan Qaboos University (Dept. of Petroleum and Chemical Engineering) was instrumental in providing us with guidelines for understanding Arabic as it was understood before renaissance took place in Europe and subsequently took over world education system.

REFERENCES

Abou-Kassem, J.H., Farouq Ali, S.M., and Islam, M.R., (2006) *Petroleum Reservoir Simulation: A Basic Approach*, Gulf Publishing Company, Houston, TX, USA, 480 pp.

Abou-Kassem, J.H., Zatzman, G.M, and Islam, M.R., (2007) Newtonian Mechanism and the Non-Linear Chaos of Nature: Preliminary Investigations in the Mathematics of Intangibles, *J. Nature Sci. and Sust. Tech.*, in press.

Armstrong, K., (1994) *A History of God, Ballantine Books*, Random House, 496 p.

Bajcsy, M., Zibrov, A.S., and Lukin, M.D., (2003) Stationary Pulses of Light in an Atomic Medium, *Nature*, vol. 426, Dec. 11, pp. 638–641.

Bear, J., (1972) *Dynamics of Fluids in Porous Media*, American Elsevier Publishing Co., New York, 761 pp.

Becker, K., Becker, M. and Schwarz, J.H., (2007) *String Theory and M-Theory: A Modern Introduction*, Cambridge University Press, New York, 756 pp.

Caroll, S., (2004) Why 3-Dimensions in Space Aren't Just Enough? Invited paper presented at the *Annual Meeting of the Philosophy of Science Association*, Austin, Texas

Chhetri, A.B. and Zatzman, G.M, (2008) Global Warming – A Technical Note, *Journal of Nature Science and Sustainable Technology*, vol. 2, no. 1.

Clayton, M.A. and Moffat, J.W., (1999) Dynamical Mechanism for Varying Light Velocity as a Solution to Cosmological Problems, *Physics Letters* B, vol. 460, No.3-4, pp. 263-270.

CNN, (2007) *Fisher Price Recalls 1M Toys,* Aug. 1, CNN online news.

Coriell, S.R., McFadden, G.B., Sekerka, R.F. and Boettinger W.J., (1998) Multiple Similarity Solutions for Solidification and Melting, *Journal of Crystal Growth*, vol. 191, pp. 573-585.

Cornell, E.A., (2005) What was God Thinking? *Science can't Tell, Time*, vol.166 (20), November 14.

Crugg, G.M. and Newman, D.J., (2001) Medicinals for the Millenia: The Historic Record, *Annals of the New York Academy of Sciences*, vol. 953, pp. 3-25.

Dorogovtsev, S. N., Mendes, J. F. F., and Samukhin, A. N., (2000) Structure of Growing Networks with Preferential Linking, *Phys. Rev. Lett.*, vol.85, No. 21, pp.4633-4636.

Editorial, (2007) Borlaug's Revolution, *The Wall Street Journal*, July 17, p. A16.

Fleischhauer, M., (2007) Indistinguishable from afar, *Nature,* vol. 445, pp. 605-606.

Freeth, T., Bitsakis, Y., Moussas, X., Seiradakis, J.H., Tselikas, A., Mangou, H., Zafeiropoulou, M., Hadland, R., Bate, D., Ramsey, A., Allen, M., Crawley, A., Hockley, P., Malzbender, T., Gelb, D., Ambrisco, W. and Edmunds, M.G., (2006) Decoding the Ancient Greek Astronomical Calculator Known as the Antikythera Mechanism, *Nature*, vol. 444, pp. 587-591.

Ginsberg, N.S., Garner, S.R., Hau, L.V., (2007) Coherent control of optical information with matter wave dynamics, *Nature,* vol. 445, Feb. 8, pp. 623-626.

Gleick, J., (1987) *Chaos – Making a New Science*, Penguin Books, NY, 352 pp.

Hadiths of The Prophet, (2007) Collected and Translated by Sahih Bukhari [online] Available:(http://www.usc.edu/dept/MSA/fundamentals/hadithsunnah/bukhari/001.sbt.html) [April 23, 2007]

Hau, L.V., Harris, S.E., Dutton, Z., and Behroozi, C.H., (1999) Light Speed Reduction to 17 Metres Per Second in an Ultra Cold Atomic Gas, Nature, vol. 397, pp. 594–598.

Hawking, S., (1988) *A Brief History of Time*, Bantam books, London, UK, 211 pp.

Islam, M.R. and Nandakumar, K., (1986) Multiple Solution for Buoyancy-Induced Flow in Saturated Porous Media for Large Peclet Numbers, Trans. *ASME Journal of Heat Transfer,* vol.108, No. 4, pp. 866-871.

Islam, M.R. and Nandakumar, K., (1990) Transient Convection in Saturated Porous Layers with Internal Heat Sources, *Int. J. Heat and Mass Transfer*, vol. 33, No.1, pp. 151-161.

Islam, M.R. and Chilingar, G.V., (1995) Mathematical Modeling of Three-Dimensional Microbial Transport in Porous Media, *International Journal of Science and Technology*, vol. 2, No. 2, pp. 55-64.

Islam, M.R., (2002) Emerging Technologies in Subsurface Monitoring of Petroleum Reservoirs, *Petroleum Res. J.*, vol. 13, pp. 33-46.

Islam, M.R., (2005 b) Unraveling The Mysteries of Chaos and Change: the Knowledge-Based Technology Development, Invited speech, Joint Session of Fifth International Conference on Composite Science and Technology and First International Conference on Modeling, Simulation and Applied Optimization, Sharjah, UAE, February.

Islam, M.R., 2008, Without the Science of Intangibles, the Earth is Still Flat, *Journal of Physics*, Non-linear dynamics series, 13 pp.

Islam, M.R. and Zatzman, G.M., (2006) Emulating Nature in the Information Age, 57th *Annual Session of Indian Chemical Engineering Congress*, Chemcon-2006, Ankelshwar, India, Dec. 27-30.

Islam, M.R., Mousavizadegan, H., Mustafiz, S., and Belhaj, H., (2007) *A Handbook of Knowledge-Based Reservoir Simulation*, Gulf Publishing Co., Houston, TX, to be published in Jan., 2008.

Joseph, G.G., (2000) *The Crest of the Peacock: Non-European Roots of Mathematics*, Princeton University Press, NJ, USA, 416 pp.

Kaku, M. and O'Keefe, R., (1994) *Hyperspace: a Scientific Odyssey through Parallel Universes, Time Warps, and the Tenth Dimension*, Oxford University Press, New York.

Katz, S., Chilingar, G.V., and Islam, M.R., (1995) Estimation of Reservoir Porosity and Relative Volume of Pore Filling Components Using Multiple Sources of Geophysical Data, *J. Pet. Sci. Eng*., vol. 13, No. 2, pp. 103-112.

Ketata, C., Satish, M.G. and Islam, M.R., (2006a) The Meaningful Zero, *Proceedings of the 36th International Conference on Computers and Industrial Engineering* (ICCIE), Taipei, Taiwan, R.O.C., June 20-23.

Ketata, C., Satish, M.G., and Islam, M.R., (2006b) The Meaningful Infinity, Proc. Conference on Computational Intelligence for Modelling, Control and Automation (CIMCA), *IEEE Conference, Sydney,* Australia, November 28 – December 1.

Ketata, C., Satish, M.G., and Islam, M.R., (2006c) Multiple-Solution Nature of Chaos Number-Oriented Equations, *Proc. Conference on Computational Intelligence for Modelling, Control and Automation (CIMCA), IEEE Conference*, Sydney, Australia, November 28 – December 1.

Ketata, C., Satish, M.G., and Islam, M.R., (2006d) Chaos Numbers, *Proc. Conference on Computational Intelligence for Modelling, Control and Automation (CIMCA), IEEE Conference*, Sydney, Australia, November 28 – December 1.

Ketata, C., Satish, M.G., and Islam, M.R., (2007a) Abacus-Based Calculators, *Proc. International Conference of Computational and Experimental Engineering and Sciences* (ICCES-2007), Miami, Florida, Jan.

Ketata, C., Satish, M.G., and Islam, M.R., (2007b) Dynamic Numbers for Chaotic Nature, *Proc. International Conference of Computational and Experimental Engineering and Sciences (ICCES-2007),* Miami, Florida, Jan.

Ketata, C., Satish, M.G. and Islam, M.R., (2007c) Chaos Laws of Motion, *Proc. International Conference of Computational and Experimental Engineering and Sciences* (ICCES-2007), Miami, Florida, Jan.

Khan, M.I., (2006) Towards Sustainability in Offshore Oil and Gas Operations, Ph.D. Dissertation, Faculty of Engineering, Dalhousie University, Canada, 440 pp.

Khan, M.I. and Islam, M.R., (2007a) *Handbook of Petroleum Engineering: Sustainable Operations*, Gulf Publishing Co., Houston, USA, 500 pp.

Khan, M.I. and Islam, M.R., (2007b) *True Sustainability in Technological Development and Natural Resources Management,* Nova Science Publishers, New York, USA, 381 pp.

Khan, M.M., Zatzman, G.M., Ketata, C., and Islam, M.R., (2007) Development of a Sustainability Model and Detailed Analyses with the Science of Intangibles, paper submitted for publication in *Int. J. Mat. Prod. Tech*., submitted.

Kline, M., (1972) *Mathematical Thought from Ancient to Modern Times*, Oxford University Press, New York, USA, 1238 pp.

Kvitco, V. (2007) Mathematical Disproof of Lorentz' Mathematics and Einstein's Relativity Theory, *Physics Letters A*. in press.

Liu, C., Dutton, Z., Behroozi, C.H., and Hau, L.V., (2001) Observation of Coherent Optical Information Storage in an Atomic Medium Using Halted Light Pulses, *Nature*, vol. 409, pp. 490–493.

Lu, P.J. and Steinhardt, P.J., (2007) Decagonal and Quasicrystalline Tilings in Medieval Islamic Architecture, Science (Washington, DC, United States), vol. 315, pp.1106-1110.

Mandelbrot, B., (1967) How Long is the Coast of Britain? Statistical Self-Similarity and Fractional Dimension. *Science,* New Series, Vol. 156, No. 3775, May 5, pp. 636-638.

Mittelstaedt, M., (2006) *Toxic Shock, The Globe and Mail*, May 5, A4.

Mittelstaedt, M., (2007) '*Inherently Toxic' Chemical Faces Its Future*, The Globe and Mail, April 7, pp. A10.

Mustafiz, S., (2007) Modeling Certain Phenomena during Fluid Movement in Petroleum Reservoirs, Ph.D. Dissertation, *Dept. of Civil and Resource Engineering*, Dalhousie University.

Mustafiz, S., Mousavizadegan, H., and Islam, M.R., (2007) The Knowledge Dimension: Towards Understanding the Mathematics of Intangibles, *J. Nat. Sci. Sust. Tech.*, vol. 1, No. 3.

Lu, P.J. and Steinhardt, P.J., (2007) Decagonal and Quasicrystalline Tilings in Medieval Islamic Architecture, *Science* (Washington, DC, United States), vol. 315, pp.1106-1110.

Pickover, C.A., (2004) *The Paradox of God and the Science of Omniscience*, Palgrave MacMillan, N.Y., 288 pp.

Ravndal, F., (2003) Quintessence from Extra Dimensions, The Gunnar Nordstrom *Symposium on Theoretical Physics*, Helsinki, Aug. 27-30.

Saunders, D., (2007) *Dream farms turning into nightmares*. Globe and Mail, July 7, A12.

Shapiro, R., Zatzman, G.M. and Mohiuddin, Y., (2007) Towards Understanding Disinformation. *J. Nature Science and Sustainable Technology*, vol. 1, no. 3.

Schewe, P.F. and Stein, B., (1999) Light Has Been Slowed To A Speed of 17 m/s, American Institute of Physics, *Bulletin of Physics News*, no. 415, February 18.

Tung, S., Schweikhard, V., and Cornell, E.A., (2006) Observation of Vortex Pinning on Bose-Einstein Condensates, *Phys. Rev. Lett.*, vol. 97, No. 24, pp. 240402.

UNDP, (2002) *The Arab Human Development Report*, UNDP Regional Bureau for Arab States (RBAS), 178 pp.

Vaziri, H., Zatzman, G.M. and Islam, M.R., (2007) "Preface", Handbook of Sustainable Petroleum Engineering Operations, Gulf Publishing Co., Houston, 452 pp.

Website 1, (2007) Wikipedia site: http://en.wikipedia.org/wiki/Speed_of_light, last accessed August 5, 2007.

Website 2, (2007) Wikipedia site: http://en.wikipedia.org/wiki/Satyendranath_Bose, last accessed August 5, 2007.

Website 3, (2008), http://ca.news.yahoo.com/s/capress/080702/science/science_solar_System, accessed July 7, 2008.

Wessen, P.S., Liu, H., and Scahra, S.S., (2000) The Big Bang as Higher-Dimensional Shock Wave, Astron. *Astrophys.*, vol. 358, pp.425-427.

Wessen, P.S., (2002) Five Dimensional Realtivity and Two Times, *Phys. Lett.,* B538, pp.159-163.

Zatzman, G.M., (2007) Paradigm Shift Needs a Shift in Standards, from Artificial to Natural – A Technical Note, *J. Nature Science and Sustainable Technology*, vol. 1, no. 4.

Zatzman, G.M. and Islam, M.R., (2007a) Truth, Consequences and Intentions: The Study of Natural and Anti-Natural Starting Points and Their Implications, *J. Nature Science and Sustainable Technology*, vol. 1, No. 2, pp. 1-38.

Zatzman, G.M. and Islam, M.R., (2007b) *Economics of Intangible*, Nova Science Publishers, New York, p.393.

In: Focus on Information, Intelligence and Knowledge ISBN 978-1-61209-979-8
Editor: Rafiq Islam 2011 Nova Science Publishers, Inc.

Chapter 2

EVALUATION AID-TOOL FOR IMPACTS ASSESSMENT OF URBAN MOBILITY

Hichem Omrani [a, c]***, Anjali Awasthi*** [b]***,***
Luminita Ion-Boussier [c] ***and Philippe Trigano*** [a]

[a]UMR CNRS 6599 HEUDIASYC, Université de Technologie de Compiègne, U.T.C, BP 20529 - 60205 Compiègne Cedex, France
[b]Centre for Operations Excellence, University of British Columbia, 209-2053 Main Mall, Vancouver, CANADA – V6T1Z2
[c]Ecole d'Ingénieur en Génie des Systèmes Industriels, EIGSI, 26 rue de Vaux de Foletier, 17041 La Rochelle cedex1, France

ABSTRACT

For environment impacts assessment, several methods have been proposed in literature, such as Life cycle analysis, Multi criteria analysis and Cost benefit analysis. Recently, other methods combining Multi Criteria Decision Analysis (MCDA) and Artificial Intelligence (AI) have been explored to develop enhanced methodologies for knowledge based decision support. In this paper, a hybrid approach is based on multi criteria analysis and evidence theory is presented for impacts assessment of urban mobility. Various criteria are evaluated frequently using complex data obtained from several information sources. A global evaluation on criteria set is done by the Choquet integral aggregation taking into account the interaction between criteria. Finally, each evaluation process is illustrated by using cases studies.

Keywords: Evidential Reasoning, Multi Criteria Analysis, Evidence Theory, Environmental Impacts Assessment, Urban mobility.

1. Introduction

An increase in the urban mobility and its impacts required serious attention in the last few years because of the significant relationships between the characteristics of an urban traffic system and the quality of the urban environment. In many urban areas, rapidly increasing traffic volumes, lower vehicle operating speeds, type of fuel used and the aging of vehicles have resulted in excessive gas emissions, generation of noise pollution and deterioration in living conditions of citizens. In response, the decision makers associated with traffic engineers and urban specialists, elaborated various projects for decreasing negative consequences of urban traffic (massive development of the public transport, introduction of fleets with clean vehicles, parking policy in the downtowns, etc). At the same time, any new project is prone to financial risks and its implementation must not affect the quality of the accessibility for people and goods. In these contexts, the local authorities and transportation community are strongly interested in methods that can evaluate impacts of any projects on urban traffic.

This work included here was carried in SUCCESS project (Smaller Urban Communities in CIVITAS for Environmentally Sustainable Solutions) of the "Urban Community" of La Rochelle (France) supported by Preston (UK) and Ploiesti (Romania) in response to the European Commission's call for proposals of specific program for research, technological development and demonstration.

The aim of this project is to adapt innovative methods and techniques to estimate the environmental impacts of transport activities and components. The results are a methodology and evaluation aid tool for local authorities/managers to compare environment friendly actions and their consequences (estimated value of the various environmental criteria), pointing out the relations between them (synergies or reverse between variables). Particular attention is the flexibility of this tool which gives the project manager possibility to integrate particularities of his city.

This paper is structured as follows. Section 2 gives a short overview about the most applied approaches to evaluate the environmental and socio-economic criteria. Section 3 underlines main problems to adapt one of them in our context for rigorous evaluation of efficiency of a transport project. In the fourth section we present the proposed approach. Finally, the conclusion underlines the benefits of the proposed approach and a description of the potential for further research in this area.

2. Overview

In literature, the most commonly applied approaches to evaluate the environmental and socio-economic criteria are:

1. Life Cycle Analysis (LCA): Originally developed for industrial processes, the use of LCA (Goedkoop, 2000) (Guine, 2002) to evaluate the environmental impact of transport system is growing. Its central concept is to combine, in a small number of criteria, the polluting emissions and resources used during the life course of a product. This method has been the subject of considerable efforts to standardise the

impacts assessment and the results interpretation. However, it does not take into consideration for example social aspects, so this method does not relate to our application.

2. Cost-Benefit Analysis (CBA) and Cost-Effectiveness Analysis (CEA): The cost-benefit analysis is based on taking into account the monetary equivalent of all the positive and negative impacts of the company project. When the advantages of a project are not quantifiable in a monetary way or when the realization degree of the result to reach is given, we speak about cost effectiveness analysis. This type of study aims to minimize the costs necessary to the achievement of a given objective and not for maximizing the advantages. With CBA and CEA approaches (Kunreuther, 2003), it is extremely difficult to estimate directly external and social costs (e.g. air pollution, noise pollution, accidents, congestions and fuel costs). For example, air pollution costs, related to transportation, is difficult to estimate because several other economy sectors also generate pollution. In this case the environmental costs should be apportioned according to each sector emission.
3. Environmental impact assessment (EIA): The aim of this method is to assess the environmental impacts of a new localised pollution source, such as an industry or highway, and its surroundings (Bond, 2001) (Fischer, 2002) (Jay, 2001) (Wood and Ahmad, 2002) (Wood, 2002). Applied to transport, EIA has been used to study the environmental impact of some practices. This method is standardized and consists of several stages from the recording of the emissions to decision-making by the authorities. The three aspects of sustainability namely environmental, economic and social are taken into account, as one tries to evaluate the impact of a new activity on the environment, the population and the attractiveness of the neighbourhood of the site.
4. Multi-Criteria Decision Analysis (MCDA) methods: A wide set of MCDA methods exist: *Multi-Attribute Value Function Theory (MAVT), Multi Attribute Utility Function Theory (MAUT), Analytic Hierarchy Process (AHP)* and *ELECTRE methods,* etc. Decision analysis provides both tools and a framework for the integration of tools from other disciplines to create an overall strategy for decision support. Several applications have been developed especially for management environment (Beinat, 2001) (Chang and Chen, 1991) (Chen, 2003) (Dempster, 1968), etc. by selecting one alternative among several. Generally, in a multi-criteria problem there is no solution optimizing all the criteria at the same time therefore compromise solutions must be found. However, we should note that when different conflicting evaluation criteria are taken into consideration, a multi-criteria problem is mathematically ill defined (Little and Rubin, 2002) (Simos, 1990). Decision analysis looks at the paradigm in which an individual decision maker (or decision group) contemplates a choice of action in an uncertain environment. In MCDA methods, the selection is facilitated by evaluating each choice on the criteria set. The criteria must be measurable; even if the measurement is performed only at the nominal scale (yes/no; present/absent) and their outcomes must be measured for every decision alternative. Criterion outcomes provide the basis for choices comparison and consequently facilitate the selection of one satisfactory choice. Criterion outcomes of decision alternatives are collected in a table (called decision matrix or decision table) comprising of a set of columns and rows. MCDA methods (Delgado, 1993)

(Delgado, 1992) (Herrera and Verdegay, 1993) (Little and Rubin, 2002) (Yager, 1992) (Yager, 1993), especially AHP and ELECTRE, are generally used for ranking several alternatives under criteria set, by using one expert or several experts and by using a numeric same scale or linguistic terms.

Recently, others methods combining MCDA and AI have been explored to develop enhanced methodologies for knowledge based decision support system. By combining MCDA with fuzzy logic theory (Zadeh, 1965) (Zadeh, 1986), new methods have been developed like, the most useful method, Fuzzy AHP (Simos, 1990), Fuzzy comprehensive assessment (Lu, 1999) (Yang, 1998). In addition, some approaches using the framework of evidence theory with MCDA methods have been proposed. In fact, Beynon (Beynon, 2002) (Beynon et al., 2000) proposed AHP with Dempster-Shafer (DS) Theory (Dempster, 1968) (Smets and Kennes, 1994). Also, an evidential reasoning (ER) approach has been developed for dealing with a complex decision problems in management (Simos, 1990) (Xu, 2005). This type of approach, using an evidential reasoning, was already used in several applications (Denoeux and Smets, 2006), but it was not extensively applied to achieve an evaluation of impacts in transport field. In our knowledge, only one research team has recently published papers for environmental analysis which use the evidence theory and multi-criteria analysis (Xu, 2005) (Yang, 2005). But in this method, only expert opinions are taken into account. The final solution could be to couple or to adapt different approaches.

3. Efficiency of a Transport Project

3.1. Main Problems for a Rigorous Evaluation

The approaches described before are generally inapt to apprehend the complexity of the systems with their environmental, social, economic and cognitive dimensions. The difficulty is to treat both inaccurate and subjective judgments (satisfaction degree of the users, spatial accessibility, security, quality of service etc.) and imprecise numerical data (cost, traffic and congestion levels, noise etc.). Furthermore, uncertainty and vagueness common in human knowledge cannot be properly catered for in such aforementioned methods.

There exist several tools for traffic analysis related to urban mobility based on these methods (i.e. (Scemama, 2004)). Generally, these tools are mono-users, study only one category of impacts like traffic system or environment and they don't deal the overall complexity of the evaluation process related to urban mobility. Users frequently accuse a rigid tool, such a black box where particularities of his city, of evaluation process can not be easily integrated.

A project manager using an evaluation aid-tool frequently bases his analysis on two steps:

- An initial evaluation, the phase which explains what is going to be measured and the actual impacts in a "do-nothing" scenario.

- An ex-post evaluation, which provides the results of the demonstration project and its impacts in close comparison with the estimated scenario to show if the project has met its objectives.

The classical process is to evaluate the criteria evolution in order to quantify impacts expected to occur as a result of the implementation of the project. In this case, the evolution of each selected criterion of the project is analysed in order to compare these two phases.

For the ex-post evaluation, a project manager can easily establish a strategy to collect data about each criterion after selection of a corresponding information source. But for the first phase, he frequently uses existing data which could be collected at different period before project implementation by different organisations having punctual objectives. In this case, several problems occur for each criterion evaluation because several information sources (i.e. experts, sensors, models, questionnaire etc.) could be available with various reliability degrees (as shown in Figure 1). In this case collected data are heterogeneous, quantitative and qualitative data (see Table 1). In addition, information sources may be different for the initial and ex-post evaluations.

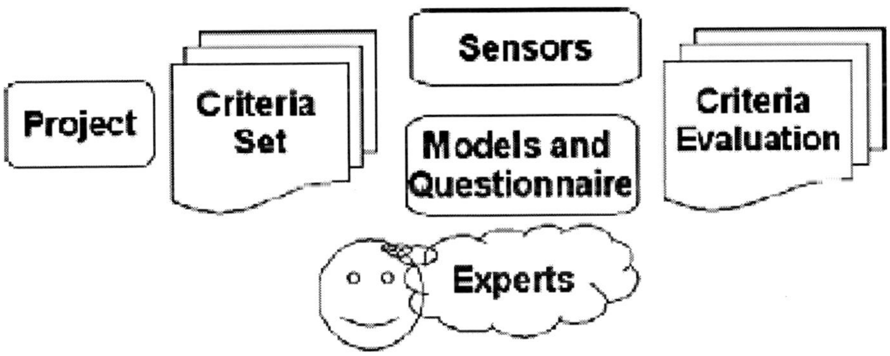

Figure 1. Information sources and data variety.

Table 1. Criteria type

Criteria	Criteria measurement (Source)	Criteria type
Pollution: NO2	measured (M; S)	Qt
Society: Security	evaluated (E;Q)	Qt
Cost	estimated (M;E)	Qt or Ql
Traffic: Congestion	measured (M;C;E)	Qt or Ql

M: Model, S: Sensor, E: Experts opinions, Q: Questionnaire, C: Counting, Qt: quantitative, Ql: qualitative.

So a first major problem is to find a common framework in order to aggregate all data for estimating the evolution of each criterion. The problem becomes more complicated when a project manager wants to study the global efficiency of the project because few criteria can be beneficial, the others can be detrimental. In addition, the criteria have different weights and the interactions between criteria can not be neglected.

3.2. Common Framework

The evidence theory (Dempster-Shafer (DS) (Smets, 1992) (Smets and Kennes, 1994) is a framework to represent quantified uncertainties based on belief functions, unrelated to any underlying probability model. We will first present an overview of principal concepts of belief theory in order to show its utility in our application.

Let Ω={A,B,C} be the domain of variables which represents a finite set of mutually exclusive and exhaustive hypotheses, called the frame of discernment. This frame varies according the application domain. In our application it represents evaluation levels of criteria (i.e. very important, important etc.).

The central element of the belief theory is the basic belief assignment (bba), denoted m from 2^{Ω} to [0; 1] verifying:

$$\begin{cases} m: 2^{\Omega} \rightarrow [0,1] \\ \sum_{A \in 2^{\Omega}} m(A) = 1 \end{cases} \tag{1}$$

with $A \subset \Omega$, $m(A)$ represents the belief that one is willing to commit exactly to A, given a certain piece of evidence. The subsets A of Ω such as $m(A) > 0$ are called the focal elements of m.

No we show how the bba is established for each information source (i.e., questionnaire, experimental data, etc). When several information sources exist for the same criterion, data fusion is done using Dempster-Shafer Theory (DST). Let us denote two masses distributions m1 and m2 from two sources. Then, mass obtained by orthogonal rule is given by (Eq. 2):

$$(m_1 \oplus m_2)(A) = \sum_{B,C \subseteq \Omega, B \cap C = A} m_1(B) \times m_2(C) \tag{2}$$

where: $m_1 \oplus m_2$ is the bba representing the combined impact of the two pieces of evidence and Ω is the frame of discernment.

Conflict between sources can be important and better manner to explain it is the reliability of information sources. Before fusion process, the technique of discounting allows us to take in consideration the reliability of the information source. The bba is discounted into a new less information bba denoted by mα (Smets, 1992):

$$\begin{cases} m^{\alpha}(A) = \alpha \times m(A) \quad for\ A \subset \Omega \\ m^{\alpha}(\Omega) = (1-\alpha) + \alpha \times m(\Omega) \end{cases} \tag{3}$$

Generally, the coefficient α is arbitrary defined in a range scale to 0 from 1. We will show in the next paragraph how project manager will handle it, in agreement with project particularities.

After fusion process masses corresponding to singletons, indecision or ignorance are obtained for each criterion evaluated by several sources. Beliefs can be transformed into a probability measure denoted by BetP (Smets, 1992) (Smets and Kennes, 1994). The function building this probability is called the "pignistic" transformation and it is defined as Eq. (4).

$$BetP(A) = \sum_{B \subseteq \Omega} \frac{|A \cap B|}{|B|} \times \frac{m(B)}{1 - m(\varnothing)}, \quad \forall A \subseteq \Omega \tag{4}$$

where |B| denotes the number of elements in the set B and m(Ø) is the mass allowed to the conflict between sources. We will present the application of "pignistic" probabilities distribution in the section 4.

3.3. Mass Assignment and Source Reliability

When a criterion is evaluated with a single information source for both phases of evaluation (initial and ex-post evaluations), the tool user can directly compare the results of the evolution of the criterion (i.e., experimental measurements of noise level). But in order to aggregate the results for several criteria in the global evaluation process of a project, a common framework is necessary for selected criteria set and their collected data. This operation becomes essential if information sources differ from two phases (i.e., experimental measurements for noise level for the first phase and questionnaire based on acoustic level perception for the second one). In addition, if several information sources exist for an evaluation phase, data fusion, under the framework of belief theory, must be done by using the orthogonal sum of Dempster-Schafer (see Eq. 2).

a) Single Information Source

The initialization methods for the mass function assignment are various and depend on the field application like classification, identification, detection, decision making etc. We present here, as below, a mean for calculating mass structure for each information source (Omrani et al., 2007).

Evaluators: They give their opinions according to a range scale from 0 to 10 (from the worst to the best). Under this scale, fuzzy variables are defined (i.e. five terms, see Figure 2, which indicate the perceived measurement related to each criterion, like acceptance, accessibility etc.). For the construction of the mass function (noted by *m* in Eq. (1)), we use the theory developed in (Radon, 2004). The results of mass assignment are given according to Figure 3.

The formula for mass assignment from evaluator opinions is shown below. Let *f* be a fuzzy subset of a finite universe Ω such that the range of the membership function (μf) of *f*, is {y1,y2,…,yn} where yi > yi+1. Then the mass assignment of *f*, denoted mf, is a probability distribution on 2Ω satisfying:

$$\begin{cases} m_f(F_i) = y_i - y_{i+1} & for\ i = 1,\ldots,n-1 \\ m_f(F_n) = y_n & and \\ m_f(\varnothing) = 1 - y_1 \end{cases} \tag{5}$$

where $F_i = \{x \in \Omega; \mu_f(x) \geq y_i\}$ for $i = 1, \ldots, n$ and $\{F_i; i = 1, \ldots, n\}$ as referred to as the focal element of m_f.

For evaluation and fusion steps, we begin by defining the frame of discernment. Let Ω be the frame of discernment (which corresponds to the evaluation levels) where Ω={Ω1, Ω2, Ω3, Ω4, Ω5}={A.D, P.D, N.S nor D.S, P.S, A.S}={Absolutely Dissatisfied, Partly Dissatisfied, Neither Satisfied nor Dissatisfied, Partly Satisfied, Absolutely Satisfied} as given according to Figure 2. For example, if an evaluator gives an evaluation equal to 6 on a scale of 0 to 10, then this last value will be transformed into qualitative value alike "Neither Satisfied nor Dissatisfied" (N.S nor D.S) and "Partly Satisfied" (P.S) with the appropriate degree 0.5 for " Neither Satisfied nor Dissatisfied" and 0.5 for "Partly Satisfied" (see Figure 2).

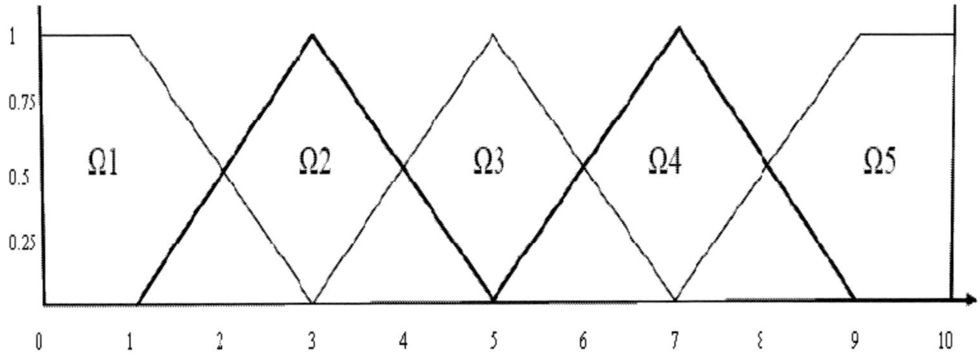

Figure 2. Fuzzy sets corresponding to the meanings of the words related to the frame of discernment. From left to right Appropriatess degrees for the frame of discernment Ω.

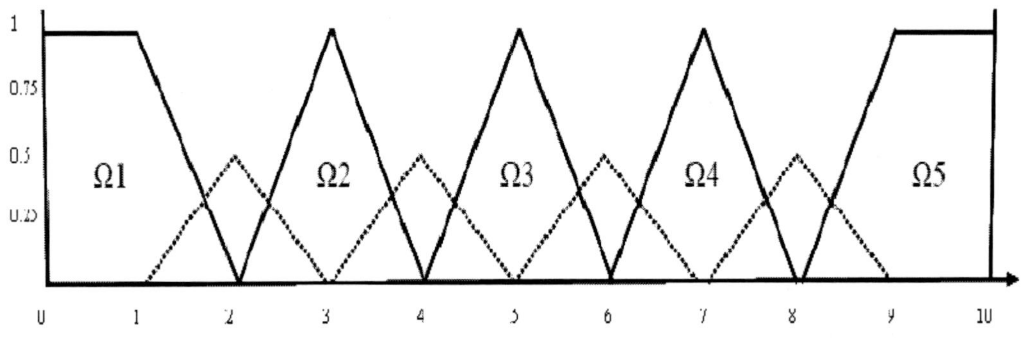

Figure 3. Mass assignment for varying x from 0 to 10, m(A.S), m(P.S), m(N.S nor D.S), m(P.D) and m(A.D) are shown from left to right with solid line, m(Ωi, Ωj) and m(Ø) are shown with a dashed line.

For the above example from an evaluator, we obtain for the criterion satisfaction degree the mass game as follows: m(N.S nor D.S, P.S) = 0.5 and m(Ø) = 0.5 (see Figure 4).

Questionnaire: The mass assignment is based on frequentist analysis of an attribute in the measurement sample. For example, for the criterion satisfaction degree, we obtain from a questionnaire, the mass game given in Figure 4.

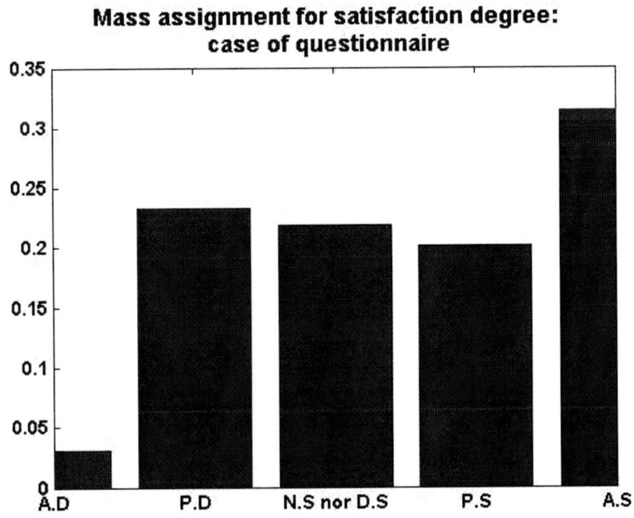

Figure 4. Mass assignment for satisfaction degree (case of questionnaire).

Sensors: Information obtained by experimental measurements is treated using fuzzy logic theory. The values to be fuzzified are averages of experimental results for a period (such as CO concentration for atmospheric pollution) or coupled criteria such as $N \times v$ for the traffic flow level (N: number of vehicles, v: average of speeds). Triangular functions are used for it and the range scale is estimated by using maximum value (e.g. for the flow level, N_{max} corresponds to maximal number of vehicles which could be parked in a section of the road and v_{max} is the legal speed; see an example in Figure 5). The membership functions obtained are directly converted in bba.

b) Several Information Sources

Now suppose that for the initial evaluation phase, several information sources exist for a criterion evaluation. Each information source has a reliability degree depending on:

- data quality (i.e. aberrant or missing data).
- source pertinence (e.g. perception of noise level by questionnaire can be considered more adequate than an acoustic model).
- periods to collect the data from various sources.

It is possible to propose a general coefficient to discount the bba of an information source due of the data quality; but the pertinence degree of a source frequently depends on data collection process (a questionnaire for the noise perception loses its "high" reliability if the number of respondents is statistically not significant). For the last point, when an initial evaluation process is done by the project manager, frequently data issued from information sources were not collected at the same period. So even if an information source is strongly pertinent, the reliability of information is affected if data were collected several month (or years) ago. In addition, the reliability information in time strongly depends on city particularities (for a city where traffic evolution, households activities had not considerably changed during several years, an "old" data base can be considered like reliable).

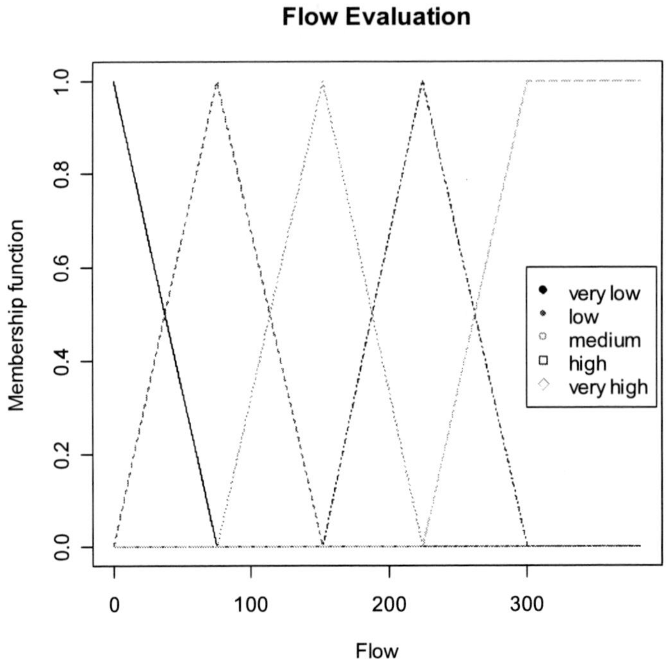

Figure 5. Mass assignment for traffic flow.

For these reasons, the research of a general analytic form of the reliability coefficient does not seems to be an adequate solution. Indeed, we propose an approach which allows a project manager to elaborate himself a model for the reliability for each information source.

This method is based on stated-preferences (marketing technique called "Liouviere" method (Louviere, 1988) where a respondent (project manager in our case) evaluates the quality of a service for hypothetical scenarios.

In order to obtain the maximum information with a minimum number of questions, we implemented an orthogonal factorial array plan which is a subset of a complete array. Figure 6 (in the left) shows an example where (A, B, C, D, E, F, G) are tested variables and the values 1 and 2 are the levels of variables. The horizontal line of the array is a hypothetical scenario to be evaluated.

In a scenario (or question), deliberately one or more variables are changed the same time. Varying several variables simultaneously may have interactive effects on the studied response (when the effect of a variable depends on the level of another, an interaction exists). Corresponding linear graphs allows us to affect columns for variables and interactions. The fractional arrays reduce the number of necessary combinations and are obtained by using orthogonality theories after the "sacrifice" of several interaction effects. The columns of the array presented in Figure 6 could be used as follow: A, B, D, G for variables and C, F, E for interactions.

In our case, the variables are respectively: A: aberrant data, B: missing data, D: pertinence level, G: period for collection process. The levels values are arbitrarily defined (i.e., A_1: important, B_1: important, D_1: low, G_1: old data). By using the linear graph estimated interactions are AB, AD and BD as shown in Figure 6 (in the right). For each scenario, the project manager evaluates the reliability of a source by using a linear range that he can define (e.g. from 1 to 5).

In order to evaluate all existing scenarios, a matrix model based on Analysis of Means (ANOM) is used (Sisson, 2000). The determinist part of the model can be schematically represented as follows:

$$\alpha = \alpha_{mean} + (a_1 \quad a_2)A + (b_1 \quad b_2)B + \ldots \begin{pmatrix} a_1b_1 & a_1b_2 \\ a_2b_1 & a_2b_2 \end{pmatrix} \ldots \tag{6}$$

where:

a_i : matrix elements representing the mean effect of variable A at level i

a_ib_j : matrix elements for the mean effect of the AB interaction when A is at level i and B at level j

$$\alpha = \alpha_{mean}\left(A_i\right) - \alpha_{mean}$$
$$a_i b_j = \alpha_{mean}\left(A_i, B_j\right) - \alpha_{mean} - a_i - b_j \tag{7}$$

with:

α_{mean} : mean value of all reliabilities

$\alpha_{mean}(A_i)$: mean value of reliabilities for combinations where A is at level i

$\alpha_{mean}(A_i, B_j)$: mean value of reliabilities when A is at level i and B at level j.

For a scenario which was not evaluated (e.g. A_2, B_1, D_1, G_1), the model is applied as follows:

$$\alpha = \alpha_{mean} + a_2 + b_1 + d_1 + g_1 + a_2b_1 + a_2d_1 + b_1d_1 \tag{8}$$

Let us take the same example, now with the satisfaction degree. Suppose that for the initial evaluation process this criterion was measured by two information sources (questionnaire and expert's opinion) and each source gives us a mass structure (as shown before). Corresponding models allow us to estimate $\alpha_{questionnaire} = 0.85$ and $\alpha_{expert} = 0.6$.

Then we combined the two masses structure related to the criterion degree of satisfaction (given in Table 5) in order to take into account the two information sources and their reliability coefficients. The results of fusion are given in Table 2.

The evidence theory will be used in different steps of our approach. Firstly, in order to combine the evaluations from heterogeneous information sources for a given criterion. Secondly, to manage indecision and ignorance related to the evaluators' opinions in order to compute criteria weights and interaction indexes between criteria.

Factors and interactions

Test	A	B	C	D	E	F	G
1	1	1	1	1	1	1	1
2	1	1	1	2	2	2	2
3	1	2	2	1	1	2	2
4	1	2	2	2	2	1	1
5	2	1	2	1	2	1	2
6	2	1	2	2	1	2	1
7	2	2	1	1	2	2	1
8	2	2	1	2	1	1	2

Linear graph

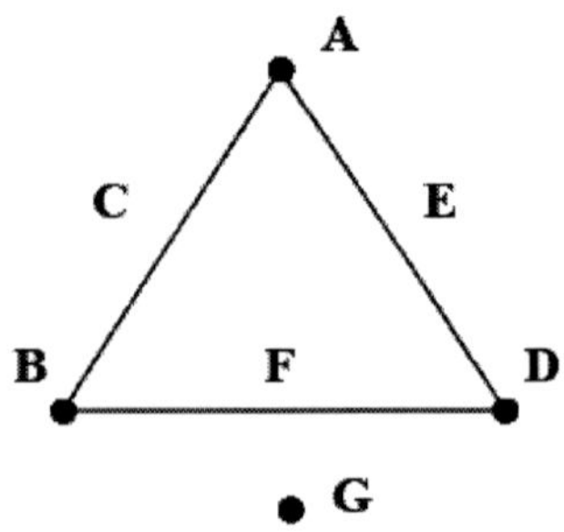

Figure 6. Factorial array (in the left); linear graph (in the right).

A Hybrid Approach to Evaluate the Efficiency of a Transport Project

The proposed approach is entitled "HA4IA" (Hybrid Approach for Impacts Assessment). In order to give flexibility in evaluation-aid tool, we implemented two evaluation steps:

- Evaluation of criteria evolution, which consists in pertinent criteria selection and aggregation of information obtained if several information sources exist.
- Evaluation of global efficiency of a project, where criteria weights and interactions between criteria are taken into account.

Table 2.Result of fusion related to the criterion degree of satisfaction ("Sa.")

	Expert's opinions	Questionnaire
Mass structure related to "Sa."	m(N.S nor D.S, P.S)=0.5 m(Ø)=0.5	m(A.S)=0.314; m(P.S)=0.202 m(N.S nor D.S)=0.219 m(P.D)=0.233; m(A.D)=0.031
Reliability by source	Less reliable (0.6)	Very reliable (0.85)
Discounting mass structure (Eq. 3)	m(N.S nor D.S, P.S)=0.3 m(Ø)=0.3 m(Ω)=0.4	m(A.S)=0.266; m(P.S)=0.171 m(N.S nor D.S)=0.186 m(P.D)=0.198; m(A.D)=0.026 m(Ω)=0.15
Fusion result with Normalisation (Eq. 2)	m(A.S)=0.207; m(P.S)= 0.191; m(N.S nor D.S)= 0.207 m(P.D)=0.138; m(A.D)=0.018; m(Ω)=0.106 (for conflict)	
BetP (Eq. 4)	m(A.S)=0.228; m(P.S)= 0.212; m(N.S nor D.S)= 0.228 m(P.D)=0.159; m(A.D)=0.039	
Utility allowed to satisfaction degree (Sa) (Eq. 14)	U(Sa)=0.54	

4.1. Evaluation of Criteria Evolution

4.1.1. An Approach for Estimating a List of Pertinent Criteria

A list of criteria usually employed in social, transport, environment, economy categories is suggested by the evaluation aid-tool after our synthesis of references in these fields. This predefined criteria set $\{C_1, C_2, \ldots, Cn\}$ is relevant for the impacts assessment of a project (e.g. traffic flow, noise, congestion rate, atmosphere pollution, satisfaction degrees for service users) but the tool user can complete or modify this list.

Table 3. Opinions of evaluators

Evaluators / Criteria	E1	E2	E3	E4	E5	E6
Aq.	1	2	3	3	3	UNK
Ac.	1	{1,2}	3	3	3	1

Table 4. Pignistic probability

Criteria	1	2	3	4
Aq.	5/24	5/24	13/24	1/24
Ac.	5/12	1/12	6/12	0

We distinguished two cases:

- Adequate criteria are identified by the project manager but a classification must be done in order to select a list of the most pertinent criteria.
- Adequate criteria are unknown and ill defined and a classification must be also done.

a) Well Defined Adequate Criteria

When pertinent criteria are well defined, the judgements of evaluators are used. Let $\{E_i, i=1,\ldots,p\}$ represent a group of experts and $\{C_k, k=1,..,n\}$ be a group of criteria which will be classified by estimating their weights $\{w_k, k=1,\ldots,n\}$. We define a set of pertinence degrees that the evaluators use for giving their opinion with respect to the pertinence for each criterion (i.e. Very Pertinent, Less Pertinent etc.). The tool user has the possibility to choose the granularity number.

Indecision and ignorance relative to the judgement of experts are also taken into account. An expert can be indecisive, for example when he is indifferent between the levels of evaluation {i.e. Pertinent and Very Pertinent}. He can also be in a situation of total ignorance when he ignores the importance of a criterion for the evaluation of a project (i.e. he does not know if the criterion is really important for the evaluation).

The evaluation levels will constitute the frame of discernment such as Ω = {Not Pertinent (1), Less Pertinent (2)...). In the belief theory, when a decision must be taken, a prudent transformation known by "pignistic probability" is applied. We note also that there exist other rule like the plausibility and the credibility measures. The difference between "pignistic" transformation and the others is that "pignistic" transformation is a prudent rule. In our case, for the estimation of criteria weights, all the game of "pignistic" probability is used.

The estimation of the criteria weights, with *n* levels of pertinence degrees, is given by the following equation:

$$\begin{cases} \omega_i = \sum_{i=1}^{n} BetP(i) \times V(i) \\ BetP(i) = Freq(i) + \sum_{j \neq i, |j| \geq 2}^{n} \frac{Freq(j)}{|j|} \end{cases} \quad (9)$$

where:

V (i): Level of degree of importance.
$i \in \Omega = \{1, \ldots, n\}$; with n levels of pertinence degrees (noted by i) from not pertinent to very pertinent
Freq(i): frequency of appearance of pertinence degrees i, with i varies from 1 to n.
| j | : cardinality of the set "*j*"
BetP(i): the "pignistic probability" (Smets, 1992), it consists to manage the indecision and ignorance, by distributing the later for singleton degree of importance (e.g. {1},…, and {n})

For the function *V(i)*, we choose a linear function (linear mapping between the degree of importance of a criterion and its score) that reflects the degree of importance of a criterion. If we dispose *n* levels of pertinence then this function is given by the following equation:

$$V(i) = 1/(n-1) * (i-1) \quad (10)$$

After the calculation of weights with the equation cited before (Eq. 9), normalisation is necessary in order to have the sum of weights equal to 1.

$$\omega_i^* = \frac{\omega_i}{\sum_{i=1}^{n} \omega_i} \quad (11)$$

The approach is applicable to one or a group of evaluators.

Illustration: Suppose that a project must be evaluated and two criteria were considered adequate by the project manager: Air quality (Aq.), Accessibility (Ac). Let Ω = {Not Pertinent (1), Less Pertinent (2), Pertinent (3), Very Pertinent (4)} be this set of pertinence degree. Let us suppose that we have the data game according to the opinion of 6 evaluators (noted by E). The criterion is unknown, where the evaluator does not have any knowledge on its degree of importance for the evaluation of the project. In this case, he can reply by UNK "I do not know". After the computation of the frequency of apparition according to the pertinent degrees, the computation of the "pignistic probability" is done (Table 4). The 'pignistic transformation' (Smets, 1992) (Smets and Kennes, 1994) allows us to distribute the ignorance and the indecision (see Table 3) on the singletons of pertinence degrees.

The example presented before, allows us to attribute weights for a list of pertinent criteria predefined by the project manager. After applying Eq. 5, the criterion weight for air quality (Aq.) is 0.56 and Accessibility (Ac.) is 0.44.

b) Ill Defined Adequate Criteria

Let us suppose now, that the project manager ignores the adequate criteria for evaluating the project. In this case, the project manager examines an exhaustive list of criteria which can be ill defined (i.e. not all criteria are relevant for a project assessment) with the aid of a set of experts.

In this case, we extend the proposed approach (presented in Eqs. 9, 10 and 11) in order to reject certain criteria judged non pertinent before the estimation of their weights by using the "pignistic risk" (Smets and Kennes, 1994).

Illustration: Let $A=\Omega=\{1, 2, 3, 4\}$ be a set of pertinence degrees, $C: A \times \Omega \rightarrow R$ be a cost function and m be a mass function defined on Ω. For choosing the pertinence degree relative to each criterion on the basis of evaluator's opinion, we use the 'minimum of pignistic risk' (Eq. 12) where the pignistic probability is computed according (Eq. 9).

$$R_{BetP}(\alpha) = \sum_{\omega \in \Omega} C(\alpha, \omega)\, BetP(\omega) \qquad (12)$$

where $C(\alpha,\omega)$ denotes the loss incurred if pertinence degree $\alpha \in A$ is selected and $Bet(\omega)$ is the pignistic probability as defined in Eq. 9.

Cost function:

$$\begin{aligned} &C(1,2)=C(2,3)=1/3\\ &C(1,3)=C(2,4)=2/3\\ &C(1,4)=1\\ &C(1,1)= C(2,2)= C(3,3)= C(4,4)=0 \end{aligned} \qquad (13)$$

With the cost function defined above (Eq. 13), we obtain the "pignistic" risk related to each pertinence degree according to Table 5.

Table 5. Pignistic Risk

Criteria	1	2	3	4
Aq.	0.472	*0.277*	0.388	0.527
Ac.	0.361	*0.305*	0.305	0.638

For the game of "pignistic" probabilities (Table 3), on applying the rule of 'minimum of pignistic risk' with the cost function (Eq. 13), we chose the degree of importance related to each criterion. In our case study, we found that the two criteria "Air quality" (Aq.) and "Accessibility" (Ac.) are considered 'less pertinent (noted by 2)'. If we admit that a criterion is retained if it is judged 'not pertinent (noted by 1)', then the two criteria 'Aq.' and 'Ac.' are retained because they are respectively judged less pertinent (noted by 2, see Table 4) for the project evaluation, and their weights are estimated as indicated before (see Eq. 9).

4.1.2. Indicator of Criteria Evolution

For a complete evaluation of criteria evolution an indicator must be defined in order to establish how criteria changed between two phases. After mass assignments for each information source, a utility function is computed for each criterion by using the "pignistic" probability (denoted by BetP) and a score function (denoted by u) as given in Eq. (14). The results provided to the project manager are in numerical form starting from the computation of a utility (u_i) for the criterion (C_i), defined by Eq. (14).

$$u_i = \sum_{k=1}^{p} u(H_k) \times BetP(H_k) \tag{14}$$

where: $u(H_k)$ represents the utility of an evaluation level H_k, $u(H_{k+1}) \geq u(H_k)$ if H_{k+1} is preferred to H_k, and $BetP(H_k)$ represents the "pignistic" probability related to each evaluation level H_k.

After computing the pignistic probability according to Eq. (4), we determine a score related to the satisfaction degree (Eq. (14)) by using a linear function on the evaluation levels (e.g. u(A.D, P.D, N.D nor D.S, P.S, A.S)=(0, 0.25, 0.5, 0.75,1).).Table 5, presented above, corresponds to initial evaluation phase where the value of utility related to the satisfaction degree (Sa.) is equal to 0.54.

In general, such utility is used to compare criteria state during several implementation phases of a given project, in order to underline if there is a benefit or decrease on each criterion. In this way, the user of tool has the possibility to exploit all existing data before project implementation, as well as to change the information source for the post-evaluation phase.

4.2. Evaluation of Global Efficiency of a Project

A global evaluation of project efficiency needs the aggregation of all results concerning the criteria evaluation, taking into account their weights and possible interactions between criteria.

4.2.1. An Approach for the Estimation of Criteria Interaction

We note that the interaction indices (noted by I) take their values in the interval [-1,1]. A negative value is significant of a redundancy between the criteria. Two criteria C_1 and C_2 are redundant if the satisfaction of the one of the criteria is sufficient to have a significant effect on the total evaluation. A positive value means that there are complementarities between the two criteria. The simultaneous satisfaction of the criteria C_1 and C_2 is required to have an effect on the global evaluation process.

The Choquet integral (Grabisch, 1996) (Sugeno, 1974) is a general tool for dealing with multiple criteria decision making taking into account the interaction between criteria. Besides, a difficulty which has posed when we want to apply the Choquet integral model is the estimating of these interaction indexes is not easy. (Grabisch, 1996) has proposed the concept of k-additive measure and some methods for computing interaction indexes between criteria. These methods deal with classification problems (i.e. candidates ranking etc.) and they are not adequate in our work. Our approach is based on belief theory and an example is used to better explain its application.

Let $\{E_i, i=1,...,p\}$ be a set of experts and $\{C_k, k=1,..,n\}$ be a set of criteria having weights evaluated (by Eqs. 5,7 and 8) whose index interaction we want to determine $\{Iij, i,j=1,...,n\}$. We define a set of interaction importance degrees that the experts use for giving their opinion. This set of interaction importance degrees can be defined using 5 levels as follow:

Ω={High Negative Interaction (-2), Less positive Interaction (-1), Not Interaction (0), Less positive Interaction (+1), High positive Interaction (+2)}.

As for the estimation of criteria weights, we deal with the indecision and the ignorance relative to the judgement of the evaluators. The estimation of the criteria interactions with 5 levels of interactions degrees is given by the following equation:

$$\begin{cases} I_{ij} = \sum_{k=1}^{5} BetP(k) \times V(k) \\ BetP(k) = Freq(k) + \sum_{l \neq k, |l| \geq 2}^{5} \frac{Freq(l)}{|l|} \end{cases} \tag{15}$$

where:

$V(k)$: level of interaction degree.

$k \in \Omega$ = {-2,-1,0,+1,+2}= {high negative interaction, less negative interaction, no interaction, less positive interaction, high positive interaction}.

Freq(k): frequency of appearance of importance degrees of interaction, $k \in \Omega$

$|l|$: cardinality of the set $l \in \Omega$

BetP(k): the "pignistic probability" related to the interaction degree $k \in \Omega$.

We have chosen, for the function V, a linear function that reflects the interaction importance degree of a pair of criteria. If we dispose n levels of interaction importance degree then this function is given by the following equation:

$$V(k) = \frac{2}{(n-1)} \times k \tag{16}$$

Now consider that we evaluate a project with 3 criteria of evaluation (air quality (Aq.), accessibility (Ac.) and traffic flow (Tf.)) and only 2 evaluators (E1, E2). Let Ω be a set of 5 levels of interaction degrees such as: Ω={-2,-1,0,1,2}={high negative interaction, less negative interaction, no interaction, less positive interaction, high positive interaction}. Let V be the function which reflects the interaction importance degree of a pair of criteria such as:

V{-2,-1,0,1,2}={-1,-0.5,0,0.5,1} (Eq. 16). Suppose that we have the data game according to the opinion of 2 evaluators as given in Tables 6 and 7.

Table 6. Matrix of confusion according to evaluator (E1)

Criteria	Aq.	Ac.	Tf.
Aq.		0	-2
Ac.	---		{1,2}
Tf.	---	---	

Table 7. Matrix of confusion according to evaluator (E2)

Criteria	Aq.	Ac.	Tf.
Aq.		UNK	{-2,-1}
Ac.	---		1
Tf.	---	---	

The computation of the frequency of apparition, according to the interaction degrees is given according to Table 8, as shown below. After the computation of "pignistic probability" as shown in Table 9, the interaction indexes between criteria are given in Table 10. We conclude that there is no interaction between Air quality and Accessibility besides there is a synergy (positive interaction) between the pair (Accessibility, Traffic) and redundancy (negative interaction) between the pair (Air quality and Traffic).

Table 8. Frequency of apparition

levels / Criteria	-2	-1	0	1	2	{-2,-1}	{1,2}	UNK
{Aq., Ac.}	0	0	0.5	0	0	0	0	0.5
{Aq., Tf.}	0.5	0	0	0	0	0.5	0	0
{Ac., Tf.}	0	0	0	0.5	0	0	0.5	0

Table 9. Pignistic probability

levels / Criteria	-2	-1	0	1	2
{Aq., Ac.}	0.1	0.1	0.6	0.1	0.1
{Aq., Tf.}	0.75	0.25	0	0	0
{Ac., Tf.}	0	0	0	0.75	0.25

Table 10. Interaction indexes

Criteria	Interaction index
{Aq., Ac.}	0
{Aq., Tf.}	-0.875
{Ac., Tf.}	0.625

4.2.2. Indicator for Efficiency of a Transport Project

For a global evaluation, a simple method of aggregation is used based on the multi-attribute utility theory (MAUT) techniques.

The utility function is not necessary additive and as in our application, criteria set can be in interaction (in synergy or in redundancy). So, to improve the evaluation process, a method of aggregation based on the Choquet integral (Grabisch, 1996) (Sugeno, 1974) is applied in order to take into account the interactions between criteria. The Choquet integral is an extension of the weighted average.

This method of aggregation considers at the same time the weight of each criterion and the interaction between them and models in a comprehensible way the redundancy and the complementarities of the criteria. The criteria aggregation by the 2-additive form of the Choquet integral (Grabisch, 1996) is written as follow:

$$u = \sum_{i=1}^{n} \omega_i \times u_i - \frac{1}{2}\sum_{s=1}^{n} \sum_{t=s+1}^{n} I_{st} \times |u_s - u_t| \quad (17)$$

with: ω_i is the weight of the criterion C_i, u_i is the criterion evaluation defined by Eq. (14) and I_{st} is the interaction between the criteria C_s and C_t which are computed according the proposed approach mentioned before.

It is possible to establish an evaluation of project (noted by *P*) efficiency during for example two implementation phases (noted by *ph1* (before implementation) and *ph2* (after implementation)) as the following index (noted by E in Eq. 18):

$$E(ph1 \rightarrow ph2) = \frac{u(ph2) - u(ph1)}{u(ph1)} \quad (18)$$

The estimation of project efficiency (*E*) is equal to the relative deviation between the utilities functions *u* at phase (*ph1*) and u at phase (*ph2*).

Table 11. Initial evaluation (u1) and Post-evaluation (u2)

Category	Criteria (C)	Initial evaluation (u1)	Post-evaluation (u2)	Criteria weights (w)	Interaction indexes (I)
Environment	C1. Air quality (Aq.)	0.245	0.2	0.2	I(Un.,Sa.)=+0.1
Society	C2: Users number (Un.)	0.140	0.15	0.4	I(Cg.,Aq.)=-0.10
	C3: Satisfaction (Sa.)	0.47	0.48	0.25	I(Cg., Un.)=+0.05
Transport	C4: Congestion (Cg.)	0.145	0.17	0.15	I(Aq., Sa.)=+0.15

Illustration: Park and Ride project in CIVITAS program (briefly described in section 1). The set of criteria used for initial evaluation phase and post-ex evaluation process are {Aq., Un., Sa., Cg.}={Air quality, Users number, Satisfaction degree and Congestion}. After applying the proposed approach for criteria weights and interaction indexes estimation before mentioned we summarize the whole evaluations according to Table 11. At last a global evaluation is computed taking into account the criteria weights and interaction indexes between criteria (see Table 12).

Table 12. Global evaluation with and without interaction

Global evaluation (without interaction)		Global evaluation (with interaction)	
Initial evaluation (u1)	Post-evaluation (u2)	Initial evaluation (u1)	Post-evaluation (u2)
0.2442	0.2455	0.2157	0.2090

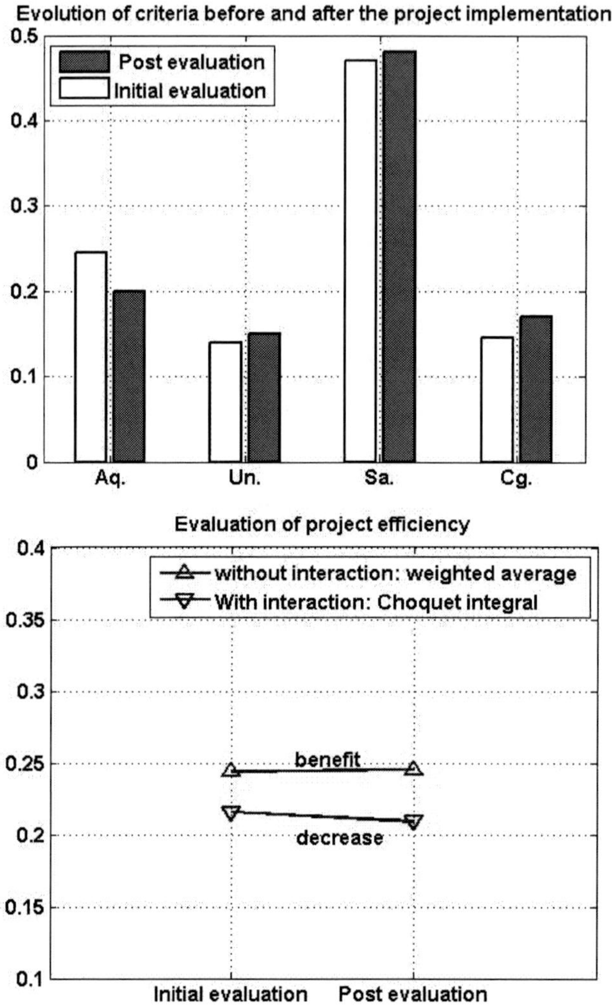

Figure 7. Evaluation of criterion and project efficiency.

Figure 7 shows both kinds of information that the tool user can exploit: the evaluation of each criteria evolution as well as the efficiency of a project. Figure 7 shows also the interest to take into account the interactions between criteria to evaluate the project efficiency.

Table 13. Approaches combining belief theory with multicriteria analysis

Approach	Source	Operator used	Aggregation	reliability Source	Alternative
AHP-Dempster (Beynon, 2000)	1 source : experts	Dempster -Shafer	Additive : Weighted average	No	Several
ER approach (Yang et al., 2005)	1 source : experts	Dempster -Shafer	Additive : Weighted average	No	Several
HA4IA (Omrani et al., 2007)	Several sources	Dempster -Shafer	Non additive : Choquet integral	Yes	One or Several

Finally, I underline that in this paper, we have proposed a hybrid approach for impacts assessment based on muti criteria analysis and belief theory. In table 13, we perform a comparison between the proposed approach and two others recently proposed approaches. From this comparison, we highlight the benefit of our approach which is more general by using several information sources, non additive operator (Choquet integral) during aggregation and take into account sources reliabilities. The proposed approach is applied in this paper for park and ride evaluation but also it can be applied in several application using multiple criteria and several information sources.

Conclusion and Future Work

Mobility in cities has been increasing in the last few years and with it, the awakening on the environmental or socio-economic harmful impacts, without a real control of the development and management of the urban transport system. In this paper we have proposed a methodology for the evaluation of the impacts related to the urban mobility under the framework of multi criteria analysis and belief theory. Information coming from different sources, even in the case of the fuzzy data and uncertainty is combined in an effective way. This approach also integrates criteria weights estimation, evaluation of interaction between criteria and reliability of information sources. We estimate that this kind of approach could be successfully used for several MCDM applications. In addition to the theoretical proposition an interactive Web-tool for a decision-making aid will be implemented for data collection, information treatment, evaluation process and allowing the visualization of knowledge and evaluation resulting from various information sources. The evaluation results will be represented with graphics techniques or via GIS (Global Information System). But the most appreciated points of this tool concerns its flexibility, multi-users participation, the interaction between project manager and the tool in different steps of evaluation process as well as the possibility to elaborate models (reliability of sources, weights) in agreement with the specificities of a city and the particularities of data collection process.

This approach will be extended by taking mass assignments by unit or by interval or fuzzy. A supplementary step will be developed to suggest strategies to project manager to obtain multi-levels optimisation taking into account characteristics of project to be implemented.

Acknowledgments

This work is funded by the "Conseil Général" (General Council) of La Rochelle, France. It is carried out under the framework of SUCCESS which is a European project funded by the European Commission under the CIVITAS program (Contract no.: 513785). There are three participants in SUCCESS project; La Rochelle- France, Ploiesti-Romania and Preston-UK. The authors are thankful to Hervé Le Berre (Engineer at CdA, La Rochelle-France) and Tom Parker (Project Evaluation Manager of SUCCESS, Transport and Travel Research in UK) for valuable discussions and to anonymous referees for their valuable suggestions.

References

Beinat, (2001) '*Multi-criteria analysis for environmental management*'. Journal of Multi-Criteria Decision Analysis, 10:51.

Beynon MJ, (2002) '*DS/AHP method: a mathematical analysis, including an understanding of uncertainty*', European Journal of Operational Research 140: 149-165.

Beynon MJ, Curry B, Morgan PH, (2000) '*The Dempster Shafer theory of evidence: an alternative approach to multi criteria decision modelling*'. OMEGA 28(1): 37-50.

Bond, R., Curran, J., Kirkpatrick, C., and Lee, N. (2001) '*Integrated Impact Assessment for Sustainable Development: A Case Study Approach*'. World Development 29(6)1011-1024.

Buckley JJ., (1985) '*Fuzzy hierarchical analysi's. Fuzzy Sets and Systems* 28(1): 233-47.

Chang P., Chen Y., (1991) '*Fuzzy Number in Business Conditions Monitoring Indicators: Fuzzy Set Methodologies in Economic Condition, in: Fuzzy Engineering toward Human Friendly Systems*', Proceedings of the international Fuzzy Engineering Symposium 91, Yokohama, Japan: 109:1100.

Chen M., Tzeng G., Liu D., *Multi-Criteria Task Assignment in Workflow Management Systems*, IEEE, Proceedings of the 36th Hawaii International Conference on System Sciences, (2003).

Denoeux T., Smets P. (2006) '*Classification using Belief Functions: the Relationship between the Case-based and Model-based Approaches*', IEEE Transactions on Systems, Man and Cybernetics B , Vol. 36, Issue 6, Pages 1395-1406.

Delgado M., Verdegay J.L, and Vila M.A, (1993) '*On aggregation Operations of Linguistic Labels*', Int. J. of Intelligent System 8: 351-370.

Delgado M., Verdegay J.L, and Vila M.A, (1992) '*Linguistic Decision Making Models*', Int. J. of Intelligent System 7: 479-492.

Dempster A.P., (1968) '*A generalisation of Bayesian inference*', Journal of the Royal Statistical Society, 205-247.

Fischer, T., Wood, C.M. and Jones, C.E. (2002) '*Policy, plan and programme environmental assessment in England, the Netherlands and Germany: practice and prospects*'. Environment and Planning B(29): 159-172.

Goedkoop M.J, Spriemsma R., (2000) '*the Eco-indicator, a damage oriented method for Life Cycle Impact Assessment. Methodology Report and Methodology*', *second edition*, 82-132. report (INERTS).

Grabisch M. (1996) '*The application of fuzzy integrals in multicriteria decision making*'. European J. of Operational Research 89, 445-456.

Guine J.B, (2002) '*Handbook on Life Cycle Assessment. An operational guide to the ISO standard*'. London: Kluwer Academic, 704.

Herrera, F., and Verdegay., J.L., (1993) '*Linguistic Assessments in Group Decision*', Proc. of First European Congress on Fuzzy and Intelligent Technologies, Aachen, 941-948.

Jay, S. and Handley, J. (2001) '*The Application of Environmental Impact Assessment to Land Reclamation Practice*'. Journal of Environmental Planning and Management 44(6): 765-782.

Kunreuther H., Grossi P., Seeber N., Smyth A., (2003) '*A Framework for Evaluating the Cost-Effectiveness of Mitigation Measures*', Columbia University, Etats-Unis d'Amrique.

Louviere, J.J. (1988) '*An Experimental Design Approach to the Development of Conjoint Based Choice Simulation Systems With an Application to Forecasting Future Retirement Migration Destination Choices.*' In R. G. Gollege and H. Timmermans (eds.) Behavioural Modelling Approaches in Geography and Planning. London: Croom-Helm.

Lu, R.S., Lo, S.L., Hu, J.Y., (1999) '*Analysis of reservoir water quality using fuzzy synthetic evaluation*'. Stochastic Environmental Research and Risk Assessment 13, 327–336.

Omrani, H., Ion-Boussier, L., Trigano, P., (2007) '*A New Approach for Impacts Assessment of Urban Mobility*', WSEAS Transactions on Information Science and Applications, vol. 4, num. 3, pp. 439-444.

Radon N.J, (2004) '*Fuzzy and Random Set Based Introduction Algorithms*', Thesis report, Université de Bristol, département d'ingénierie mathématique.

Roy, B., (1985) '*Méthodologie multicritère d'aide la décision*', Economica, Paris, or Multi criteria methodology for decision aiding, Kluwer, Dordrecht.

Scemama G. et Carles O., (2004) 'Claire-siti, *public road transport network management control: a unified approach.* In 12th IEE Int. Conf. on Road Transport Information and Control (RTIC'04), London (UK).

Shafer G., (1976) '*A Mathematical Theory of Evidence*', Princeton University Press, Princeton.

Sisson J-C., (2000) '*Pratique des plans d'expériences fractionnels.* Approche Taguchi-Box", rapport interne, CACEMI, CNAM.

Simos J., (1990) '*Evaluer l'impact sur l'environnement, une approche originale par l'analyse multicritère et la négociation*', PPUR, Lausanne, Suisse.

Smets P., (1992) '*The transferable belief model for experts' judgments and reliability problems. In Analysis and Management of uncertainty: Theory and Applications*', Elsevier Science Publishers B.V.

Smets P. and Kennes R., (1994) '*The Transferable Belief Model. Artificial Intelligence*', 66:191–243.

Tao, Y., Xinmiao, Y., (1998) '*Fuzzy comprehensive assessment, fuzzy clustering analysis and its application for urban traffic environment quality evaluation*'. Transportation Research-D 3 (1), 51–57.

Sugeno M. (1974) '*Theory of fuzzy integrals and its applications*'. PhD thesis, Tokyo Inst. of Technology.

Wood, C. and Ahmad, B. (2002) '*A comparative evaluation of the EIA systems in Egypt, Turkeyy and Tunisia*'. Environmental Impact Assessment Review 22: 213-234.

Wood, C. (2002) '*Environmental Impact Assessment: a Comparative Review'.* 2nd edition. Prentice Hall, UK, 405pp.

Xu, D. L., Yang. J. B. and Wang Y. M., (2005) '*The ER approach for multi attribute decision analysis under interval uncertainties'*, European Journal of Operational Research.

Yager R.R, (1993) '*Families of OWA operators'*, Fuzzy Sets and Systems, 59,125-148.

Yager R.R, (1992) '*Applications and extensions of OWA aggregation operators in multi criteria decision making'*, Int. Journal of Man-Machine Studies, 37, 103-132.

Yang J. B., Wang Y. M, Xu D. L. and Chin K. S., (2005) '*The evidential reasoning approach for MCDA under both probabilistic and fuzzy uncertainties'*, European Journal of Operational Research.

Yang T, Yang X. (1998). *Fuzzy comprehensive assessment, fuzzy clustering analysis and its application for urban traffic environment quality evaluation.* Transportation Research D, vol. 3, No. 1, pp. 51-57.

Zadeh LA. (1965) '*Fuzzy set. Information and Control',* 8(3) 338-53.

Zadeh L.A. (1986) '*A Simple View of Dempster-Shafer Theory of Evidence and its Implications for the Role of Combination'*, The AI Magazine.

In: Focus on Information, Intelligence and Knowledge
Editor: Rafiq Islam

ISBN 978-1-61209-979-8
2011 Nova Science Publishers, Inc.

Chapter 3

ON LINE PID CONTROLLER TUNING THROUGH BIO-MIMICRY OF BACTERIA FORAGING

Rajesh Kumar*, Anupam Kumar and Devendra Sharma
Department of Electrical Engineering,
Malaviya National Institute of Technology, Jaipur, India

ABSTRACT

In this chapter we present on line tuning of PID controller through bio-mimicry of bacteria foraging. Genetic Algorithm is currently a standard for optimization. Bio-mimicry is a new technique proposed for optimization by Kevin M. Passino of Ohio State University, Ohio. A comparison of famous Genetic Algorithm with bio-mimicry of bacteria foraging has been made. This comparison is done by comparing their application to the problem of tuning a PID controller. In this chapter we observe that bio-mimicry is equally applicable to the problem of control system design (here tuning PID controller). The results of bio-mimicry algorithm are better than to that of genetic algorithm.

Keywords: PID, On line tuning, Genetic Algorithm, Bacteria Foraging.

1. INTRODUCTION

The problem of PID tuning has been widely studied in control system literature. The problem requires optimum tuning of PID parameters using a suitable optimization algorithm. In fact control engineers have vastly explored all proposed method to optimally tune PID controllers for various control systems. Of course the most conventional of these methods is the classical Zieglar Nicholas method. Presently Genetic Algorithm (GA) is becoming an industry and research standard for optimization problems.[6,9] The technique of GA has been vastly applied for tuning PID controllers.

* Email: rk.rlab@gmail.com

The most important requirement of any tuning method is its ability to tune the controller rapidly with least computation complexity. This is an unavoidable constraint for real time or on-line tuning of a controller.

Bio-mimicry of bacteria foraging is a new optimization algorithm proposed by Passino. This techniques model the swarming of e-coli bacteria as they move about in natural environment in search for food. Passino claims that nature has forged an optimal nutrient acquiring strategy in these bacteria. This strategy enables them to gain maximum energy in nutrient rich environment and survive in hazardous environment. In his work Passino has proposed that this optimal bacterial behavior can be modeled into an optimization algorithm suited to computation.

In this chapter we check the suitability of GA for on-line tuning of PID controllers. Then we apply the bacteria foraging algorithm to the same problem of tuning PID controller. We observe that the bacteria foraging algorithm responds equally well to the problem. The computation time is nearly same and the results point to a global optimum. Further the programming model for the foraging algorithm is straightforward and easy to construct from its "pseudo- code".

2. PID TUNING

The following figure shows a simple control system with a PID controller. We consider the plant to have a linear transfer function of order one and two. The problem now remains to tune the parameters of PID controller that would maintain the plant at the desired output.[1,2,4,7]

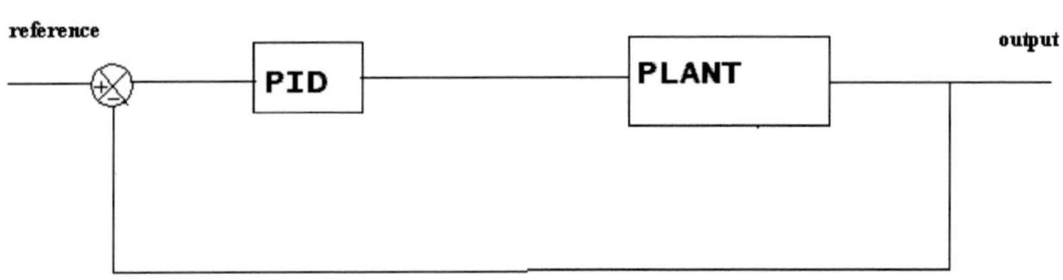

Figure 1. System with PID controller.

A PID controller shown in fig (1) has the following transfer function:

$$G(s) = K_d \times s + K_p + K_i / s \tag{1}$$

Here K_p, K_d and K_i are the coefficients of proportional (P), differential (D) and integral (I) gains for the controller respectively. It is clear that a PID controller introduces a transfer function with one pole at origin and two zeros that can be located anywhere in the s plane by carefully choosing the values of gain parameters. In general therefore we note that PID control is very useful for reducing steady state error and improve the transient response of the system.

Now the tuning problem is defined as finding the value of K_p, K_d and K_i such that the plant provides desired output. Mathematically be desired output mean that the plant would reach the desired stead state with least integral of absolute error (IAE). We call this value as our "objective function". We define error and IAE as follows:

$$IAE = \sum_{k} e(k) \text{ , } e(k) = ref - y(k) \qquad (2)$$

Here ref is the reference input, y(k) is the output at time k, e(k) is error for time unit k.

3. Genetic Algorithm

Genetic Algorithm (GA) was introduced by Holland and his colleagues at the University of Michigan[8]. GA models the adaptive process of natural system to design artificial systems software that retains the important mechanisms of natural systems. GA is based on the Darwinian theory of natural selection and survival of the fittest.

Typically GA aims to find the global optimum in a search space. But as opposed to normal optimization techniques GA starts with a large number of random solutions in search space. Other conventional algorithms generally start with a single and deterministic point. The initial set of random solution is known as population in GA lingua.

Further this population is not a set of real numbers but a coding of the real numbers. Generally this coding is a binary representation of the real numbers. A detailed understanding of this coding can be found in . Also GA is suited for typically for problems of a large number of variables. So the coding is therefore multi-parameter and mapped[8].

Now the operators of GA namely reproduction, crossover and mutation is applied to the (binary) coding of initial population. These operators emulate their biological counterparts. When these operators are repeated for a sufficient number of iterations (Generations) we obtain the solution that is closest to the global optimum for all practical purposes. In each generation GA will be preferential to those solutions which have a more desirable (maximum or minimum as per the situation) value to the objective function.

It must be noted that the programming model for GA is a stochastic in nature rather than being deterministic. That means reproduction, crossover and mutation are probabilistic operators. Yet these probabilistic operators are more robust than a deterministic model.

The genetic algorithm can be summarized as follows:

Algorithm 1:

```
t:=0;
initialize P(t);
evaluate P(t);
while not terminate do
P'(t)=variation[P(t)];
evaluate [P'(t)];
P(t+1) :=select [P'(t)UQ];
t:=t+1;
od
```

It must be noted that GA is mostly a mathematically tool. The objective functions as well as parameters therefore have to be utilized in a way to be simply applicable to problems of computation. Several things regarding this have to be considered. For example, the basic GA proposed by Holland was aimed only at maximizing the objective function and minimization can be achieved using various techniques. The objective function then has to be transformed as shown below. Here g(x) is the original fitness function and f(x) is the transformed fitness function.

$$f(x) = C - g(x) \quad if \ g(x) < C$$
$$= 0 \qquad otherwise \tag{3}$$

where C is the maximum of all previous generations.

Another transformation for minimization can be.

$$f(x) = 1/[1 + \{g(x)\}^2] \tag{4}$$

Another problem is mapping of the parameter. The unsigned integer parameter is mapped linearly from $[0,2^{lchrom}]$ to a specified interval $[U_{min}, U_{max}]$. The precision of mapped coding is calculated as

$$p = (U_{max} - U_{min})/(2^{lchrom} - 1) \tag{5}$$

Similarly there are various different options available for reproduction and cross-over operators viz. tournament selection, rank based selection, proportional selection etc. and multipoint crossover, single point cross-over etc. These decisions can vary the overall performance of GA[8].

For our problem of tuning the PID controller using GA the model of the plant for on-line tuning would look like following fig(2). The integral of absolute error (IAE) as defined earlier will work as objective function for GA.[6]

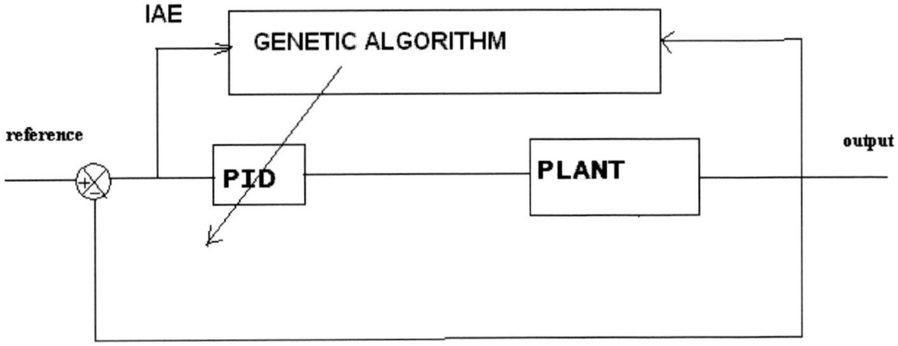

Figure 2. System with GA tuned PID.

4. Bio-Mimicry of Bacteria Foraging

Bio-mimicry of bacteria foraging is a new optimization algorithm proposed by Kevin M. Passino of Ohio State University. This algorithm models the foraging strategy e-coli bacterial swarms as they move about in search for food. Passino[10] claims that Evolution has "engineered" what is sometimes referred to as an "optimal foraging policy" in these bacteria. He goes on to claim that if this foraging policy is studied carefully it can be converted into a mathematical or programming model that can be applied to a variety of engineering problems. Here we will apply this strategy to the problem of tuning a PID controller. First we study the behaviour of e–coli bacteria and foraging theory.

Foraging theory is based on the assumption that animals search for and obtain nutrients in a way that maximizes their energy intake E per unit time T spent foraging. Hence they try to maximize a function like E/T (or they maximize their long-term average rate of energy intake). This essentially means that animals have the capability to make a decision about direction of food search and they don't about just randomly. This involves taking risks and making optimal decisions. Optimal foraging theory formulates a computational model of this foraging ability of animals. Many biologists claim that optimal foraging behaviour in animals cannot be proved in animals and the very base of the computational model is shaky. But proponents of foraging theory claim its suitability by arguing that they try to emulate optimal foraging behaviour.[10]

An important aspect of animate foraging is that they forage in groups. This gives rise to the phenomenon of social foraging. When one member of group finds a good food source it would signal about it to others by some special technique. For example ants would a chemical trail of "pheromone" on its path and animals like monkeys would signal food and danger by certain sounds.

The foraging strategy of every animal in nature is different. Here we would try to understand the foraging strategy in e-coli bacteria that are generally found in human intestine. Their behaviour is summarized below:

- E-coli bacteria live in a liquid medium and essentially have mechanism to determine whether they are in favourable or hazardous environment.
- Throughout its life time a bacterium would alternate between a swim and a tumble. A swim means a straight run in nutrient medium and a tumble means a random change in direction in search for more favourable environment. Passino [10] describes the biological actualization of swim and tumble in e-coli.
- Whenever the bacteria is in a nutrient rich medium its swim length and duration would be considerably more than compared to its swim length in a hazardous medium.
- This means that when in hazardous medium the bacteria would frequently change its direction to come out of hazardous environment. Further when in a nutrient rich medium it would still change its direction in search of more nutrients.
- This shows that the bacteria have decision making capability.
- The bacteria also perform social foraging by sending cell to cell chemical attractant

Making use of the bacterial model Passino derived an optimization algorithm. This algorithm treats the search space as the nutrition medium and a test point on the search space as the current location of the bacteria. These bacteria socially forage on the nutrient space to reach the global optimum.

We have a three-dimensional search space for our problem of tuning PID controller. The IAE objective function becomes the nutrient surface in our case. And the values of parameters collectively define the position of bacterium in the nutrient space. The plant model is as shown in following figure (3)

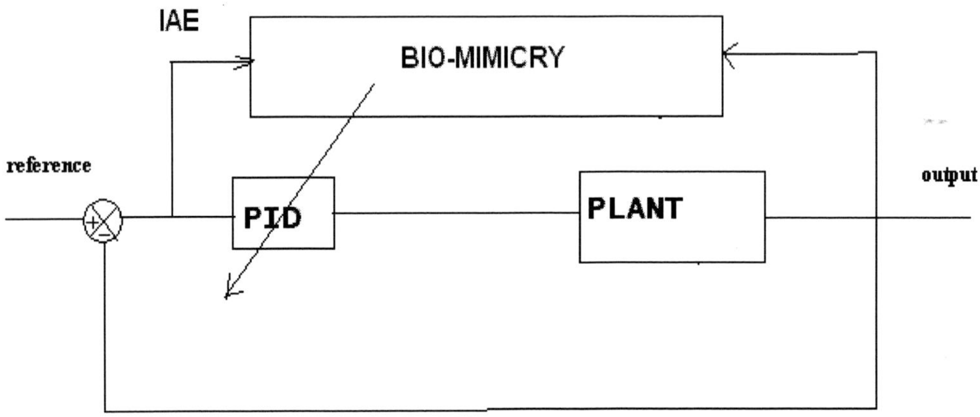

Figure 3. System with Bio-mimicry tuned PID.

The algorithm is explained as follows:

For initialization, you must choose *p, S, Nc , Ns , Nre , Ned , ped* ,and the *C(i) , i= 1,2,….,S*. Here *p* is the no. of variables, *S* is population size, *Nc* is number of chemotactic steps, *Ns* is swim length, *Nre* is the number of reproduction steps, *Ned* is the number of elimination and dispersal steps and *ped* is the probability for the same.

If swarming is also included pick the parameters of the cell-to-cell attractant functions. Also, initial values for the $\theta(i)$, *i= 1,2,…,S,* must be chosen.

The algorithm that models bacterial population chemotaxis, swarming, reproduction, elimination, and dispersal is given here (initially, j=k=l=0).

1) Elimination-dispersal loop:l=l+1
2) Reproduction loop:k k = +1
3) Chemotaxis loop:j= j+1

   a) For *i= 1, 2… S.*, take a chemotactic step for bacterium i as follows.
   b) Compute $IAE(i, j, k, l)$.

Let $IAE(i, j, k, l) = IAE(i, j, k, l) + IAE_{cc}(\theta^i, j, k, l)$.

Here

$$IAE_{cc}(\theta, j,k,l) = \sum_{i=1}^{S} [-d_{attract} \exp(-w_{attract} \sum_{m=1}^{p} (\theta_m - \theta_m{}^i)^2)]$$

$$+\sum_{i=1}^{S} [-h_{repellant} \exp(-w_{repellant} \sum_{m=1}^{p} (\theta_m - \theta_m{}^i)^2)]$$

where *w, d* and *h* are attraction and repulsion strengths.

c) Let $IAE_{last} = IAE(i, j,k,l)$ to save this value since we may find a better cost via a run.

d) Tumble: Generate a random vector $\Delta(i)$ with each element $\Delta_m(i)$, m= 1, 2… p, a random number on [−1, 1].

e) Move: Let

$$\theta^i(j+1,k,l) = \theta^i(j,k,l) + C(i)\frac{\Delta(i)}{\sqrt{\Delta^T\Delta(i)}}$$

This results in a step of size ***C(i)*** in the direction of the tumble for bacterium i.

f) f)Compute $IAE(i, j,k,l)$ .Let $IAE(i, j,k,l) = IAE(i, j,k,l)$ + $IAE_{cc}(\theta^i, j,k,l)$

g) Swim:

i) Let m = 0 (counter for swim length).
ii) While *m* <*Ns*

Let m =*m +1.*

If $IAE(i, j+1,k,l) < IAE_{last}$ (if doing better), let $IAE_{last} = IAE(i, j+1,k,l)$ and let

$$\theta^i(j+1,k,l) = \theta^i(j+1,k,l) + C(i)\frac{\Delta(i)}{\sqrt{\Delta^T\Delta(i)}}$$

and use this $\theta^i(j+1,k,l)$ to compute the new $IAE(i, j+1,k,l)$ as we did in (f).

Else, letm Ns = . This is the end of the while statement.

h) Go to next bacterium *(i + 1)* if *i≠ S* (i.e., go to b) to process the next bacterium).

4) If *j* < *Nc*, go to step 3. In this case, continue chemotaxis, since the life of the bacteria is not over.

5) Reproduction:

a) For the given k and l, and for each $i= 1,2,....,S$, let

$$IAE^{i}_{health} = \sum_{j=1}^{Nc+1} IAE(i,j,k,l)$$

be the health of bacterium i. Sort bacteria and chemotactic parameters C(i)in order of ascending cost IAE_{health}.

b) The S_r bacteria with the highest IAE_{health} values die and the other S_r bacteria with the best values split (and the copies that are made are placed at the same location as their parent).

6) If $k < Nre$, go to step 2. In this case, we have not reached the number of specified reproduction steps, so we start the next generation in the chemotactic loop.
7) Elimination-dispersal: For $i= 1, 2... S$, with probability *ped*, eliminate and disperse each bacterium (this keeps the number of bacteria in the population constant). To do this, if you eliminate a bacterium, simply disperse one to a random location on the optimization domain.
8) If $l < Ned$, then go to step 1; otherwise end.

5. Results

Case-1

For a first order system with following open loop transfer function the various results are shown.

$$G(s) = \frac{1}{s}$$

A PI form of PID controller was found sufficient to the problem. That means K_d was safely chosen to be 0.The range for K_p, and K_i was chosen to be [-100 100] for both. This necessarily means that the search space is large enough.

The following values are obtained:

Technique	Kp	Ki	IAE
GA	20.025	1.13	1.0004
Bio-Mimicry	19.995	0.9891	1.0000

The above table shows that there is not much difference in the results of either technique. The below plots show that K_p and K_i converge to similar values. The convergence rate is fairly equivalent.

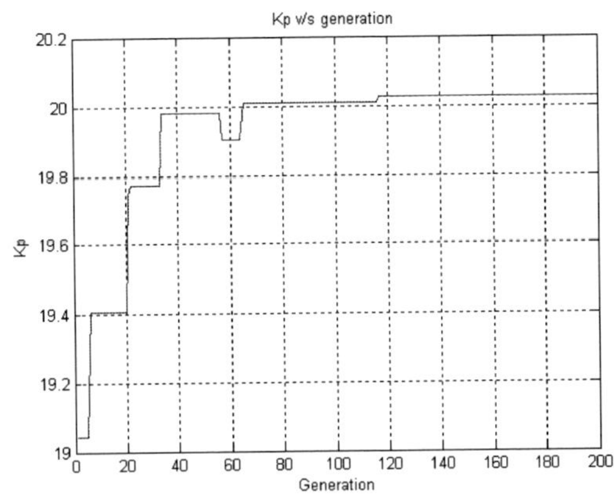

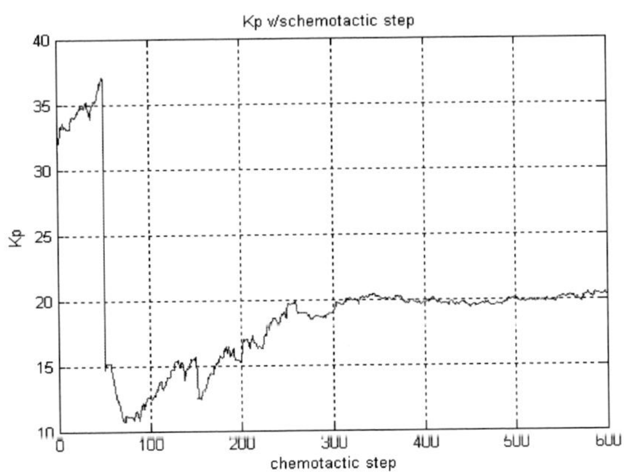

Figure 4. (Top) Kp for GA v/s generations.

Figure 5. (Bottom) Kp for Bio-mimicry v/s chemotactic steps.

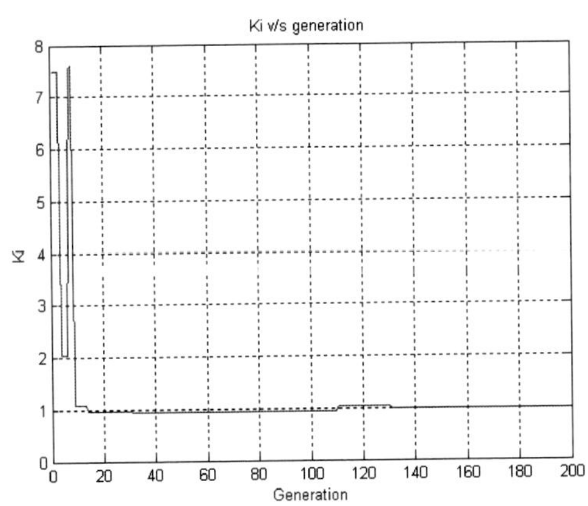

Figure 6. Ki for GA v/s generations.

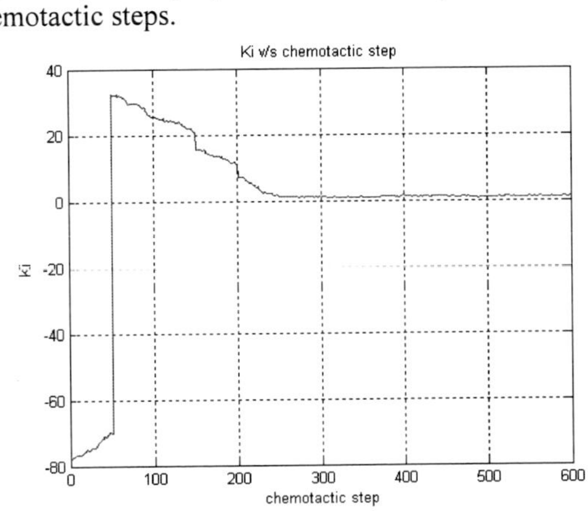

Figure 7. Ki for Bio-mimicry v/s chemotactic steps.

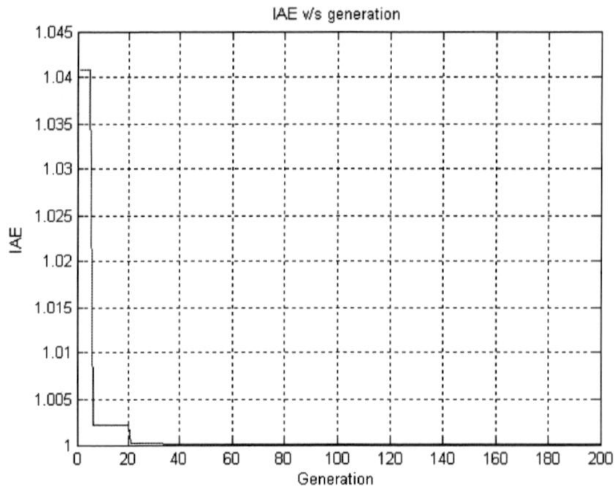

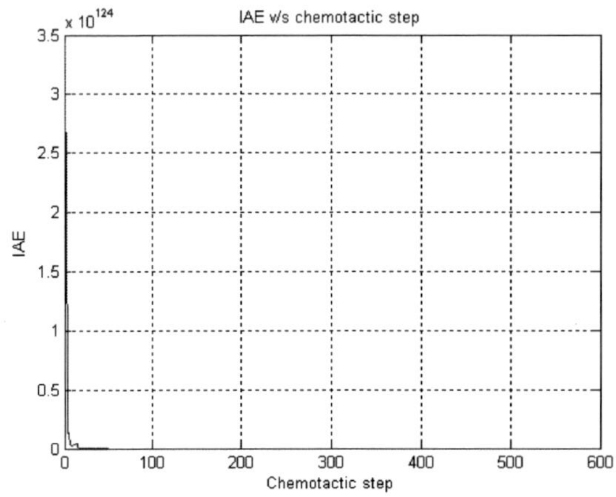

Figure 8. (Top) IAE for GA v/s generations.

Figure 9..(Bottom) IAE for Bio-Mimicry v/s chemotactic steps.

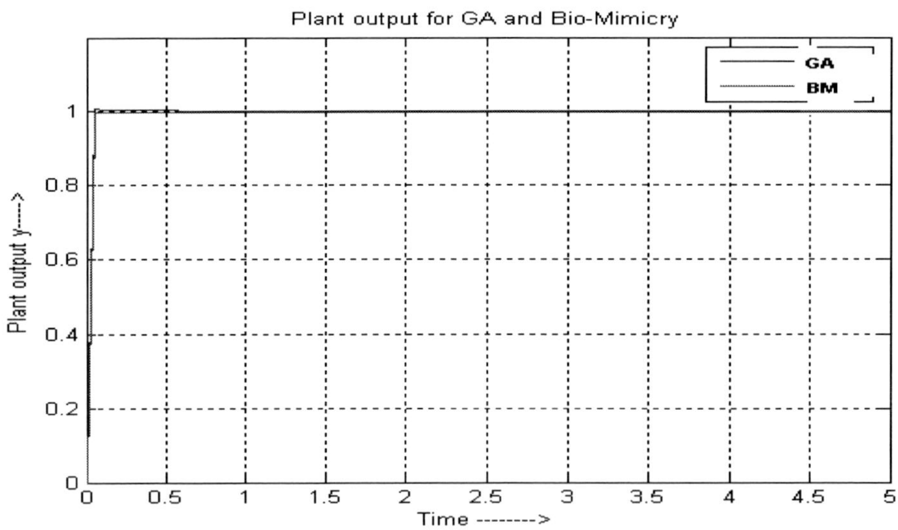

Figure 10. Plant output with GA and Bio-mimicry.

Case 2

Next the experiment is repeated for a second order system with following open loop transfer function:

$$G(s) = \frac{\omega_n^2}{s^2 + 2\delta\omega_n s}$$

Here δ is the damping factor and ω_n is the natural frequency of the system. We choose both an under-damped and over-damped system. It was found that a PD form of PID controller is sufficient for the nature of the problem and K_i was safely chosen to be 0. The range for K_p, and K_d was chosen to be [-100 100] and [-500 500] respectively. This necessarily means that the search space is large enough.

A. $\delta = 0.5$ *and* $\omega_n = 4$ *(Under-damped plant).*

The following values are obtained

Technique	Kp	Kd	IAE
GA	25.0543	625.8742	1.0253
Bio-Mimicry	24.3838	611.8771	1.0021

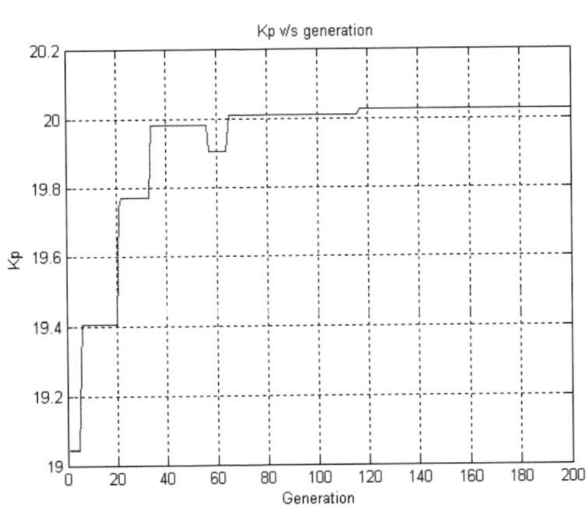

Figure 11. Kp for GA v/s generations.

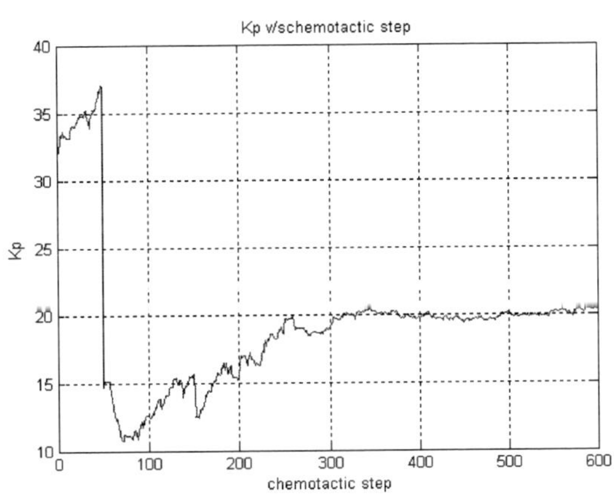

Figure 12. Ki for Bio mimicry v/s chemotactic steps.

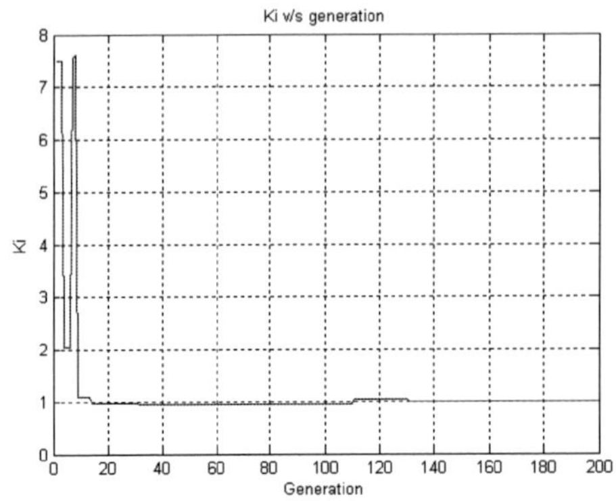

Figure 13. Kp for GA v/s generations.

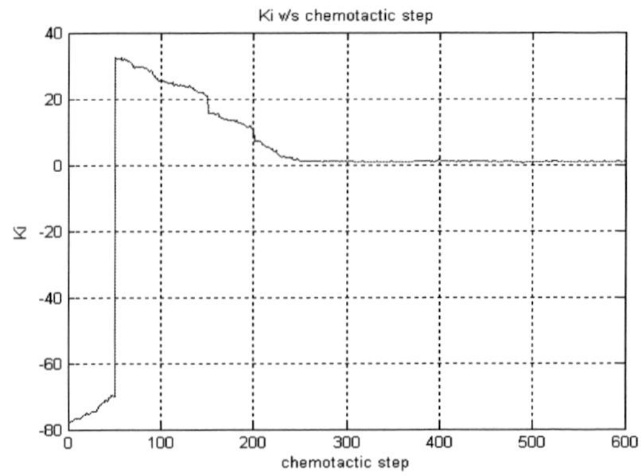

Figure 14. Ki for Bio-mimicry v/s chemotactic steps.

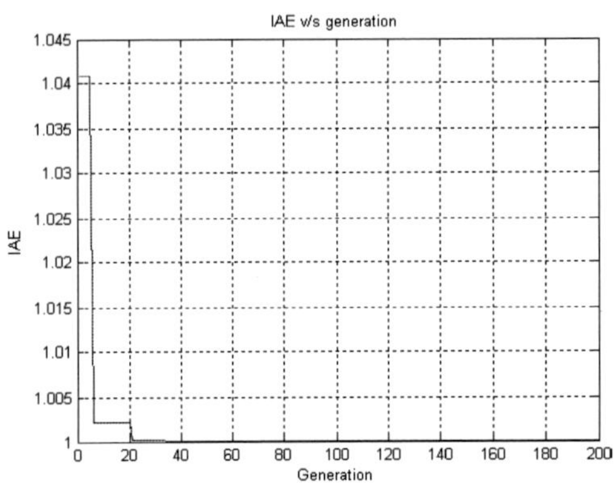

Figure 15. Kp for GA v/s generations.

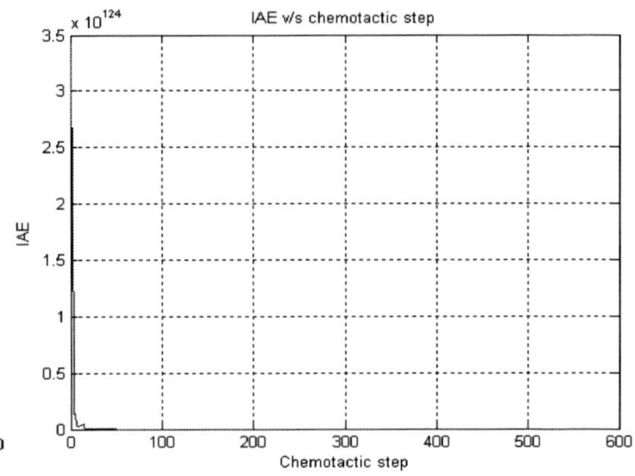

Figure 16. Ki for Bio-mimicry v/s chemotactic steps.

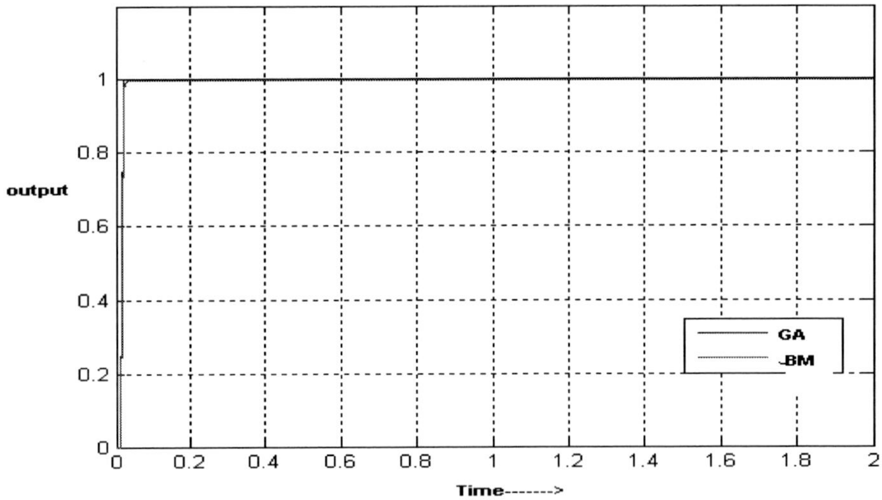

Figure 17. Plant output with GA and Bio-mimicry.

The above plots show that K_p and K_d converge to similar values. The convergence rate is fairly equivalent.

B. $\delta = 1.5$ *and* $\omega_n = 4$ *(Over-damped plant)*

The following values arc obtained

Technique	Kp	Kd	IAE
GA	75.0298	625.4969	1.0049
Bio-Mimicry	74.6732	625.0010	1.0004

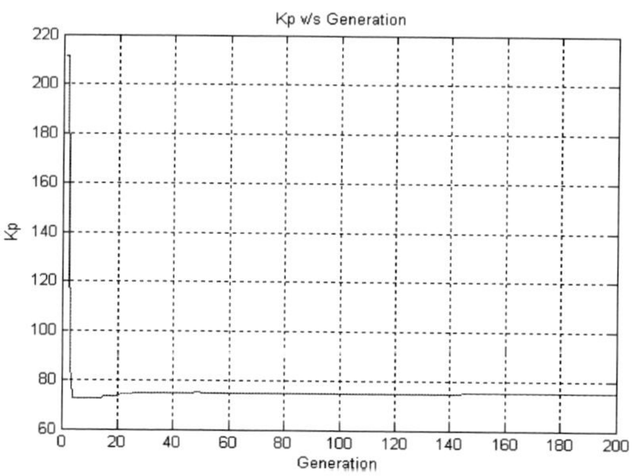

Figure 18. Kp for GA v/s generations.

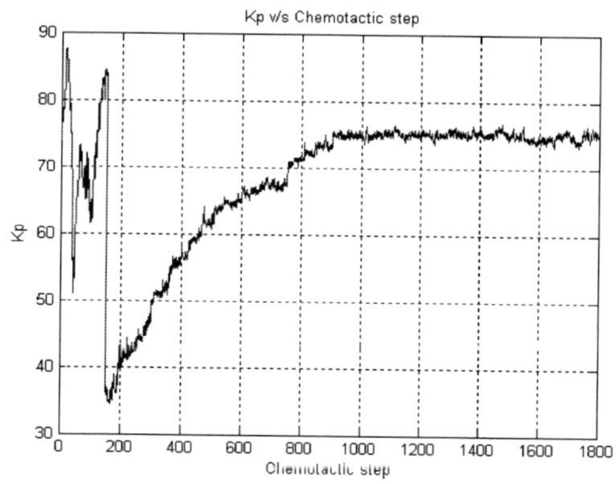

Figure 19. Ki for Bio-mimicry v/s chemotactic steps.

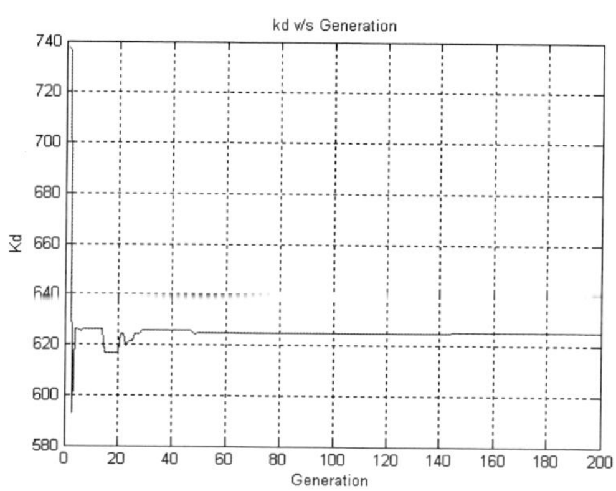

Figure 20. Kp for GA v/s generations .

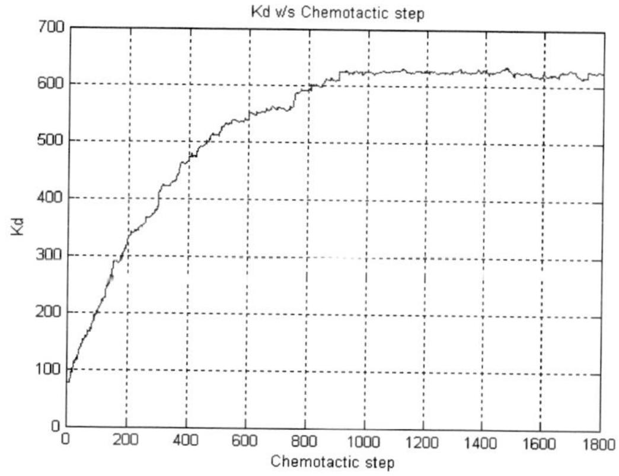

Figure 21. Ki for Bio-mimicry v/s chemotactic steps.

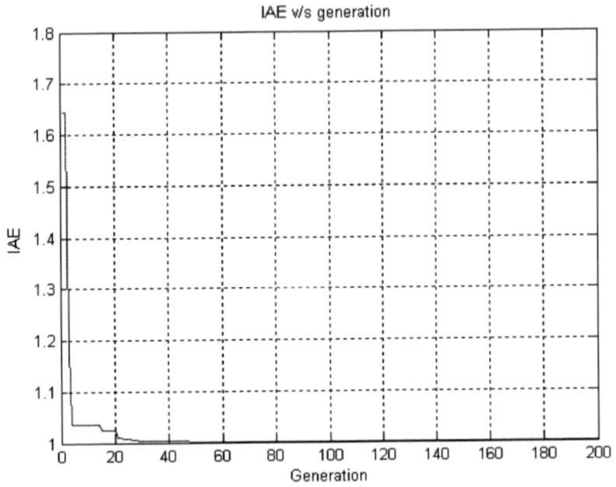

Figure 22. Kp for GA v/s generations.

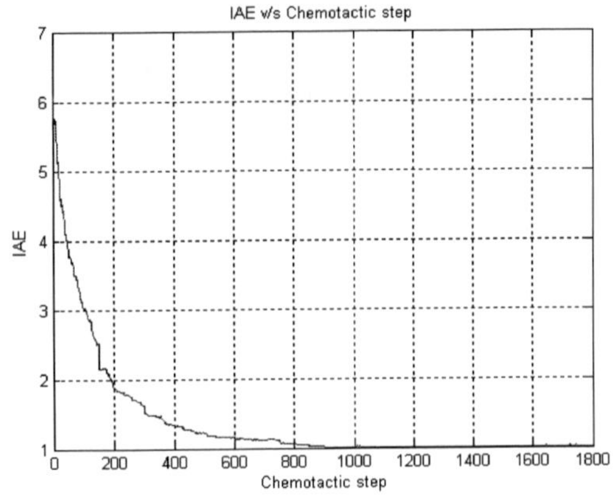

Figure 23. Ki for Bio-mimicry v/s chemotactic steps.

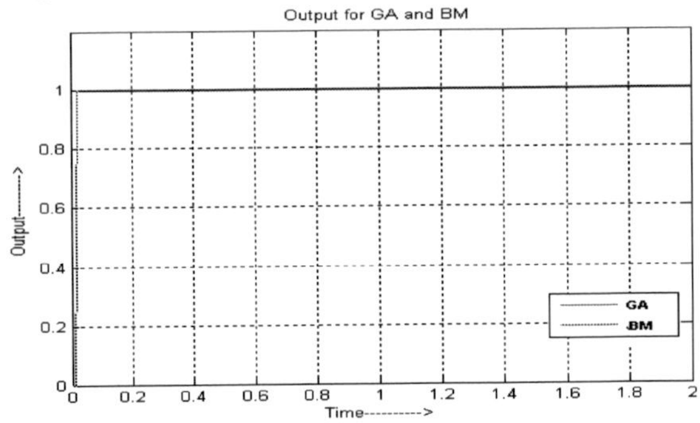

Figure 24. Plant output with GA and Bio-mimicry.

CONCLUSION

It is thus observed that the working of Bio-Mimicry algorithm have been almost equivalent to those of GA. This lets us hypothesize that Bio-Mimicry is equally applicable to PID tuning as GA. Although a thorough study of convergence rate of Bio-Mimicry algorithm wasn't done but the performance of both the algorithms was satisfactory. The results obtained through Bio-Mimicry algorithm are better than the GA in some extent. In all Bio-Mimicry seems to be a promising optimization tool. But a detailed application to vast domain of optimization problems still remains.

REFERENCES

[1] Bose B. K., "*Power Electronics and AC Drives,*" Englewood Cliffs, NJ: Prentice-Hall, 1986.

[2] Chih-Min Lin and Chun-Fei Hsu, "*Adaptive Fuzzy Sliding – Mode Control for Induction Servomotor Systems,*" IEEE Tran. On Energy Conversion, March 23, 2001.

[3] Jones, A.H. and Oliviera P.B. DE Moura, "*Genetic Auto-Tuning of PID Controllers*", IEE Conference on Genetic Algorithms in Engineering Systems: Innovations and Applications, 12-14, September, 1995.
[4] Ogata, K, "*Modern Control Engineering*" Prentice Hall of India, 2002.
[5] Salami, M. and Cain, G., "*An Adaptive PID controller Based on Genetic Algorithm Processor*" IEE Conference on Genetic Algorithms in Engineering Systems: Innovations and Applications 12-14, September, 1995.
[6] Caminos, P. and Munro, N., "*PID controllers: Recent tuning methods and design to specification.*" IEE proceedings, Online No. 20020103, 7 January, 2002.
[7] Dorf, R. and Bishop R., "*Modern Control Systems, Pearson Education Inc,*" 1998.
[8] Goldberg, D. E., "*Genetic Algorithm in Search Optimization and Machine Learning*", Addison Wesley Inc., 1989.
[9] Srivastava, R.P, "*Use of Genetic Algorithms for Optimization in Digital Control of Dynamic Systems*",ACM, 1992.
[10] Passino, K.M., "*Biomimicry of Bacterial Foraging for Distributed Optimization and Control*", IEEE Control System Magazine, June 2002.

In: Focus on Information, Intelligence and Knowledge
Editor: Rafiq Islam

ISBN 978-1-61209-979-8
2011 Nova Science Publishers, Inc.

Chapter 4

INTELLIGENT SYSTEM FOR THREATENING E-MAIL DETECTION: A TEXT MINING APPROACH

Appavu alias Balamurugan[1], Rajaram[2], Ponvidhya[3], Priyadharsini[4] and Uma maheswari[5]
[1]Department of Information Technology,
Thiagarajar College of Engineering, Madurai, India
[2]Department of Computer Science,
Thiagarajar College of Engineering, Madurai, India
[3]Department of Information Technology,
Thiagarajar College of Engineering, Madurai, India
[4]Department of Information Technology,
Thiagarajar College of Engineering, Madurai, India
[5]Department of Information Technology,
Thiagarajar College of Engineering, Madurai, India

ABSTRACT

Data mining is the quest for knowledge in databases to uncover previously unimagined relationships in the amount of data. . This paper proposes to detect malicious emails (i.e. related to terrorism, fraud, etc.). Deception theory suggests that deceptive writing is characterized by reduced frequency of first person pronouns and exclusive words and elevated frequency of negative emotion words and action verbs. We apply this model of deception to the e-mail dataset, then applied decision tree algorithm to generate the decision tree .The decision tree that is generated is used to test the e-mail as threatening or not. In particular we are interested in detecting fraudulent and possibly criminal activities from e-mail corpora.

Keywords: data mining, deceptive theory, decision tree based classification.

* Corresponding author: E-mail: sbit@tce.edu

1. Introduction

Data mining has attracted considerable attention from database practitioners and researchers because of its applicability in many areas such as decision support, market strategy, financial forecast, etc. Combining techniques from the fields like statistics, machine learning, databases, etc., Data mining helps in extracting useful and invaluable information from database. Detecting unusual communication patterns in various means and channels of communications, represents an important class of application directly relevant to security informatics. The e-mail has become one of today's standard means of communication. The large percentage of the total traffic over the internet is the e-mail. The e-mail data is also growing rapidly, creating needs for automated analysis. So, to detect crime, a spectrum of techniques should be applied to discover and identify patterns and make predictions. Data mining has emerged to address problems of understanding ever-growing volumes of information for structured data, finding patterns within data that are used to develop useful knowledge.

As individuals increase their usage of electronic communication, there has been research into detecting deception in these new forms of communication. Models of deception assume that deception leaves a footprint. Work done by various researches suggests that deceptive writing is characterized by reduced frequency of first-person pronouns and exclusive words and elevated frequency of negative emotion words and action verbs. We apply this model of deception to the e-mail dataset and preprocess the email body and to train the system we used ID3 (Iterative Dichotomiser3) algorithm to generate a decision tree that categorize the e-mail as threatening or not.

Contemporary world requires more security. Research works pay more attention to security based issues. Concern about National security has increased significantly since the terrorist attacks on 11 September 2001. Catastrophes may continue if prior measures are not taken at the appropriate time. The CIA, FBI and other federal agencies are actively collecting domestic and foreign intelligence to prevent future attacks. These efforts have in turn motivated us to collect data and undertake this paper work as a challenge.

Data mining is a powerful tool that enables criminal investigators who may lack extensive training as data analyst to explore large databases quickly and efficiently. Computers can process thousands of instructions in seconds, saving precious time. In addition, installing and running software often costs less than hiring and training personal. Computers are also less prone to errors than human investigators. So, this system helps and supports the investigators.

This paper is organized as follows: Section 2 defines problem statement and related work in this area and section 3 describes the proposed work. System implementation presented in section 4. Section 5 discuss about experimental topology. Section 6 describes Performance measure. Finally, Section 7 concludes the paper and point out some potential future work.

2. Problem Statement and Related Work

It's hard to remember what our lives were like, without e-mail. Ranking up with the web as one of the most useful features of the Internet, billions of messages are sent each year.

Though e-mail was originally developed for sending simple text messages, it has become more robust in the last few years. So, it is one possible source of data from which potential problem can be detected. Thus the problem is to find a system that identifies the deception in communication through emails.

P.S. Kaila and Skillicon developed a method based on the singular value decomposition to detect unusual and deceptive communication in e-mail. The problem with this approach is that it does not deal with incomplete data in an efficient and elegant way and can not incorporate new data incrementally without having to reprocess the entire matrix.

Y.Yang compared a cross–experiment between 14 classification methods, including decision tree, naive bayesian, neural networks, linear square fit and rocchio. KNN is one of the top performers, and it performs well in scaling up to very large and noisy classification problems. Seongwook Youn and Dennis McLeod compared a cross experiment between four classification methods and decision tree is one of the top performer. Appavu alias Balamurugan and Rajaram presents association rule mining for suspicious e-mail detection but the accuracy of the classifier is relatively low. Appavu alias Balamurugan and Rajaram compared a cross experiment between four classification methods including decision tree, naive bayes, SVM and ID3 for suspicious e-mail detection and experimental result showed ID3 perform better than other classifiers.

3. PROPOSED WORK

3.1. Classifier Construction Framework

The general framework of the classifier constructions is shown in Figure 1. Data pre-processing, building classifier and classifier testing are three major stages and each stage contains several sub stages. Classification is the process of finding a set of models that describe and distinguish data classes and concepts, for the purpose of being able to use the model to predict the class label of objects whose class label is unknown.

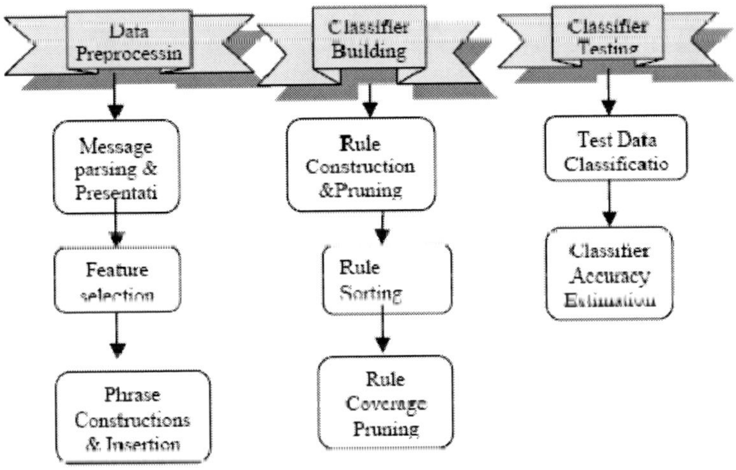

Figure 1. Classifier construction framework.

Figure 2 shows a general framework of the classification process. In the example *object* corresponds to e-mail messages and *object class label* correspond to e-mail category. Every e-mail message contains many features such as keywords, tense and sender details that are used to predict the category of message. Classification is a two step process: 1. Build classification model using training data. Every object of the data must be pre-classified i.e. its class label must be known.

Email	Tense	Bomb	Blast	Terrorist	Attack	Threaten	Class
1	past	y	y	y	y	n	informative
2	past	n	n	y	y	y	informative
3	present	y	y	y	y	n	alert
4	future	n	y	n	y	y	alert
5	past	n	n	n	n	n	normal
6	present	y	y	y	n	n	alert
7	past	n	n	n	n	y	informative
8	past	y	y	y	y	y	informative
9	future	n	y	n	y	y	alert
10	future	y	n	y	n	y	alert

Figure 2. A Sample Training Data set.

In Figure 2, features keyword, and tense are used to build a model that replies how these attributes determine the category of each message. 2. The model generated in the preceding step is tested by assigning class labels to data objects in a test dataset. The test data is different from the training data. Every element of test data is also pre-classified in advance. The accuracy of the classification model is determined by comparing true class labels in the testing set with those assigned by the model.

Example of Threatening and Normal E-Mail

Example of classifying Threatening into Alert and informative email

The informative e-mails provides us with the information about the past historical criminal activities by enhancing some common sense to us such as in the example shown above we came to know that these types of e-mail will never have any consequences in future. The alert e-mails were identified using the deceptive theory and the future tense verbs used in the e-mails. By which the security enforcing methods can be strengthened. Also we can prevent the occurrences of future attacks.

3.2. Database Used in Experiment

The Microsoft Access data base is used to store the e-mail messages. The data set contains the folder information for each of the threatening and normal e-mails. Each message present in the folders contains the sender and receiver e-mail address, date and time, subject, body and text and some other e-mail specific technical details. We created MS Access database for the dataset to store the e-mail message, our database contains two tables. The first table contains the information of the e-mail message, the sender, subject, text and other information. The second table contains the recipient's information .It contains the email address of the recipient and the type (To, CC, BCC) in which message was sent to the recipient.

3.3. E-Mail Preprocessing

Email pre-processing involves the process of transforming the training dataset into a representation suitable for the decision tree- ID3 (Iterative Dichotomiser 3) algorithm. The basic steps are tokenization, data cleaning, stop word removal, stemming (paice/husk) and identification of most interesting terms.

Cleaning

Two techniques are used to clean the e-mail documents. The first technique is removing special characters and symbols from the documents. Special characters include comma, *, #, $, (,) and so on. After removing the special characters from the documents the second technique, case conversion is used. Here the upper case letters in the document are converted to lower case letters. For case conversion first the document is parsed. After parsing the document, each word is retrieved from the document and they are checked. If the words are in upper case letters they are converted to lower case letters and the corresponding words are replaced in the document.

Stop Word Removal

This module is used to remove the words which are not necessary for searching such as prepositions, pronouns and auxiliary verbs. After the documents are cleaned, stop words should be removed from documents. To remove stop words a separate file maintained, which contains stop words. The words are retrieved one by one from the cleaned document. Then the retrieved word is compared with the stop words in the file. If it matches then it is removed from the document.

Stemming

Stemming is an algorithm that allows a document that contains a certain word (for example, informing) to match a query for a related term (for example, informs).There are many stemming algorithms for stemming are porters stemming algorithms, lovins stemming algorithms, paice/husk stemming algorithm and S-removal algorithm.Stemming module is used to reduce the word to its stem or root form. In stemming there are different algorithms used for the identification of the root words such as Lovin's, Porter's, Paice/Husk, S removal algorithms. We have selected the Paice/Husk algorithm for the effective stemming. This algorithm eliminates the extreme outliers and these outliers are clearly over stemmed (too many characters were removed).

Identification of Most Interesting Terms

In e-mails certain words are most frequent and are not discriminative for message contents, such as prepositions, pronouns and conjunctions. Examples of such words are "we", "that", "and"," this", etc. Some widely used conversational English words, such as \I'm", \isn't", \can't", etc are of less importance and elimination of these terms is performed in this step.

Based on the theory of deception a deceptive email will have highly emotional words and action verbs. So, such words are set as keywords and extracted from the input dataset and the most frequent but less deceptive words are eliminated in this step. Example for highly emotional words and action verbs are "lifeless","anger","kill","attack",etc. Prior to indexing and classification, a number of preprocessing steps were performed.

1. E-mails were converted to plain –text from .mbox files.
2. Headers and HTML components were removed.
3. Body of the message was extracted.
4. The messages body was tokenized in to words, stop words were removed and words were converted in to lower case.

3.4. Building the Decision Tree

We have used ID3 (Iterative Dichotomiser 3) Decision Tree algorithm to classify the records. ID3 (Iterative Dichotomiser 3) is an algorithm used to generate a decision tree.The algorithm is based on Occam's razor: it prefers smaller decision trees (simpler theories) over larger ones. However, it does not always produce the smallest tree, and is therefore a heuristic. Occam's razor is formalized using the concept of information entropy:

$$I_E(i) = -\sum^{m}_{j=1\, f(i,j)} \log f(i, j)$$

ID3 algorithm overview

ID3 (Examples, Target_Attribute, Attributes)

(1) Create a root node for the tree
(2) If all examples are positive, Return the single-node tree Root, with label = +.
(3) If all examples are negative, Return the single-node tree Root, with label = -.
(4) If number of predicting attributes is empty, then Return the single node tree Root, with label = most common value of the target attribute in the examples.
(5) Otherwise Begin

- A = The Attribute that best classifies examples.
- Decision Tree attribute for Root = A.
- For each possible value, v_i, of A,
  - Add a new tree branch below Root, corresponding to the test A = v_i.
  - Let Examples(v_i), be the subset of examples that have the value v_i for A
  - If Examples(v_i) is empty
  - Then below this new branch add a leaf node with label = most common target value in the examples

4. System Implementation

Algorithm for classification is based on construction of a tree to model classification process. When the complete tree is constructed it is applied to each message and hence does classification for each. We used 50% of the available email data set (2500 e-mails) for training the ID3 and the remaining 50%(2500 e-mails) for validation (to test the performance of the classifier) and 106 emails out of 2500 in the validation set were incorrectly classified. The accuracy rate for ID3 classifier is 95.76%.

5. Experimental Topology

5.1. Data Source

We have collected over 5000 emails through brain storming session to evaluate the system.Also an email database containing 5000 sample having diverse data like normal and suspicious email.

5.2. Training and Test data

These are the few set of emails used in the experiment and below is the tabulated result after preprocessing. We have shown some of the attributes in the row. The last column represents the outcome. Yes (Y) denotes the occurrence of the attribute in the e-mail and no (N) denotes the non-occurrence of the attribute in the e-mail.

5.3. Experimental Result

The application of data mining to the task of automatic suspicious email classification is done; experiments were carried out on a small email corpus. In order to conduct an experiment setting, different sets of 5000 emails are used. A mixture containing 2000 suspicious emails and 3000 normal emails. The system was trained with the training dataset and the information gain and the entropy were calculated. When the training process was finished, the best quality rules were taken as the final classification rules.

Some of the rules generated by the Decision tree based classifications are:

Rule 1: If Tense=Past, Attack=Yes and then mail =Informative
Rule 2: If Tense=Present, Attack=Yes, Hijack=Yes then mail=Alert
Rule 3: If Tense=Future, bomb=Yes, Assassinate=Yes then mail=Alert.
Rule 4: If Tense=Future, Attack=Yes then mail=Alert
Rule 5: If Tense=Future, Bomb=No, Assassinate=No, Weapon=No, Demolish=No and Blast=No then mail=Normal.

```
=== Evaluation on test set ===

=== Summary ===

Correctly Classified Instances        2394        95.76
Incorrectly Classified Instances       106         4.24
Kappa statistic                          0.9261
Mean absolute error                      0.0375
Root mean squared error                  0.1618
Relative absolute error                  9.6589 %
Root relative squared error             36.8006 %
Total Number of Instances             2500
```

```
=== Detailed Accuracy By Class ===

TP Rate  FP Rate  Precision  Recall  F-Measure  Class
 0.986    0.063     0.934     0.986    0.959    informative
 0.955    0.012     0.984     0.955    0.969    alert
 0.83     0.003     0.97      0.83     0.894    normal

=== Confusion Matrix ===

    a    b    c   <-- classified as
 1171   15    2 |   a = informative
   45 1028    4 |   b = alert
   38    2  195 |   c = normal
```

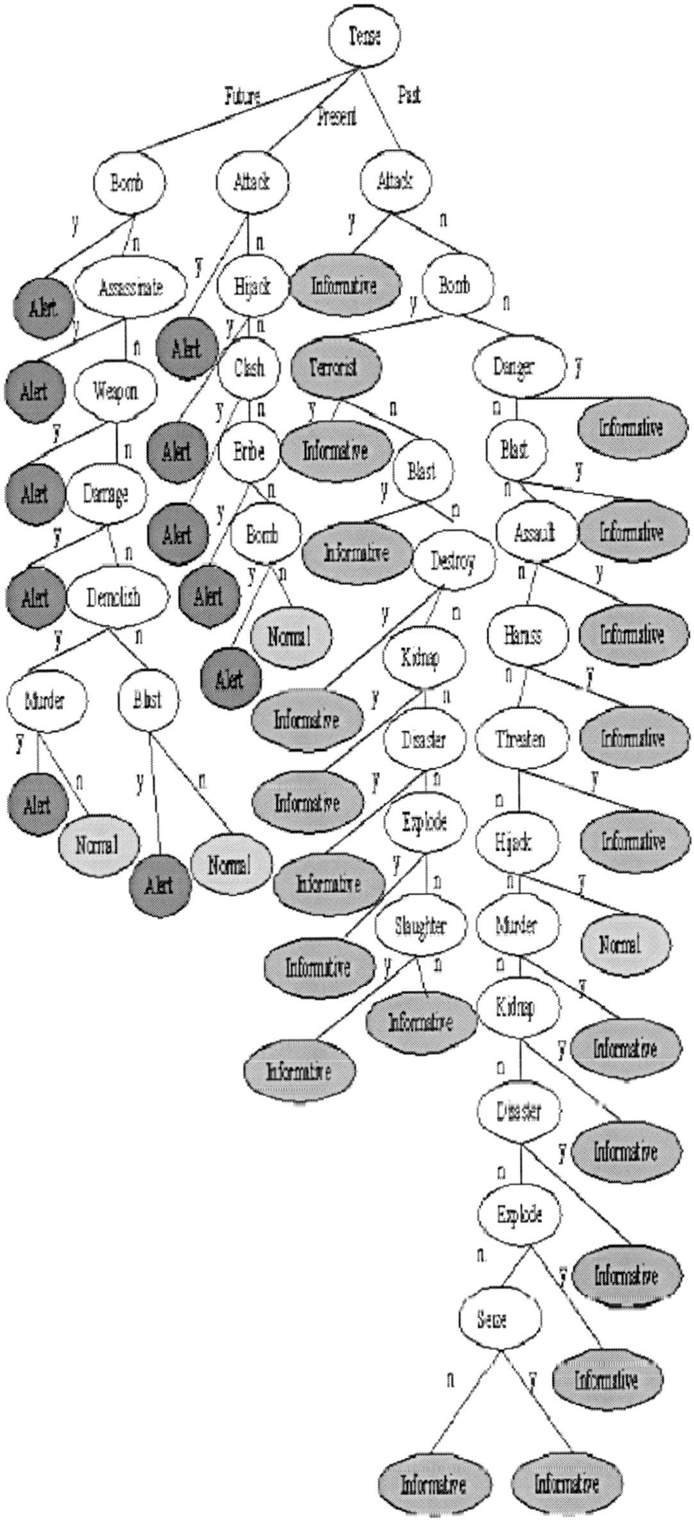

Figure 3.Proposed decision tree.

6. Performance Measure

To evaluate the classifiers on testing dataset, we defined an accuracy measure as follows.

Accuracy (%) =correctly_classified_emails/Total _emails*100

Table 1. Classification accuracy based on data size

No of record	NB	SVM	NN	J 4.8	ID3
500	97.6	98.4	98.6	98.4	98.8
1000	90.2	92.8	91.90	92.9	93.7
1500	92.26	93.54	93.46	93.9	94.66
2000	91.45	92.7	92.65	93.06	93.7
2500	94.04	87.88	95.44	93.36	95.76

Table 2 .Informative Email Precision/Recall/TP/FP Rate based on data size of 2500

	NB	NN	SVM	J48	ID3
Prec	0.925	0.932	0.926	0.926	0.934
Rec	0.98	0.986	0.986	0.986	0.986
TP Rate	0.98	0.986	0.986	0.986	0.986
FP Rate	0.072	0.065	0.072	0.072	0.063

Table 3. Normal Email Precision/Recall/TP Rate/FP Rate based on data size of 2500

	NB	NN	SVM	J48	ID3
Prec	0.976	0.964	0.854	0.788	0.97
Rec	0.698	0.804	0.865	0.791	0.83
TP Rate	0.698	0.804	0.878	0.791	0.83
FP Rate	0.02	0.003	0.890	0.022	0.003

Table 4. Alert Email Precision/Recall/TP/FP Rate based on data size of 2500

	NB	NN	SVM	J48	ID3
Prec	0.953	0.979	0.831	0.978	0.984
Rec	0.950	0.953	0.953	0.907	0.955
TP Rate	0.950	0.953	0.953	0.907	0.955
FP Rate	0.036	0.015	0.147	0.015	0.012

An experiment measuring the performance of ID3 with other classifiers such as NB, NN, SVM was conducted using dataset of different sizes. For example, in case of 2500 dataset, accuracy was 95.76% using ID3 classifier. According to the experimental study on the testing datasets, good classification result order in the experiment was ID3 classifier. Also, precision and recall were used as the metrics for evaluating the performance of each email classification approach.

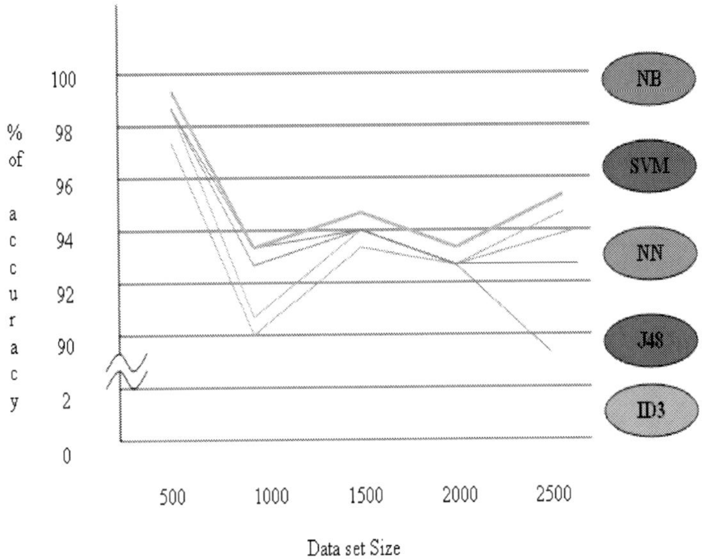

Figure 4. Performance evaluation.

A problem we faced when trying to test out new ideas dealing with e-mail systems was an inherent limitation of the available data, because we only have access to our own data, our results and experiments no doubt reflect some bias. Much of the work published in the email classification domain also suffers from the fact that it tries to reach general conclusion using very small data sets.

Acknowledgments

We are grateful for the computing resources provided by IT Department at Thiagarajar College of Engineering, Madurai.

References

Appavu alias Balamurugan,S. and Rajaram,R (2007) 'A Novel Data mining Approach to detect unusual and deceptive communication in email text', *Proceedings of National Conference on Advanced Computing,MIT,Chennai,Febravary.*

Appavu alias Balamurugan,S. and Rajaram,R (2007) 'Association rule mining for suspicious email detection :A Data mining Approach', *Proceedings of International Conference on Intelligence and Security Informatics ,New Jersey,USA,May.*

Appavu alias Balamurugan.S. and Rajaram,R (2007) 'Data mining Techniques for Suspicious email Detection:A Comparative Study', *Proceedings of European Conference on Data mining,Portugal,June.*

G. Wang, H. Chen, H. Attabakhsh, 'Automatically Detecting Deceptive Criminal Identity'.

Ian H.Written and Eibe Frank, 'Data Mining, Practical Machine Learning Tools and Techniques'.

J. Han, M. Kamber. 'Data Mining Concepts and Techniques'. *Morgan Kaufmann Publishers.*

J. Tang and H. Li, Y. Cao and Z.Tang (2005) 'Email Data cleaning', *Proceedings of International Conference on knowledge Discovery in Database Chicago, USA.*

P.S.Keila and D.B.Skillicorn, 'Detecting unusual and deceptive communication in e-mail ', *Technical reports, June 2005.*

Seongwook Youn and Dennis McLeod, 'A Comparative Study for Email Classification'.

Y.Yang and J.Pedersen. (1997) 'A Comparative study on Feature selection in Text Categorization', *Proc.of the International Conference on Machine Learning, pp.412-420.*

Y.Yang. (1999) 'An Evaluation of Statistical Approaches to Text Categorization', *Journal of Information Retrieval, Vol. 1, No.1/2,,pp.67-88.*

In: Focus on Information, Intelligence and Knowledge
Editor: Rafiq Islam

ISBN 978-1-61209-979-8
2011 Nova Science Publishers, Inc.

Chapter 5

INFRASTRUCTURE OF A CONTEXT ADAPTIVE AND PERVASIVE MULTIMODAL MULTIMEDIA COMPUTING SYSTEM

Manolo Dulva Hina*[*,1,2], Amar Ramdane-Cherif[3], Chakib Tadj[4] and Nicole Levy[5]

[1]Université du Québec, École de technologie supérieure, Montréal, Canada
[2]PRISM Laboratory, Université de Versailles-Saint-Quentin-en-Yvelines, France
[3]PRISM Laboratory, Université de Versailles-Saint-Quentin-en-Yvelines,
45 avenue des États-Unis, 78035 Versailles Cedex, France
[4]Université du Québec, École de technologie supérieure,
1100, rue Notre-Dame Ouest, Montréal, Québec H3C 1K3 Canada
[5]PRISM Laboratory, Université de Versailles-Saint-Quentin-en-Yvelines,
45 avenue des États-Unis, 78035 Versailles Cedex, France

ABSTRACT

The aim of pervasive multimodal multimedia computing is to realize anytime, anywhere computing using various modes of human-computer interaction. The current state-of-the-art pervasive systems and solutions, however, do not include applications that are related to pervasive multimodality. Also, the current multimodal interfaces were designed with pre-defined modes of human-machine interaction that were not chosen based on the given context of the user, of his environment and of his computing system. This paper addresses these weaknesses by proposing a pervasive multimodal multimedia computing system in which its modalities are chosen based on their suitability to the given interaction context. This same system chooses a multimodal (or unimodal) interface based on the given context, available media devices and user preferences. This paper discusses the challenges in designing the infrastructure of such computing system and illustrates how we addressed those challenges. This work is our contribution to the ongoing research that aims at realizing pervasive multimodality.

* Corresponding author: E-mail: manolo-dulva.hina.1@ens.etsmtl.ca

Keywords: multimodal multimedia computing; pervasive computing; human-computer interaction; context-aware system.

1. Introduction

Pervasive computing (also known as *ubiquitous computing*) aims at providing anytime, anywhere computing to a user working on a computing task using various software applications. This has been made possible because the infrastructure of pervasive computing (Satyanarayanan 2001) (Pahlavan and Krishnamurthy 2002), (Satyanarayanan 1990; Satyanarayanan 1996) does exist, an infrastructure that allows both wired and wireless computing and communications. *Pervasive multimodal multimedia computing*, on the other hand, aims to provide the infrastructure that would realize anytime, anywhere computing using various modes of human-computer interaction. Context awareness is an integral characteristic of pervasive computing systems. Context-awareness implies that the system is capable of adapting its operations to the most current context without explicit user intervention. Some context-aware systems have been developed to deliver pervasive healthcare, education, and communication, just to cite a few. Noticeably, however, there is something missing in the current state-of-the-art pervasive applications – one that would permit pervasive multimodal multimedia computing.

A *multimodal system*, in the context of human-computer interface, refers to the fusion of two (or more) input modes – such as speech, pen, gesture, gaze and head and body movements (Oviatt 2002). In contrast, a *unimodal recognition system* or *interface* involves only a single recognition-based technology, such as speech, pen and vision. Some of the current multimodal interfaces do fusion of speech and pen inputs, speech and lip movements, speech and manual gesturing, and gaze tracking and manual input. This implies, however, that the modes of human-machine interaction are already pre-defined from the very beginning. In most of these interfaces, there is an assumption that the setting is ideal (i.e. that there is barely a change in environment's context) and that the user is stationary. For mobile computing, a new conflicting requirement arises – that is, mobility requires that the computing terminal be light and small yet the system is required to deliver more advanced multimedia features. Such requirement suggests that keypads should possibly shrink or even vanish. To this end, the suitability of manual input modality also shrinks while the others – specifically, the vocal input modality – augments. Hence, the necessity for a wireless user interface (Ringland and Scahill 2003) in which speech is the mode for data input interaction.

Most of the multimodal interfaces are not suitable to mobile users. Most of their modes for data input are pre-defined from the beginning and not selected based on their suitability to the environment's context. The drawback of such non-adaptive system is that if a context parameter changes (e.g. the environment becomes noisy) then the effectiveness of the interface is compromised (e.g. speech, as mode for data input, is not effective in a noisy workplace). To this end, we believe that an ideal pervasive system for multimodal application must have a wide-range selection of multimodal interfaces (aside from the regular unimodal interfaces) and at any given time, one particular interface is selected based on its suitability to the context of the user, of his environment and of computing system (henceforth called the *interaction context*).

Modality, in the context of multimodal multimedia computing, refers to the mode of human-computer interaction for data input and output. Given the current state-of-the-art systems and solutions, we have noted that the infrastructure to realize pervasive multimodality is missing. Such infrastructure is important as it is meant to be the backbone that (i) *implements either stationary or mobile computing*, (ii) *allows the invocation of modalities based on context suitability and availability of supporting media devices*, and (iii) *appropriately selects a unimodal or multimodal interface based on the given interaction context*. This paper, therefore, is intended to present the design of such infrastructure, the challenges involved in the design and our proposed solutions to address these challenges.

Apart from this introductory section, the rest of this paper is structured as follows. Section 2 discusses the works related to pervasive computing, multimodal system and the shortcomings of the current state-of-the-art systems and solutions, and presents our idea of addressing them. In section 3, we list down software engineering challenges related to this work and how we address them, and in the process presents our contribution. In section 4, we explain the concepts of context and its relationship with modality and media devices. In section 5, we explain our proposed method for selecting appropriate modalities for the user interaction interface. Sample cases are cited in section 6 and the architectural framework of our proposed system is explained in section 7. Finally, we conclude this paper in section 8 and provide future works that we intend to do.

2. Related Work

Pervasive computing (aka ubiquitous computing) (Weiser 1991; Vasilakos and Pedrycz 2006) realizes anytime, anywhere computing; In doing so, a user's productivity increases as he can continue working on an interrupted computing task whenever and wherever he wishes. Context awareness, along with context management, heterogeneity, scalability, mobility, transparent user interaction, dependability and security are some software infrastructural issues for ubiquitous computing (Costa, Yamin et al. 2008). Several applications of pervasive computing have been developed and implemented, among them are one for pervasive *healthcare* (Varshney 2003), *education* (Garlan, Siewiorek et al. 2002) and *communication* (Vallejos, Desmet et al.), just to cite a few. Missing, however, in the current state-of-the-art pervasive applications is the one that is related to multimodal multimedia computing.

In human-computer interface, multimodality refers to the fusion of two (or more) modes for data input. Since Bolt's original "*Put that there*" concept demonstration (Bouhuys 1995), which processed speech and manual pointing during object manipulation, some significant achievements in multimodal interface have surfaced, such as the one that combines *speech and pen* (Oviatt 2000), *speech and gestures* (Oviatt and Cohen 2000), *speech and lips movements* (Rubin, Vatikiotis-Bateson et al. 1998), *gaze and speech* (Zhang, Imamiya et al. 2004), *speech and mouse* (Djenidi, Ramdane-Cherif et al. 2002), *interface for Internet* (Dong, Xiao et al. 2000), and *wireless user interface* (Ringland and Scahill 2003). But *why build multimodal interface*? It is because it supports more transparent, flexible, efficient and expressive means of human-computer interaction. Multimodal interfaces are expected to be easier to learn and use, and are expected to accommodate more adverse user conditions than in the past (Oviatt 2002). The drawback to the current state-of-the-art multimodal interfaces, however, is that they are all designed with pre-defined modes for data input and without consideration to the varying conditions in the user's workplace, such as a workplace that becomes noisy. Most of these existing systems are also meant for users who are in stationary locations, and hence would become ineffective the moment the user becomes mobile.

Given the limitations cited above, we then envision a pervasive multimodal multimedia computing system. This new computing paradigm's infrastructure is characterized by the following features: (1) it is adaptive to the given interaction context – that is, the modalities for data input and output between the user and the machine are chosen based on their suitability to the given context, (2) that the modalities of interaction are chosen because they can be supported by available media devices, (3) that the chosen user interface is selected based on its suitability to the given context, the availability of supporting media devices and of user's preferences, (4) that the infrastructure supports both stationary and mobile computing, and (5) that the infrastructure itself is autonomic – specifically, it is self-optimizing, self-adaptive, self-configurable, self-optimizing (Horn 2001; Kephart and Chess 2001; Salehie and Tahvildari 2005). Due to space constraints, however, the design of the system's infrastructure as presented in this paper demonstrates only its *self-adaptive* features.

This work is our contribution to the ongoing research in making anytime, anywhere computing using the most suitable form of human-computer interaction possible.

3. Requirements Analysis and Contribution

Here, we list down some software engineering challenges by posing specific technical challenges that need to be addressed. By answering these challenges, we do explain our novel contribution to the software engineering domain.

Our goal is to model a pervasive multimodal multimedia computing system. The design of such a system needs to address some key requirements cited below:

Requirement 1: Provide a generic representation of context. Provide a methodology that allows the incremental definition of context (i.e. add, delete, modify a context parameter).

Requirement 2: Provide the relationship between modality and context. Given that the application domain is multimodality, what parameters constitute the user, environment and system contexts? On what basis a specific modality is considered suitable to a context parameter and to the overall interaction context?

Requirement 3: Given a modality that is suitable to the given interaction context, provide a mechanism that chooses its supporting media devices. Then, given the modality and media devices selections, provide the mechanism that will determine the appropriate (unimodal or multimodal) user interface. What factors should be considered in the selection of a user interface?

The technical challenges are addressed by our proposed solutions given below:

Proposed solution to requirement 1: The term context, in this work, refers to *interaction context* (*IC*) which is the combined contexts of the user, his environment and his computing system. We provide a mathematical model that defines each parameter of interaction context and a virtual machine model for its implementation.

Proposed solution to requirement 2: The modalities for human-machine interaction are *manual, visual* and *vocal*, both to input and output data (details in next section). All context parameters are related to its domain of application, which in this case is multimodality. The relationship to consider is whether a specific modality is suitable to each *IC* parameter and if so, to what extent.

Proposed solution to requirement 3: We establish a relationship between modality and media devices. Given that any selected modality is deemed appropriate to the given context, it also follows that the selected media devices supporting the modality are also suitable to the given context. The interface of user-machine interaction is either unimodal or multimodal. When more than one interface is found suitable, then another factor to consider is the user's preference vis-à-vis user interface. Hence, we propose a priority ranking being assigned to the user's preference. The same priority ranking applies to the user's preferred modality and preferred media devices. The selection of user's interface, therefore, is based on appropriately selected modalities, available media devices and user's preferences.

4. Context, Multimodality and Media Devices

Here, we define context and provide its mathematical representation. We also illustrate how we can implement an incremental definition of context. Then, we derive the relationships that exist between context and multimodality and between multimodality and media devices.

4.1. Context Definition and Representation

In chronological order, the early definition of context includes that of (Schilit and Theimer 1994; Oviatt 2002) in which context means the answers to the questions "*Where are you?*", "*With whom are you?*", and "*Which resources are in proximity with you?*". Schilit defined context as the changes in the physical, user and computational environments. This idea is taken afterwards by Pascoe (Pascoe 1998) and later on by Dey (Dey, Salber et al. 1999). Brown considered context as "*the user's location, the identity of the people surrounding the user, as well as the time, the season, the temperature, etc.*" (Brown, Bovey et al. 1997). Ryan defined context as the environment, the identity and location of the user as well as the time involved (Ryan, Pascoe et al. 1997). Ward viewed context as the possible environment states of an application (Ward, Jones et al. 1997). In Pascoe's definition, he added the pertinence of the notion of state: "*Context is a subset of physical and conceptual states having an interest to a particular entity*". Dey specified the notion of an entity: "*Context is any information that can be used to characterize the situation of an entity. An entity is a person, place or object that is considered relevant to the interaction between a user and an application, including the user and application themselves*" (Dey 2001). This definition became the basis for Rey and Coutaz to coin the term interaction context: "*Interaction context is a combination of situations. Given a user U engaged in an activity A, then the interaction context at time t is the composition of situations between time t_0 and t in the conduct of A by U*" (Rey and Coutaz 2004).

We adopted the notion of "*interaction context*", but define it in the following manner: An interaction context, $IC = \{IC_1, IC_2, \ldots, IC_{max}\}$, is a set of all possible interaction contexts. At any given time, a user has a specific interaction context *i* denoted as IC_i, $1 \le i \le max$, which is composed of variables that are present in the conduct of the user's activity. Each variable is a function of the application domain which, in this work, is multimodality. Formally, an IC is a tuple composed of a specific user context (UC), environment context (EC) and system context (SC). An instance of IC is given as:

$$IC_i = UC_k \otimes EC_l \otimes SC_m \tag{1}$$

where $1 \le k \le max_k$, $1 \le l \le max_l$, and $1 \le m \le max_m$, and max_k, max_l and max_m = maximum number of possible user contexts, environment contexts and system contexts, respectively. The Cartesian product (symbol: ⊗) denotes that IC yields a specific combination of UC, EC and SC at any given time.

The user context UC itself is composed of parameters that describe the state of the user during the conduct of his activity. A specific user context k is a tuple composed of ($ICParam_{k1}$, $ICParam_{k2}$, … $ICParam_{kmaxk}$) and is given by:

$$UC_k = \bigotimes_{x=1}^{max_k} ICParam_{kx} \tag{2}$$

where $ICParam_{kx}$ = parameter of UC_k, k = the number of UC parameters, $k \in 1 .. max_k$. Using similar convention as that of UC, a specific instance of environment context EC_l and a specific instance of system context SC_m can be specified as follows:

$$EC_l = \bigotimes_{y=1}^{max_l} ICParam_{ly} \tag{3}$$

$$SC_m = \bigotimes_{z=1}^{max_m} ICParam_{mz} \tag{4}$$

4.2. Incremental Definition of Interaction Context

As stated, an instance of *IC* is composed of specific instances of *UC, EC*, and *SC*, which themselves are composed of one or more parameters. To realize the incremental definition of *IC*, each of these parameters is introduced into the system, one at a time.

In our work, a virtual machine is designed to add, modify or delete one context parameter at a time, making *IC* parameters a reflection of the system's dynamic needs. A *virtual machine* (VM) is software that creates a virtualized environment on computer platform so that the end user can operate the software. *Virtualization* is the process of presenting a group or subset of computing resources so that they can be accessed collectively in a more beneficially manner than their original configuration. In effect, a VM is an *abstract computer*; it accepts input, has algorithms and steps to solve the problem related to the input, and yields an output. The steps taken by the VM are its "*instructions set*" which is a collection of functions that the machine is capable of undertaking. A *layered VM* is a group of VM's wherein interaction takes place only between adjacent layers. *Layering* is a design choice to limit the propagation of errors within the concerned layer only during a modification of its functionality. Generally, in layered VM, the top layer refers to the interface that interacts with the end users while the bottom layer interacts with the hardware. Hence, Layer 0 is the bottom layer composed of sensors that generate some raw data representing the value needed by the topmost VM layer.

Figure 1 shows the functionality of such "*machine*". In general, the transfer of instruction command is top-down (steps 1 to 4). At Layer 0, the raw data corresponding to the *IC* parameters are collected for sampling purposes. The sampled data are then collated and interpreted, and the interpretation is forwarded to different layers bottom-up (steps 5 to 8).

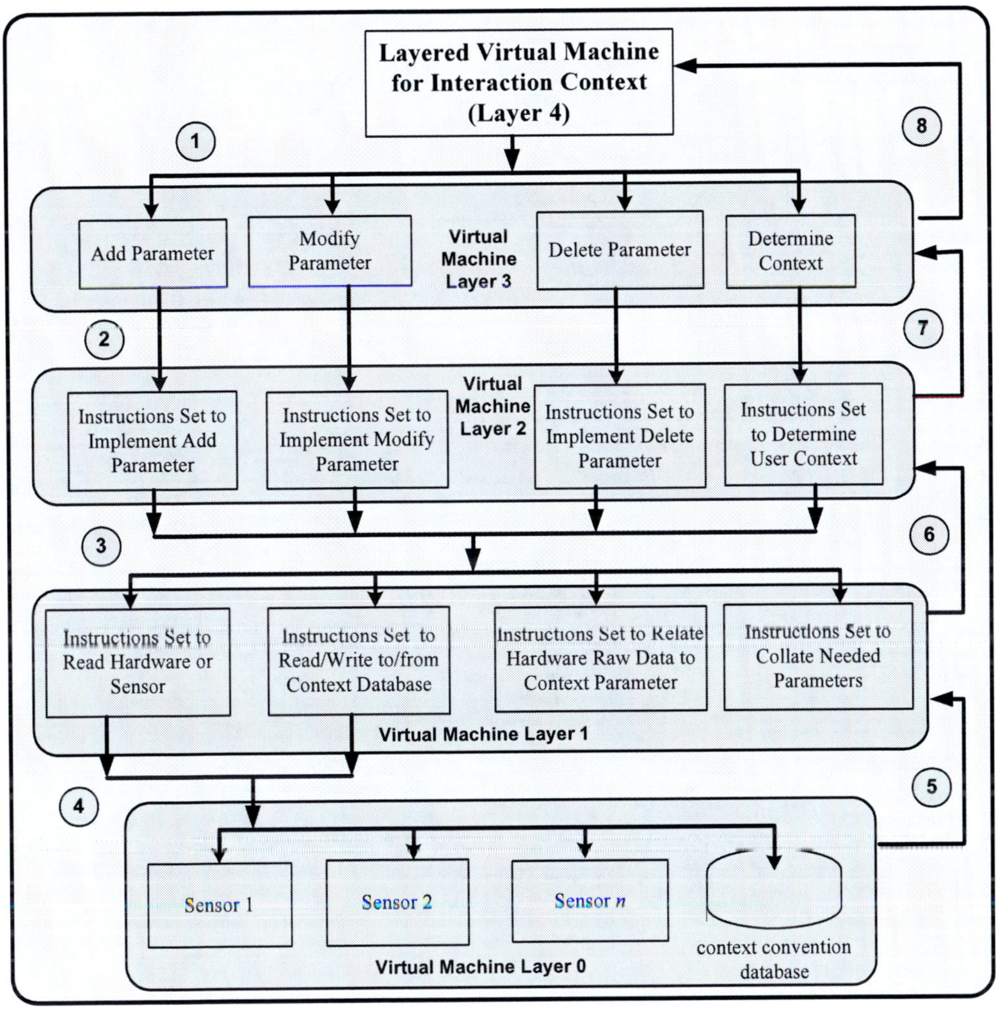

Figure 1. The design of a layered virtual machine for incremental interaction context.

The VM Layer 4 acts as the human-machine interface; its "*instruction set*" are the four functions found in Layer 3. The "*add parameter*", "*modify parameter*", and "*delete parameter*" are basic commands that manipulate the sensor-based context parameters while "*determine context*" yields the values of currently-defined parameters. VM Layer 2 is a "*library of functions*" that collectively supports Layer 3 instructions while Layer 1 is another "*library of functions*" that acts as a link between Layer 2 and Layer 0.

4.2.1. Adding a Context Parameter

Consider using VM to add a specimen context parameter: the "*noise level*". See the design of VM's user interface in Figure 2.

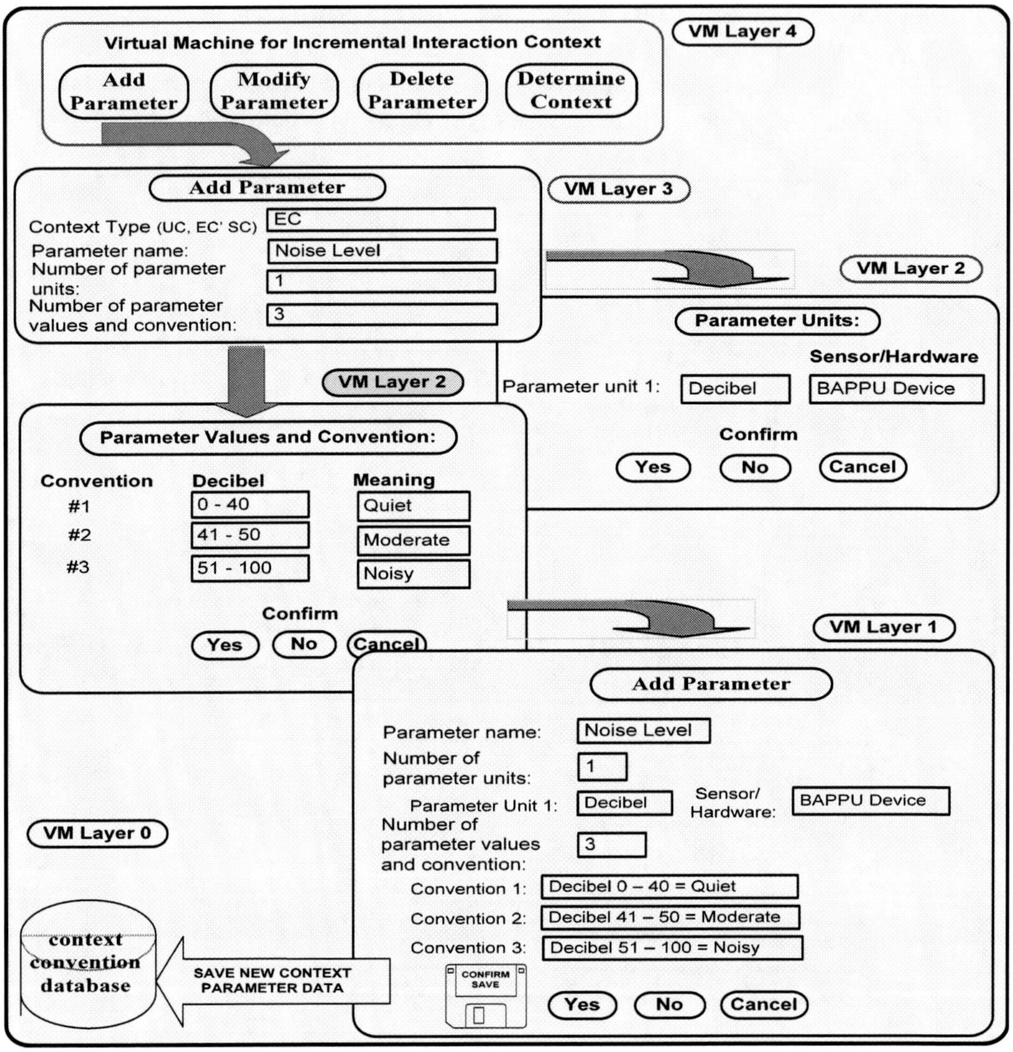

Figure 2. The interactions among layers to add a new (specimen only) context parameter: "Noise Level".

As shown in the diagram, upon invoking the VM user interface (i.e. Layer 4), the user chooses the "*Add Parameter*" menu. A window opens up which transfers the execution control to Layer 3. Then data entry takes place. To realize adding a new context parameter, at least four data entry functions must exist, namely: (i) *getting context type of the parameter*, (ii) *getting name of the parameter*, (iii) *getting the parameter's number of units*, and (iv) *getting number of parameter values and conventions*. In Layer 3, the user inputs "*Noise level*" as parameter name, itself an EC parameter, "*1*" as parameter unit, and "*3*" as parameter values and conventions. When done, two new windows open up, one window at a time, that brings up the functionalities of Layer 2. For each parameter's unit, the VM receives input for the parameter's unit name and the sensor (or hardware) that supplies its raw data. As shown, the unit for "Noise level" is specified as "*decibel*" and the BAPPU noise measuring device (http://www.bappu.com/) (or any sensor for that matter that measures noise and supplies data to the computer) as the sensor supplying the data. When done, another Layer 2 window opens up for data entry of "*Parameter values and conventions*". In the diagram, the user specifies the value (range of decibels) that he considered is equivalent to "*quiet*", "*moderate*" and "*noisy*". When done, a window for Layer 1 opens up to save the newly-added parameter information. This function interacts directly with the hardware (i.e. the context convention database).

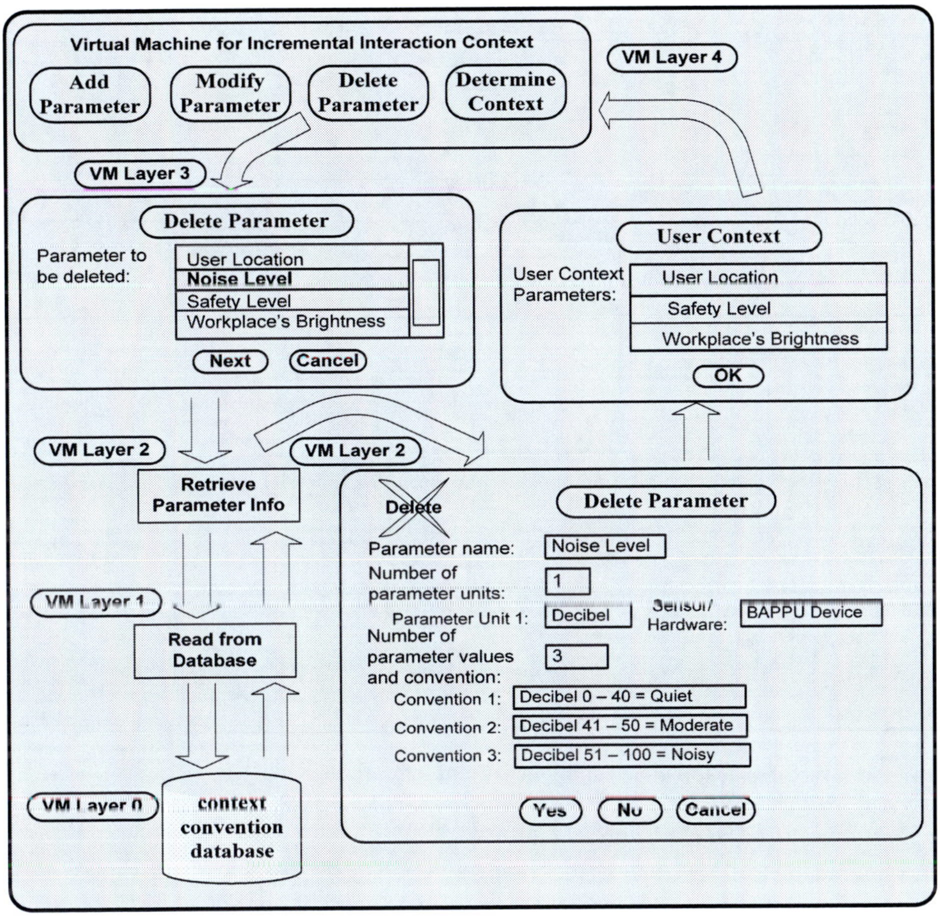

Figure 3. The VM layers interaction to realize "deleting a user context parameter".

4.2.2. Modifying and Deleting a Context Parameter

The VM layers interaction involved in "*Modify parameter*" is almost identical to that of "*Delete Parameter*" function. The only thing extra in the former is a procedure that allows user to select the context parameter that should be modified. Other than that, everything else is the same. The processes involved in "*Delete Parameter*" menu are shown in Figure 3.

Upon menu selection, the execution control goes to Layer 3, demanding the user to specify the parameter for deletion (e.g. "*Noise level*" is chosen for deletion). Upon confirmation, the information about the parameter for deletion is extracted and read from database (transfer of control from Layer 2 to Layer 1 then to Layer 0). When the information for deletion is read, the control goes back to Layer 2 where such information is presented and a re-confirmation of its deletion is required. When parameter deletion is done, the control goes back to Layer 3 which presents the updated list of context parameters. An "*OK*" button click transfers the control back to Layer 4.

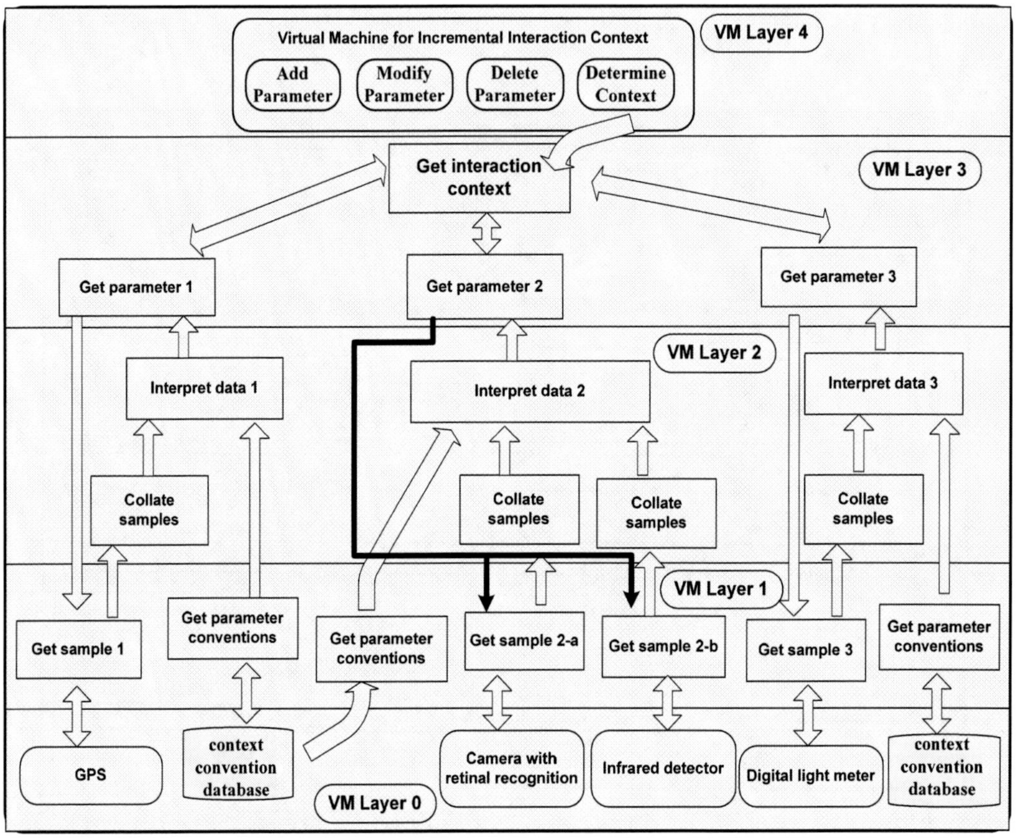

Figure 4. VM layers interaction in detecting the current interaction context.

4.2.3. Capturing the User's Current Context

The interactions of VM layers to "*Determine Context*" are shown in Figure 4. This is simulated using three specimen context parameters, namely (i) *the user location*, (ii) *the safety level*, and (iii) *the workplace's brightness*. When the user opts for this menu, the VM execution control goes to Layer 3. The function "*get user context*" creates threads equal to the number of parameters. Hence, this process produces thread "*get parameter 1*", assigned to detect user location, thread "*get parameter 2*" assigned to get the user's safety level, and the thread "*get parameter 3*" for the user's workplace's brightness (i.e. light intensity). The concepts involved are identical for each thread. Consider the case of "*user location*". The thread passes control to Layer 1 where a function takes sample data from a sensor (e.g. global positioning system, GPS, at http://www.rayming.com), attached to the user computer's USB port or value transmitted to the computer via wireless communication. In the VM design, the user can specify the number of raw data that need to be sampled and in what frequency (*n* samples per *m* unit of time). These *n* samples are then collated, normalized and interpreted.

For example, a specimen GPS data of 5 samples, taken 1 sample per minute, is shown in Figure 5. The data are then normalized (averaged), hence, the user's computer is located at 14°11' latitude and -120°57' longitude. Then, this value is interpreted using the convention values for user location parameter. Table 1 shows the format of the convention values of the specimen parameters. (Recall that the convention value of a parameter is created during the "*Add Parameter*" process.) Using Table 1-a, the interpretation identifies if the user (who uses the computer equipped with a GPS) is at home, at work or on the go.

Sample Track Data

Format: DMM M/D/Y H:M:S -5.00 hrs
Datum [104|: WGS 84
T 05/30/2006 14:46:08 47°11.839' -120°57.156'
T 05/30/2006 14:46:08 47°11.843' -120°57.192'
T 05/30/2006 14:46:08 47°11.845' -120°57.205'
T 05/30/2006 14:46:08 47°11.894' -120°57.203'
T 05/30/2006 14:46:08 47°12.001' -120°57.199'

T = Track point
DMM = Degree-Minutes.Minutes
M/D/Y H:M:S = timestamp from year to second
-5.00 hrs = time setback from GMT. Eastern standard time (New York, Toronto, Montréal are 5 hours back from GMT)
47° = latitude in degrees 11.839 = latitude in minutes
-120° = longitude, 'W120 degrees and 57.156 seconds

Figure 5. Sample GPS data gathered from Garmin GPSIII+.

Specimen parameter 2 *(the workplace's safety level)* is a function of (i) the person sitting in front of the computer, and (ii) the presence of other people in the user's workplace. A camera with retinal recognition (http://www.informatik. uniaugsburg.de/~kimjongh/ biometrics/retinal.pdf) may be used to identify the person sitting in the user's seat. The identification process would yield three values: (1) *User* – if the legitimate user is detected, (2) *Other* – if another person is detected, and (3) *Empty* – if no one is detected. Also, an infrared detector (http://www.globalsources.com/manufacturers /InfraredDetector.html) may be used to identify the presence of other person in front or in either side of the user. The identification process would yield two values: (1) *Image* – if at least one person is detected, and (2) *No Image* – if nobody is detected. (Note that the image and pattern recognition is not the subject of this work; hence, the detection process is not elucidated further in this paper.). The VM takes n = 5 samples, normalizes them and compares the result against the convention values in Table 1-b. The interpretation yields a result indicating if user's workplace is *safe*, *sensitive* or *risky*. This specimen parameter is useful for people working on sensitive data (e.g. bank manager) but can be irritating to a person working with teammates (e.g. students working on a project). Hence, this specimen parameter can be added or deleted on the user's discretion.

Table 1. Sample conventions of the specimen sensor-based context parameters

1.a – Convention format for user location

Convention No.	Latitude	Longitude	Meaning
1	<value11>	<value12>	At home
2	<value12>	<value22>	At work
3	Not <value11> and not <value12>	Not <value12> and not <value22>	On the go

1.b – Convention format for safety level in user's workplace

Convention No.	Detected in user's seat	Detected in user's workplace	Meaning
2	User	Image	Sensitive
1	User	No Image	Safe
3	Empty	Image	Sensitive
1	Empty	No Image	Safe
3	Other	Image	Risky
3	Other	No Image	Risky

1.c – Convention format for light intensity in user's workplace

Convention No.	Foot-Candle	Meaning
1	<value-range1>	Bright
2	<value-range2>	Moderate
3	<value-range3>	Dark

The third specimen parameter (i.e. *workplace's brightness*) detects the workplace's light intensity. Here, we can assume that a sensor measuring the light's intensity (http://www.gossen-photo.de/english/lichtmess_produkte.html) is attached to the computer's USB port. Its measurement unit, the *foot-candle*, is the number of "*lumens*" falling on a square foot of an inch; lumen is a unit of light used to rate the output of a bulb. For example, we may assume the following conventions in a user's workplace: (i) 0 – 9 foot candles = *dark*, (ii) 10 – 20 foot-candles = *moderate*, and (iii) 21 – 100 foot-candles = *bright*. The processes involved in sampling, collating and interpreting sensor data for parameter 3 is identical with the other 2 parameters mentioned above. Given the specimen parameters, when "*determine context*" is done, the output indicates (1) *if the user is at home, at work or on the go*, (2) *if user's workplace is safe, sensitive or risky*, and (3) *if the workplace's light intensity is bright, moderate or dark.*

4.3. Context Storage and Dissemination

In general, if a system is to obtain an accurate representation of the user's interaction context, then the system must be introduced to the most number of possible context parameters. As a context parameter is added to the system, the VM's context convention database forms a tree-like IC structure, as shown in generic format in Figure 6. Every new IC parameter is first classified as either UC or EC or SC parameter and is then appended as a branch of UC or EC or SC. Then, the conventions of the parameter are identified.

For the IC information to be propagated in a pervasive system, the data representation used is XML Schema which is based on XML (Ross and Lightman 2005). Figure 7(Left) illustrates the general XML format of a context parameter (i.e. name, units, source of raw data, and conventions) and Figure 7(Right) shows the various snapshots of windows involved in adding a parameter in the VM as implemented using Java programming language.

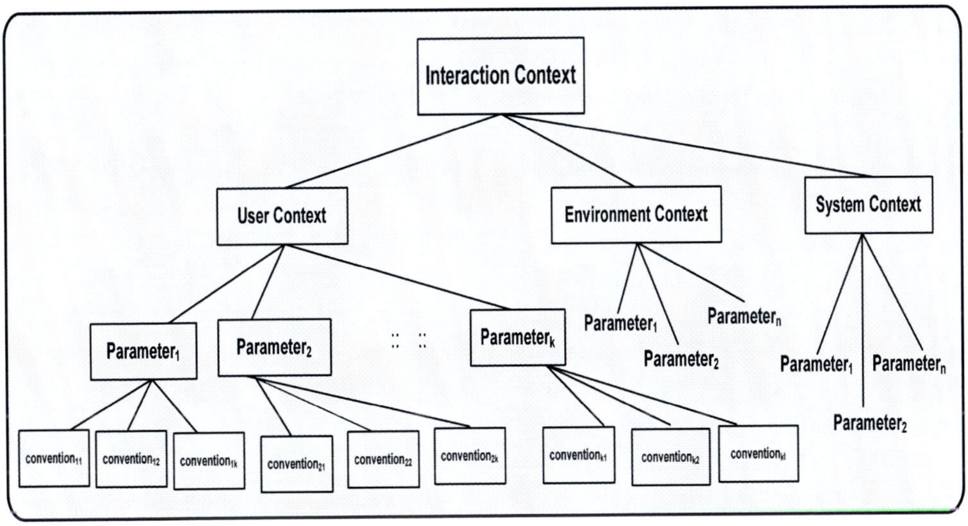

Figure 6. The structure of stored IC parameters.

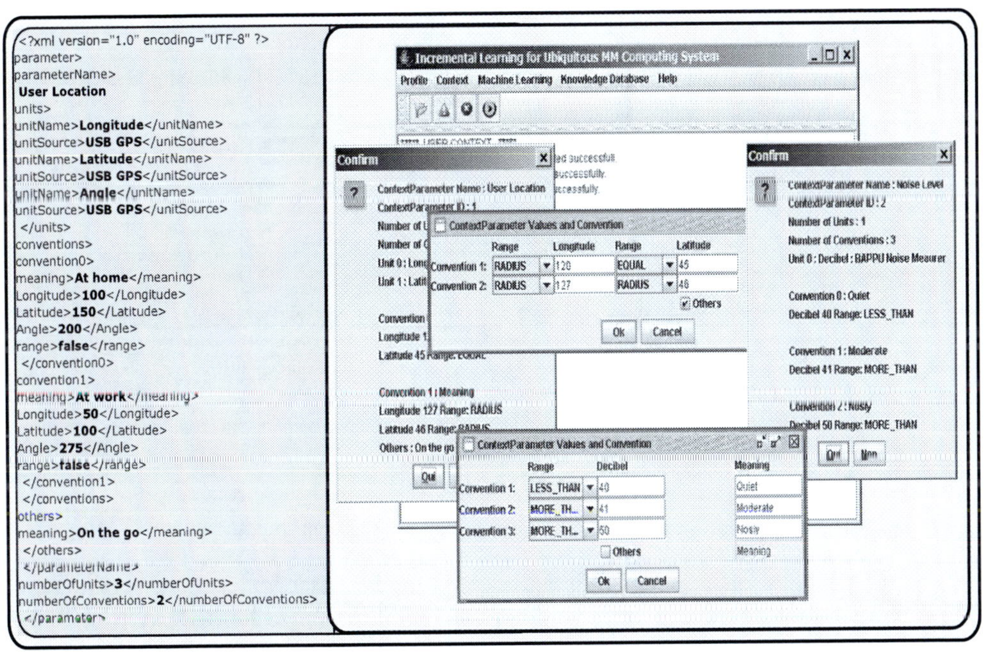

Figure 7. (Left) Sample context parameter in XML, (Right) snapshots of windows in add.

4.4. Measuring a Modality's Context Suitability

Multimodality refers to the selection of modalities based on its suitability to the given IC. Here, modality refers to the logical interaction structure (i.e. the mode for data input and output between a user and computer). A modality may only be realized if there is/are media devices that would support it. In this work, media refers to a set of physical interaction devices (plus some software supporting the physical devices). Using natural language processing as basis, we group modalities as follows: (i) Visual Input (VI_{in}), (ii) Vocal Input (VO_{in}), (iii) Manual/Tactile Input (M_{in}), (iv) Visual Output (VI_{out}), (v) Vocal Output (VO_{out}), and (vi) Manual/Tactile Output (M_{out}). Multimodality is possible if there is at least one modality for data input and at least one modality for data output.

Using Z language specification (Lightfoot 2001), let there be a set of input modalities and output modalities, as given by INPUTMODE ::= VI_{in} | VO_{in} | M_{in} and OUTPUTMODE ::= VI_{out} | VO_{out} | M_{out}. Let the relationship multimodality be a set of pairs of input and output modalities, as denoted by multimodality: P (INPUTMODE ⊗ OUTPUTMODE) where P = power set which is the set of all subsets denotes power set and ⊗ = Cartesian product. At any given time, we can test if multimodality is possible by getting an instance of input and output modalities. Assume that x: P INPUTMODE, y: P OUTPUTMODE. Multimodality is possible if x and y forms a pair within the relationship multimodality and that neither x nor y is an empty set, that is, Possible((x,y)) $\Leftrightarrow$ (x,y) $\in$ multimodality $\wedge x \neq \varnothing \wedge y \neq \varnothing$.

Accordingly, media devices themselves are grouped as follows: (i) Visual Input Media (VIM), (ii) Visual Output Media (VOM), (iii) Oral Input Media (OIM), (iv) Hearing Output Media (HOM), (v) Touch Input Media (TIM) (vi) Manual Input Media (MIM), and (vii) Touch Output Media (TIM). The relationships that map modalities with media group and then the media group with media devices are shown in Figure 8.

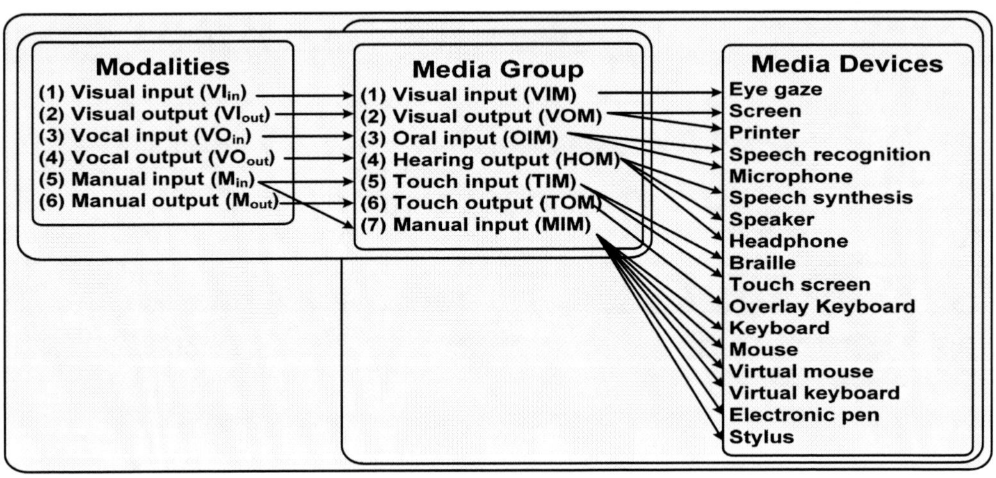

Figure 8. The relationship among modalities, media group and physical media devices.

To build a relationship between modalities and media devices, let there be a function g_1 that maps a modality to a media group, given by g_1: MODALITY → MEDIAGROUP. This is shown in Figure 8. There is often a case, however, when two or more media devices both belong to one media group. In such a case, devices selection is determined through their priority rankings. Hence, let there be another function g_2 that maps a media group to a media device and its priority rank, denoted by g_2: MEDIAGROUP → (MEDIADEVICE, $£_1$) where $£_1$ denotes an integer value greater than zero. Sample elements of these functions are:

g_1 = {(VI_{in},VIM), (VI_{out},VOM), (VO_{in},OIM), (VO_{out},HOM), (M_{in},TIM), (M_{in},MIM), (M_{out},TOM)}
g_2 = {(VIM, (eye gaze,1)), (VOM, (screen,1)), (VOM, (printer,1)), (OIM, (speech recognition,1)), (OIM, (microphone,1)), (HOM, (speech synthesis,1)), (HOM, (speaker,2)), (HOM, headphone,1)), etc.}

A modality's suitability to IC is equal to its collective suitability to the IC's individual parameters. Suitability measure is not binary, not just suitable or not suitable, because there are some cases wherein the extent of suitability lies in between. Hence, our suitability measures are *high*, *medium*, *low* and *inappropriate*. High suitability means that the modality in consideration is the preferred mode for computing; medium suitability means the modality is simply an alternative mode for computing, hence, its absence is not considered as an error but its presence means added convenience to the user. Low suitability means the modality's effectiveness is negligible and is the last recourse when everything else fails. Inappropriateness recommends that the modality should not be used at all.

If the collective IC is composed of n parameters, then the modality in consideration has n suitability scores. We then adopt the following conventions:

1. A modality's suitability to an IC parameter is one of the following: H (high), M (medium), L (low), and I (inappropriate). Mathematically, H = 1.00, M = 0.75, L = 0.50, and I = 0.
2. The modality's suitability score to an IC is given by:

$$\text{SuitabilityScore}_{\text{modality}} = \sqrt[n]{\prod_{i=1}^{n} \text{context\_parameter}_i} \quad (5)$$

where *i* = parameter index and *n* – total number of parameters. Given the calculated value, a modality's IC suitability is given by:

$$\text{FinalSuitability}_{\text{modality}} = \begin{cases} \text{H if SuitabilityScore}_{\text{modality}} = 1.00 \\ \text{M if } 0.75 \leq \text{SuitabilityScore}_{\text{modality}} < 1.00 \\ \text{L if } 0.50 \leq \text{SuitabilityScore}_{\text{modality}} < 0.75 \\ \text{I if SuitabilityScore}_{\text{modality}} < 0.50 \end{cases} \quad (6)$$

4.5. Selecting Context-Appropriate Modalities

Figure 9 shows the algorithm for determining the suitability of modalities to a given IC and if multimodality is possible (i.e. section 4.4). Checking that multimodality is possible is done by determining that not all of input modalities (i.e. specified by indexes 1, 2 and 3) are scored "inappropriate", and so does for output modalities (i.e. specified by indexes 4, 5 and 6). The optimal input modality is chosen from the group of input modalities and is one with the highest IC suitability score. The same principle applies to the selection of the optimal output modality. Subject to the availability of media devices, an optimal modality is ought to be implemented; all others are considered optional. In the absence of supporting media devices, an alternative modality is chosen and is one that has the next highest score.

```
//Initialization
// Assumption: index i = 1 to 6 represent modalities
//  VI_in, VO_in,, M_in, VI_out, VO_out, and  M_out respectively
 for i = 1 to modality_max
    modality[i] = Null
end for
 //Evaluate IC suitability of individual modality
for i = 1 to modality_max do
    // Calculate modality's IC suitability score
    temp := 1.0;
    for j = 1 to parameter_max do
       // read suitability level of a modality with respect to parameter i
       case suitabilityLevel(j) of
            suitabilityLevel(j) = High:            score = 1.00
            suitabilityLevel(j) = Medium:          score = 0.75
            suitabilityLevel(j) = Low:             score = 0.50
            suitabilityLevel(j) = Inappropriate: score = 0.0
       end case
       temp = temp * score;
    end for
    finalScore := temp ↑ (1/parameter_max)
    case finalScore of
         finalScore = 1.00:            Suitability = High
         0.75 ≤ finalScore <  1.00:  Suitability = Medium
         0.50 ≤ finalScore <  0.75:  Suitability = Low
         finalScore <  0.50:           Suitability = Inappropriate
    end case
   modality[i] = Suitability
end for
//check if multimodality is possible
if ((modality[1] != Inappropriate) OR (modality[2] != Inappropriate) OR
   (modality[3] != Inappropriate)) AND ((modality[4] != Inappropriate) OR
   (modality[5] != Inappropriate) OR (modality[6] != Inappropriate)) then
  // implement the chosen modalities
  // choose the optimal modality for data input and output
   optimalInputModality := largest(modality[1], modality[2], modality[3])
   optimalOutputModality := largest(modality[4], modality[5], modality[6])
 else
  //multimodality is not possible
end if
```

Figure 9. Algorithm to determine a modality's suitability to IC and if multimodality possible.

Again, any alternative modality must be supported by available media devices. This process is repeated until the system is able to find a replacement modality that can be supported by currently available media devices.

4.6. Selecting Media Devices Supporting Modalities

When multimodality is possible and optimal modalities have been chosen, then supporting media devices are checked for availability. Using function g_1, the media group that supports the chosen modality can be identified. Given that MODALITY = {VI_{in}, VO_{in}, M_{in}, VI_{out}, VO_{out}, M_{out}} and MEDIAGROUP = {VIM, OIM, MIM, TIM, VOM, HOM, TOM} and that g_1: MODALITY → MEDIAGROUP, then formally, using Z language, we can specify that for every p which is a selected modality, there corresponds a media group q, wherein neither p nor q is an empty set, such that the ordered pair (p, q) is a member of set g_1, that is $\forall$ p: MODALITY; $\exists$ q: MEDIAGROUP $|$ $x \neq \emptyset \wedge y \neq \emptyset \bullet (p, q) \in g_1$.

By using function g_2, the top-ranked media devices that belong to the specified media group can also be identified. Given function g_2, a media device d, priorities p_1 and p_2 of type £1, then the specification for finding the top-ranked device within the media group m is that the mapping between m and media device d with a priority ranking of d_1 does exist in function g_2 and that the numerical value of device's priority p_1 is less than p_2, (i.e. the lesser the numerical value, the higher is its priority ranking), that is, $\exists$ m: MEDIAGROUP; $\forall$ d: MEDIADEVICE; $\exists$ p_1: £1; $\forall$ p_2: £1 $|$ $d \bullet m \rightarrow (d, p_1) \in g_2 \wedge (p_1 < p_2)$.

Table 2. A sample media devices priority table (MDPT)

Media Group	Media Devices				
	Priority = 1	Priority = 2	Priority = 3	::	Priority = *n*
Visual Input	Eye Gaze				
Oral Input	Microphone, Speech Recognition				
Touch Input	Touch Screen	Braille Terminal			
Manual Input	Mouse, Keyboard	Virtual Mouse, Virtual Keyboard	Electronic pen	Stylus	Braille
Visual Output	Screen	Printer	Electronic projector		
Hearing Output	Speaker	Headphone, Speech Synthesis			
Touch Output	Braille	Overlay Keyboard			

Let there be a media devices priority table (MDPT) (see Table 2) which tabulates all media groups, and its set of media devices arranged by priority ranking. T = {T_1, T_2… T_{max_table}} is the set of MDPT's. The elements of table $T_n \in$ T, where n ∈ 1 .. max_table, are similar to elements of function g_2. No two MDPT's are identical. To create a new table, at least one of its elements is different from all other tables that have already been defined. The priority ranking of a media device may be different in each MDPT. In general, it is possible that two or more different context scenarios may be assigned to one common MDPT.

When a new media device d_{new} is added or introduced to the system for the *first time*, the device is associated to a media group and is given a priority ranking r by the user. What happen to the rankings of other devices d_i, ($i \in 1 .. n$, and n = number of media devices) which are in the same media group as d_{new} in the MDPT? Two things may happen, depending on the user's selection. The first possibility is after having the new device's priority Priority(d_{new}) set to r then the priority of the other device i, ($1 \leq i \leq n$) denoted Priority(d_i), remains the same.

The second possibility is the priority rankings of all media devices (d_i) ranked r or lower are adjusted such that their new priority rankings are one lower than their previous rankings. Formally, this is specified as: $\forall$ i, $\exists$ r: £$_1$; $\forall$ d_i, $\exists$ d_{new}: MEDIADEVICE | (Priority(d_{new}) = r $\wedge$ Priority(d_i) $\geq$ r) $\Rightarrow$ Priority(d_i)' = Priority(d_i) + 1.

5. Modalities in User Interaction Interface

Here, we wish to determine the selections of modality to be used in the user interaction interface, given that it is already known that multimodality is possible for implementation.

5.1. Media Groups and Media Devices

In general, the association between media group and media devices can be specified as:

$$\text{VIM} = \text{eye gaze} \vee \text{gesture interpreter} \tag{7}$$

$$\text{OIM} = \text{speech recognition} \wedge \text{microphone} \tag{8}$$

$$\text{MIM} = \text{keyboard} \vee \text{Braille} \vee \text{pen} \tag{9}$$

$$\text{TIM} = \text{mouse} \vee \text{virtual mouse} \vee \text{touch screen} \tag{10}$$

$$\text{VOM} = \text{terminal screen} \vee \text{printer} \tag{11}$$

$$\text{HOM} = \text{speech synthesis} \wedge (\text{speaker} \vee \text{headset}) \tag{12}$$

$$\text{TOM} = \text{tactile keyboard} \vee \text{Braille} \tag{13}$$

Note that the relationships cited above list down only limited number of commonly-used media devices. That said, these relationships can be modified easily and accordingly to include other media devices.

Given that INPUT_MEDIA_GROUP = {VIM, OIM, TIM, MIM}, then the power set (i.e. the set of all subsets) of this group is given by P(INPUT_MEDIA_GROUP) = {{VIM}, {OIM}, {TIM}, {MIM}, {VIM, OIM}, {VIM, TIM}, {VIM, MIM}, {VIM, OIM, TIM}, {VIM, OIM, MIM}, {VIM, TIM, MIM}, {VIM, OIM, TIM, MIM}, {OIM, TIM}, {OIM, MIM}, {OIM, TIM, MIM}, {TIM, MIM}, {}}. These results indicate that as far as human-machine interaction interface is concerned, there can be four types of user interface. Note that, by definition, an interface is a function of input modalities only.

Hence, the possible types of human-machine interaction interfaces are:

- *unimodal interface* – media devices (and supporting software) belonging to VIM, OIM, TIM and MIM can be used, but there is no fusion of data generated by one media with the data generated by another media (ex. speech, pen, vision)
- *bimodal interface* – there are 6 possible combinations of fusion of data generated by two media devices – that of {VIM, OIM}, {VIM, TIM}, {VIM, MIM}, {OIM, TIM}, {OIM, MIM}, and {TIM, MIM}. The current state-of-the-art multimodal interfaces fall in this category.
- *trimodal interface* – this interface allows the combination of data that are generated by three media devices into a new meaningful data; there are 4 possible selections, namely: {VIM, OIM, TIM}, {VIM, OIM, MIM}, {VIM, TIM, MIM}, and {OIM, TIM, MIM}
- *quadmodal interface* – this one would combine all types of input media altogether, {VIM, OIM, TIM, MIM}.

As far as research advancement (i.e. year 2008) is concerned, a user interface can only be unimodal or bimodal. There is no evidence that a trimodal interface, let alone a quadmodal one, exists, at least not yet.

5.2. Selection of User Interface Modalities

Given that a unimodal or bimodal user interface is possible, then the system, in consultation with the user, decides the most suitable user interface. We believe that the selection of user interface that suits the user should be based on (i) the modalities and media groups that suit the given context (ii) the availability of media devices (and their supporting software) that would support the chosen modalities, and (iii) the availability of the preferred interface system or middleware within the user's computing system, and (iv) the user's preference on these interface as given by their priority rankings.

In order to determine whether the system will implement a unimodal or multimodal interface, let there be a *human-machine interaction interface priority table* (HMIIPT). This table contains important information related to the user's preferences, such as: (i) the priority ranking of multimodal and unimodal interface, (ii) the priority ranking of modalities within the interface, and (iii) the priority ranking of media devices that support a modality. See Table 3 for a sample HMIIPT.

Suppose that the user prefers a unimodal interface over a multimodal (actually, bimodal) one. Using HMIIPT, the system then determines the ranking assigned to each of the input modalities. Then the priority ranking of media devices supporting a preferred modality is taken from the media devices priority table (see Table 1). For multimodal interface, the user is also consulted in the priority rankings of all modality combinations/fusion. In the same manner, the priority of media devices supporting the multimodal fusion is also indicated in HMIIPT.

The selection process for optimal user interface modality uses the following functions to determine the score of each user interface mode. We take the result yielding the highest score as the optimal user interface modality:

$$\text{User Interface Modality Score} = \text{User Interface Priority} \times \text{Modality Priority} \times \text{Media Devices Priority} \quad (14)$$

$$\text{User Interface Priority} = (m + 1 - p) / m \quad (15)$$

$$\text{Modality Priority} = (q + 1 - p) / q \quad (16)$$

$$\text{Media Device}_i \text{ Priority} = 1 - \sum_{1}^{i-1} (1/n) \quad (17)$$

such that for user interface priority, the variable m = number of types of user interface (i.e. unimodal, bimodal, etc. available in HMIIPT) and p = priority ranking as obtained from HMIIPT, $1 \le p \le m$. For modality's priority, the variable q = number of modality selections (i.e. available in HMIIPT) and p = priority ranking of the specified modality as obtained from HMIIPT, $1 \le p \le q$. For media devices priority, n = available media devices supporting the chosen modality and media group (see Equations (7) through (13)). Also, given the ith device, where $1 \le i \le n$, then d_i = priority ranking obtained from MDPT and n = number of media devices supporting the same modality as the ith device.

Table 3. A sample human-machine interaction interface priority table (HMIIPT)

User Interface Modality by Priority			
Interaction Interface Priority	**Priority of Modalities in Unimodal Interface**	**Priority of Modalities in Bimodal Interface**	**Priority of Media Devices in Bimodal Interface**
1. Bimodal Interface 2. Unimodal Interface	1. MIM 2. OIM 3. TIM 4. VIM	1. VIM and OIM	1. eye gaze - speech, 2. gesture - speech
		2. VIM and TIM	1. eye gaze - mouse, 2. eye gaze - Virtual mouse, 3. gesture - mouse, etc.
	Priority of Media Devices in Unimodal Interface	3. VIM and MIM	1. eye gaze - keyboard, 2. eye gaze - Braille, 3. gesture - keyboard, etc.
		4. OIM and TIM	1. speech - mouse, speech - touch screen, 3. speech - virtual mouse, etc.
	Same Priority Ranking as MDPT	5. OIM and MIM	1. speech - keyboard, 2. speech - pen, 3. speech - joystick, etc.
		6. TIM and MIM	1. mouse - keyboard, 2. mouse - Braille, 3. mouse - pen, etc.

6. Sample Cases

Here, we simulate sample cases and accordingly apply the principles discussed in the previous sections.

6.1. Sample Interaction Context and Its Corresponding Modalities Selection

Suppose that we are given the following interaction context: (i) *user context*: user location = at home, user handicap = none, (ii) *environment context:* noise level = quiet, safety factor = safe, (iii) *system context:* computing device = PDA. Suppose that the context convention database contains the conventions and suitability score of different modalities as shown in Tables 4 to 8. What will be the optimal modality?

Table 4. User location conventions and suitability scores

(a): User location convention table using GPS values

Convention No.	Latitude	Longitude	Meaning
1	<value11>	<value12>	At home
2	<value12>	<value22>	At work
3	Not <value11> and not <value12>	Not <value12> and not <value22>	On the go

(b): Modality selection based on user location

Type of Modality	User location = At home	User location = At work	User location = On the go
Visual Input	H	H	L
Visual Output	H	H	H
Vocal Input	H	H	H
Vocal Output	H	H	H
Manual Input	H	H	H
Manual Output	H	H	H

Table 5. User disability conventions and suitability scores

(a): User Profile/Disability Convention

Convention No.	User Profile
1	Regular User
2	Deaf
3	Mute
4	Visually-Impaired
5	Manually-Impaired

(b): Modality selection based on user profile/handicap

Type of Modality	User = Regular user	User = Deaf	User = Mute	User = Visually-Impaired	User = Manually Impaired
Visual Input	H	H	H	I	H
Visual Output	H	H	H	I	H
Vocal Input	H	M	I	H	H
Vocal Output	H	I	M	H	H
Manual Input	H	H	H	H	I
Manual Output	H	H	H	H	I

Table 6. Workplace safety conventions and suitability scores

(a): The safety/risk factor convention table

Convention No.	Detected in user's seat	Detected in user's workplace	Meaning
2	User	Image	Sensitive
1	User	No Image	Safe
3	Empty	Image	Sensitive
1	Empty	No Image	Safe
3	Other	Image	Risky
3	Other	No Image	Risky

(b): Modality selection based on workplace's safety level

Type of Modality	Safety level = Safe	Safety level = Sensitive	Safety level = Risky
Visual Input	H	M	I
Visual Output	H	M	I
Vocal Input	H	M	I
Vocal Output	H	M	I
Manual Input	H	M	I
Manual Output	H	M	I

Table 7. Noise level conventions and suitability scores

(a): Sample noise level convention

Convention No.	Decibel	Meaning
1	Less than 41	At home
2	41 to 50	At work
3	Greater than 50	On the go

(b): Modality selection based on noise level

Type of Modality	Noise level = Quiet	Noise level = Moderate	Noise level = Noisy
Visual Input	H	H	H
Visual Output	H	H	H
Vocal Input	H	M	I
Vocal Output	H	H	M
Manual Input	H	H	H
Manual Output	H	H	H

Table 8. Computing device conventions and suitability scores

(a): Computing device convention

Convention No.	Latitude
1	PC
2	Laptop
3	PDA
3	Cellular phone

(b): Modality selection based on user's computing device

Type of Modality	Computing device = PC	Computing device = Laptop	Computing device = PDA/ Cellphone
Visual Input	H	H	L
Visual Output	H	H	H
Vocal Input	H	H	H
Vocal Output	H	H	H
Manual Input	H	H	H
Manual Output	H	H	L

The given interaction context is IC = (c_1, c_2, c_3, c_4, c_5) = (1, 1, 1, 1, 3). The calculated final suitability scores of each type of modality are given below:

Visual Input= $[(H)(H)(H)(H)(L)]^{1/5} = [(1)(1)(1)(1)(0.50)]^{1/5}$ = 0.87 = Medium suitability
Vocal Input = $[(H)(H)(H)(H)(H)]^{1/5} = [(1)(1)(1)(1)(1)]^{1/5}$ = 1 = High suitability
Manual Input = $[(H)(H)(H)(H)(H)]^{1/5} = [(1)(1)(1)(1)(1)]^{1/5}$ = 1 = High suitability
Visual Output = $[(H)(H)(H)(H)(H)]^{1/5} = [(1)(1)(1)(1)(1)]^{1/5}$ = 1 = High suitability
Vocal Output = $[(H)(H)(H)(H)(H)]^{1/5} = [(1)(1)(1)(1)(1)]^{1/5}$ = 1 = High suitability
ManualOutput=$[(H)(H)(H)(H)(L)]^{1/5}=[(1)(1)(1)(1)(0.50)]^{1/5}$= 0.87 = Medium suitability

Given this case, multimodality is possible (see section 4.4). The preferred *input modality* is either Vocal Input (VO_{in}) or Manual Input (M_{in}). The preferred *output modality* is Visual Output (VI_{out}) or Vocal Output (VO_{out}). All non-optimal modalities are considered *optional.* Using Figure 8 as visual reference, the media groups that suit the given interaction context are OIM, MIM, VOM, and HOM.

6.2. Sample Media Devices and User Interface Selection

Consider that the same user has the following media devices: OIM: speech recognition system, microphone, MIM: mouse, keyboard and electronic pen, VOM: screen and keyboard, and VOM: speech synthesis, speaker. Question: what is the most suitable human-computer interaction interface for the user?

To answer this question, we need to know the user preferences concerning media devices and user interface. Assuming that the data in the specimen MDPT and HMIIPT apply, then:

i. Bimodal Interface Priority = (2 + 1 – 1)/2 = 1
ii. Modality Priority: OIM and MIM = (6 + 1 - 5)/6 = 0.33
iii. Media Devices Priority: speech-keyboard = 1, speech-pen = 2/3
iv. Unimodal Interface Priority = (2+1-2)/2 = 0.5
v. Modality Priority: MIM = 1, OIM = 0.75
vi. Media Devices Priority: OIM: speech = 1 MIM: keyboard = 1, pen = 3/5

The calculation for bimodal interface follows:

(i) speech-keyboard = 1*0.33*1 = 0.33
(ii) speech pen = 1* 0.33*0.67 = 0.22

For, unimodal interface, the result is:

MIM: (i) keyboard = 0.5 * 0.75 * 1 = 0.375

(ii) pen = 0.5 * 0.75 * 0.6 = 0.225, and
OIM: speech = 0.5 * 0.75 * 1 = 0.375.

In this case, the system determines that the optimal user interface is a unimodal one in which the optimal input device/modality is the keyboard and speech.

7. Our Multimodal Multimedia Computing System

7.1. Architectural Framework

Our proposed system is conceived for two purposes: (1) to contribute to multimodal multimedia computing research and (2) to further advance self-adaptive computing system. To achieve the first goal, we develop the model that relates modality with user context, and associate media devices to support the implementation of the chosen modality. In the second goal, we advocate the propagation of knowledge, acquired through training, into the user's computing environment so that such knowledge can be used for system adaptation based on user's requirements and system's restrictions. The major components of our multimodal multimedia computing system are shown in Figure 10. The functionality of each component is given below:

- *The Task Manager Agent (TMA)* – manages user's profile, task and pertinent data and their deployment from a server to the user's computing device, and vice versa.
- *The Context Manager Agent (CMA)* – detects user context from sensors and user profile, and selects the modality and media apt for the context.
- *The History and Knowledge-based Agent (HKA)* – responsible for ML training and knowledge acquisition.
- *The Layered Virtual Machine for Interaction Context (LVMIC)* – detects sensor-based context and allows the incremental definition of context parameters.
- *The Environmental Manager Agent (EMA)* – detects available and functional media devices in the user's environment.

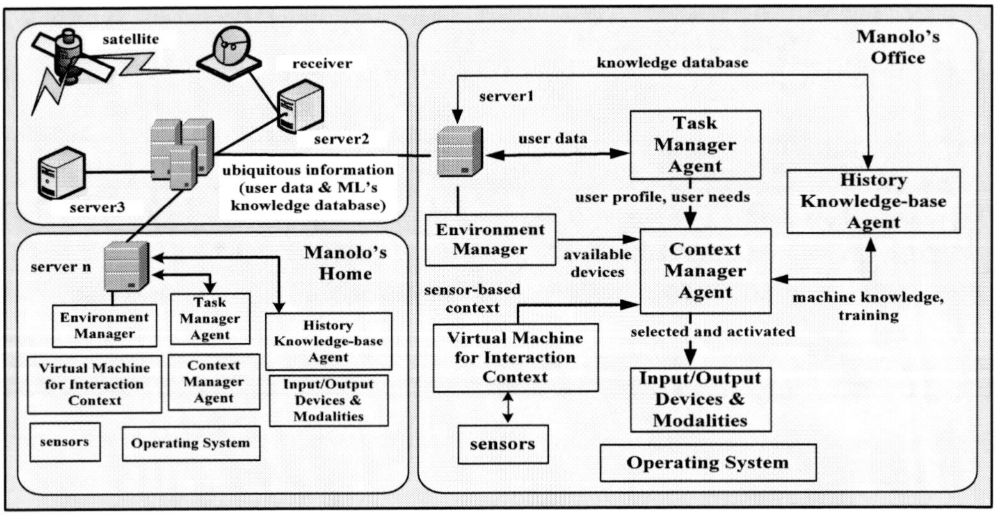

Figure 10. The architecture of a context-aware ubiquitous multimodal computing system.

In the diagram, the user (Manolo) can work at home, logs out, and still continue working on the same task at anytime and any place. Due to user's mobility, the variation in user's context and available resources is compensated by a corresponding variation in modality and media devices, and user interface selections.

7.2. Ubiquity of System Knowledge and Experience

The HKA is the component that incorporates incremental learning to the system. Its knowledge, for now, is concentrated on the system's ability to configure modalities, media devices and user interface based on the given context.

In the beginning, we assume that our system would have no knowledge whatsoever. Its initial knowledge is related to context parameter and its conventions, obtained incrementally as every relevant context parameter gets added using our VM. The system will acquire extra knowledge when HKA interacts with CMA (see Figure 11).

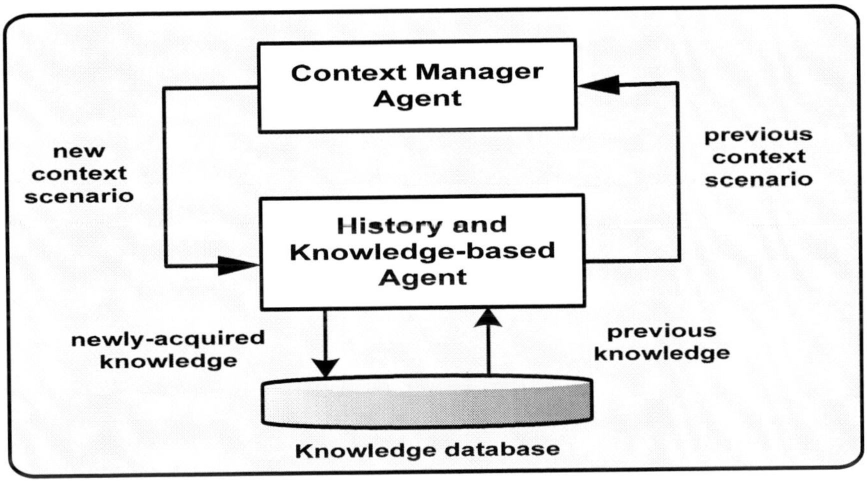

Figure 11. The History and Knowledge–based Agent at work.

Through CMA, the HKA obtains information containing the user's pre-condition scenario (i.e. the instance of interaction context), and accordingly determines the corresponding post-condition scenario (i.e. the selected modalities, media devices, and user interface). Each unique instance of interaction context (pre and post conditions) forms an entry in the knowledge database.

The system adds newly-acquired knowledge onto the database. Whenever a situation arrives that the system needs to do some decision or calculation based on the given instance of IC, the system first consults the database for any previous knowledge. If an exact match exists, then the system would simply implement the applicable set-up or post-condition scenario. If no match is found, then the system would have to do all the calculations as this is a new case. Afterwards, the result of the calculations becomes a newly acquired knowledge that is appended onto the database. Over time, the system would have enough knowledge to deal with almost all conceivable IC situations. Ideally, when the system could react automatically for almost every conceivable computing condition with minimum or no human intervention, we then say that the machine is "*intelligent*".

To implement a pervasive system, the system's knowledge database needs to be transportable from one computing environment to another. A model of migration of user's task and of machine's knowledge is already demonstrated in our previous work, as applied to visually-impaired users (Awde, Hina et al. 2007). We intend to apply the same main principle into this system.

Conclusion and Future Works

In this paper, we noted that present-day applications of pervasive system do not include that of pervasive multimodality. The current state-of-the-art multimodal interfaces themselves are not apt for pervasive application since they are conceived and developed with pre-defined modalities from the very beginning. Indeed, there is a need for a pervasive multimodal multimedia computing system that would serve both stationary and mobile users, chooses modalities (i.e. mode of human-computer interaction to input and output data) that are appropriate to the given interaction context (i.e. user context + environment context + system context), chooses media devices to support the selected modality and chooses the optimal unimodal/multimodal interface based on user's preferences. This paper enumerates some software engineering challenges in designing the system's infrastructure and explains some details on how those challenges are addressed. We then show the architectural framework of our system and explain briefly our vision on the system's incremental knowledge acquisition.

Our future works include the system's dynamic configuration of its applications whenever computing resources become scarce. Also, more knowledge acquisition and machine learning algorithms need to be developed to make our system exhibits more autonomic computing system features (i.e. self-optimization, self-protection and self-healing).

Acknowledgments

This research work has been made possible through the funding provided by the *Natural Sciences and Engineering Research Council of Canada* (NSERC), the *Décanat des études* of École de technologie supérieure and *La bourse de cotutelle* (Programme Frontenac) provided by the governments of Quebec and France.

References

Awde, A., Hina, M. D., Bellik, Y., Ramdane-Cherif, A. and Tadj, C. (2007). Task Migration in a Pervasive Multimodal Multimedia Computing System for Visually-Impaired Users. *GPC 2007 - 2nd International Conference on Grid and Pervasive Computing*, Paris, France, LNCS 4459, Springer Verlag.

Bouhuys, A. (1995). "Induction of depressed and elated mood by music influences – the Perception of Facial Emotional Expressions in Healthy Subjects." *Journal of Affective Disorders* 33: 278 - 282.

Brown, P. J., Bovey, J. D., and Chen, X. (1997). Context-Aware Applications: From the Laboratory to the Marketplace. *IEEE Personal Communications*. 4: 58 - 64.

Costa, C. A. d., Yamin, A. C., and Geyer, C.F.R. (2008). Toward a General Infrastructure for Ubiquitous Computing. IEEE Pervasive Computing, *IEEE Computer Society*. 7: 64 - 73.

Dey, A. K. (2001). "Understanding and Using Context " *Springer Personal and Ubiquitous Computing* 5(1): 4 - 7.

Dey, A. K., Salber, D., Masayasu, F., and Abowd, G.D. (1999). An Architecture to Support Context-Aware Applications. *GVU Technical Report* GIT-GVU-99-23.

Djenidi, H., Ramdane-Cherif, A., Tadj, C. and Levy, N. (2002). Dynamic Multi-agent Architecture for Multimedia Multimodal Dialogs. *IEEE Workshop on Knowledge Media Networking*.

Dong, S. H., Xiao, B., and Wang. G. P. (2000). "Multimodal user interface for internet." *Jisuanji Xuebao/Chinese Journal of Computers* 23(12): 1270-1275.

Garlan, D., Siewiorek, D., Smailagic, A., and Steenkiste, P. (2002). "Project Aura: Towards Distraction-Free Pervasive Computing." *IEEE Pervasive Computing, Special Issue on Integrated Pervasive Computing Environments* 21(2): 22 - 31.

Horn, P. (2001). Autonomic Computing: *IBM's Perspective on the State of Information Technology*, IBM Research.

Kephart, J. O. and Chess, D. M., (2001). *The Vision of Autonomic Computing*, IBM Thomas J. Watson Research Centre.

Lightfoot, D. (2001). *Formal Specification Using Z*, McMillan Press.

Oviatt, S. (2002). *Multimodal Interfaces. Handbook of Human-Computer Interaction*. J. Jacko and A. Sears. New Jersey, USA, Lawrence Erbaum.

Oviatt, S. L. (2000). "Designing the user interface for multimodal speech and gesture applications: State-of-the-art systems and research directions." *Human Computer Interaction* 15(4): 263 - 322.

Oviatt, S. L. and Cohen, P. R. (2000). "Multimodal Interfaces that Process What Comes Naturally." *Communications of the ACM* 43(3): 45 - 53.

Pahlavan, K. and Krishnamurthy, P., (2002). *Principles of Wireless Networks*.

Pascoe, J. (1998). *Generic Contextual Capabilities to Wearable Computers*. 2nd International Symposium on Wearable Computers.

Rey, G. and Coutaz, J. (2004). The Contextor Infrastructure for Context-Aware Computing. 18th European Conference on Object-Oriented Programming (ECOOP 04), *Workshop on Component-Oriented Approach to Context-Aware Systems*, Oslo, Norway.

Ringland, S. P. A. and Scahill, F. J. (2003). "Multimodality - The future of the wireless user interface." *BT Technology Journal* 21(3): 181-191.

Ross, D. A. and Lightman, A. (2005). Talking Braille: A Wireless Ubiquitous Computing Network for Orientation and Wayfinding *Proceedings of the 7th International ACM SIGACCESS Conference on Computers and Accessibility Baltimore*, MD, USA ACM Press.

Rubin, P., Vatikiotis-Bateson, E., and Benoit, C. (1998). "Audio-Visual Speech Processing." *Speech Communications* 26(1-2): 1998.

Ryan, N., Pascoe, J., and Morse, D. (1997). Enhanced Reality Fieldwork: the Context-Aware Archeological Assistant. *Exxon Computer Applications in Archeology*.

Salehie, M. and Tahvildari, L. (2005). Autonomic Computing: Emerging Trends and Open Problems. *2005 Workshop on Design and Evolution of Autonomic Application Software*, St. Louis, Missouri, USA, ACM Press.

Satyanarayanan, M. (1990). "Scalable, secure, and highly available distributed file access." *Computer* 23(5): 9-18.

Satyanarayanan, M. (1996). "Mobile information access." *IEEE Personal Communications* 3(1): 26-33.

Satyanarayanan, M. (2001). "Pervasive Computing: Vision and Challenges." *IEEE Personal Communications* 8(4): 10-17.

Schilit, B. and Theimer, M. (1994). "Disseminating Active Map Information to Mobile Host." *IEEE Network* 8(5): 22 - 32.

Vallejos, J., Desmet, B., (2007). "Pervasive Communication: The Need for Distributed Context Adaptations" ECOOP 2007 *Workshop on Object Technology for Ambient Intelligence and Pervasive Systems*, Berlin, Germany.

Varshney, U. (2003). "Pervasive Healthcare." *Computer* 36(12): 138 - 140.

Vasilakos, A. V. and Pedrycz, W. (2006). *Ambient Intelligence, Wireless Networking, Ubiquitous Computing*. USA, ArtecHouse.

Ward, A., Jones, A., and Hooper, A. (1997). "A New Location Technique for the Active Office." *IEEE Personal Communications*: 42 - 47.

Weiser, M. (1991). "The computer for the twenty-first century." *Scientific American* 265(3): 94 - 104.

Zhang, Q., Imamiya, A., Go, K. and Mao, X. (2004). *A gaze and speech multimodal interface*, Hachioji, Japan, Institute of Electrical and Electronics Engineers Inc., Piscataway, USA.

WEBSITES

Noise Measuring Device, ELK Company, at http://www.bappu.com/

Rayming Corp. USB GPS Receiver, at http://www.rayming.com

HTG Advance System's retinal recognition by a camera, at http://www.informatik. uniaugsburg.de/~ kimjongh/biometrics/retinal.pdf

Infrared Detector, at http://www.globalsources.com/manufacturers/ InfraredDetector.html

USB Digital Light Meter", at http://www.gossen-photo.de/english/ lichtmess_produkte.html

In: Focus on Information, Intelligence and Knowledge
Editor: Rafiq Islam

ISBN 978-1-61209-979-8
2011 Nova Science Publishers, Inc.

Chapter 6

RE-ENGINEERING OF PRODUCTION PROCESSES USING VALUE STREAM MAPPING TOOL. CASE STUDY: "ABC" PLANT FOR PRODUCING TISSUE AND HYGIENIC PAPER

Mohammad D. Al-Tahat*
University of Jordan, Faculty of engineering and technology,
Industrial engineering department,
Amman 11942 – Jordan

ABSTRACT

Value Stream Mapping (VSM) is a lean principle, which is used for analyzing and controlling the flow of material and information through production systems. This particular lean principle allows documenting current working methods, recording inventory levels and computing throughput time. These data are used to design a future vision of production and information flow that can be utilized as bases for lean thinking. Ultimately, this allows prioritizing the opportunities for waste elimination and time saving. The aim of this paper is to recognize the bottlenecks when producing a specific product in turn to achieve improvements through average number of jobs (WIP), throughput time, machine delay time, and product queuing time. A product was specified, production line for producing this product was monitored, and data were collected and analyzed, consequently, current state map for producing the primal products was developed and constructed, then it was discussed and brainstormed with senior engineers of the plant that produces the product, the weakness points were found in the flow of production within the system, therefore the future state map for the product was developed and presented. Finally, discussion, results, recommendations and future work were suggested and presented.

* E-mail: altahat@ju.edu.jo

Keywords: Value Stream Mapping, Kaizen, Re-engineering, Process Improvement, Lean.

1. Introduction

One of the ways to improve a production system is to use the Value Stream Mapping (VSM) as a tool for Kaizen or never-ending efforts for improvement. A value stream is a collection of all activities and actions (value added as well as non-value added) that are required to bring a product through the main flows, starting with the supplier and ending with the customer (Abdulmalek and Rajgopal, 2006). These actions consider the flow of both information and materials within the overall production line. The crucial objective of VSM is to identify all types of waste in the value stream and to take actions to try eliminating these (Rother and Shook, 1999). VSM creates a common basis for the production process, thus facilitating more thoughtful decisions to improve the value stream (McDonald et al., 2002). VSM is a pencil and paper tool, which is created using a predefined set of standardized icons, for details see (Rother and Shook, 1999; Tashtoush et al., 2001). The VSM is a physical diagram of every step involved in the process. Hence, VSM requires following a product's production path from supplier to customer and drawing a visual representation of every process in the material and information flow. As one continues to draw the maps, he will learn to see the shop floor in the way that supports lean thinking. A very brief description of the most common lean tools is given in (Monden, 1998; Feld, 2000; Nahmias, 2001; Vais A. et al. 2006.). Taking the value stream viewpoint means working on the big picture and not individual processes. VSM creates a common basis for the production process, thus facilitating more thoughtful decisions to improve the value stream (McDonald et al., 2002). Nine steps are followed when using VSM tool, these are:

(1) *Meeting* between the management and analyst to state the purpose of the VSM and to gain an accurate understanding of how the process operates.
(2) *Training* the VSM team, the team should understand the concepts of lead-time, waste, value-added activities and non value-added activities as they pertain to lean methodologies. In addition to that the VSM team must work from a standard set of definitions and icons when drawing the VSM.
(3) *Data collection.* VSM team picks a particular product and follows it through the process. Data on each activity within the value stream should be collected, such data include; cycle times; changeover or setup times; number of people involved; quantity of units processed; quantity of changeovers for a given time period; downtime; and batch sizes.
(4) *Constructing the current-state map.* When each activity has been observed and data have collected, the current-state map can be constructed. Each activity should be labeled in a process box, and the data for that activity listed below it. Arrows should be used to show the flow of the product, and triangles between the process boxes can be used to depict inventory levels of work in process before each activity. Icons and arrows showing communication flows--whether they are electronic, manual, and visible or something else--are drawn opposite the product flow, because they signal

upstream activities that will react to downstream information. Finally, a sawtooth diagram should be drawn underneath the value stream to depict process and lead-time. Once the current-state map is drawn, the process time for the value stream can be calculated by adding the times along the upper teeth of the sawtooth diagram. The total lead-time is calculated by adding all the times on the upper teeth and the lower gaps together. Often, the process time in the value stream will be less than five percent of the total lead-time.

(5) *Reviewing the current-state map.* Once the current-state map is completed, managers and analyst should be called back to review the map and validate the information. The data on the map are open for questions, and a consensus must be reached before going any further

(6) *Targeting improvement.* VSM team can begin to target improvement initiatives to eliminate and reduce bottlenecks and lead times. Brainstorming is the best method to create these initiatives. The improvements initiatives can be ranked according to the organization's operating plan. In the absence of an operating plan effect/cost approach can be used

(7) *Constructing the future-state map***.** The future-state map can be drawn once improvement initiatives have been determined. Data in the future-state map are derived from targets that will evolve as a result of implementing the improvements. In most cases, the future-state map will look similar to the current-state map, with the addition of improved metrics. If activities in the current-state map can be eliminated or combined, this should be reflected in fewer process boxes in the future-state map.

(8) *Formalizing improvement.* This step is to formalize the improvement initiatives strategy. This can take the shape of a matrix that outlines the area where the improvement will occur, describes the improvement, and states a quantifiable goal, owner and targeted completion date.

(9) *Overcoming traditional thinking.* One common pitfall to avoid during a VSM event is developing improvement initiatives that focus on the process steps where products are being transformed.

The following points are a brief of VSM advantages; Relates the manufacturing process to supply chains, distribution channels and information flows, Integrates material and information flows; Links Production Control and Scheduling (PCS) functions such as production planning and demand forecasting to production scheduling; Forms the basis for implementation of lean manufacturing by designing the production system based on the complete dock-to-dock flow time for a product family; and many others. On the other hand VSM has following drawbacks; Fails to map multiple products that do not have identical material flow maps; Fails to show the impact on Work-In-Process (WIP), order throughput and operating expenses of in-efficient material flows in the facility; Fails to factor queuing delays, sequencing rules for multiple orders, capacity constraints, etc. in any map; Lacks the capability, due to the manual mapping method, for rapid development and evaluation of multiple "what if" analyses required to prioritizing different alternatives for improving a current state map when time and/or budget constraints exist; and many others. The objective of this chapter is to use a case-based approach to demonstrate how VSM when used appropriately, can help the process industry eliminate waste, maintain better inventory control, improve product quality, and obtain better overall financial and operational control.

A plant for producing hygienic paper is used to illustrate the approach followed. Since some of the information is confidential, the plant is referred to as “ABC plant” throughout this chapter.

2. “ABC” PLANT AND PROCESS PLANNING

A large local company is used to illustrate the approach followed. Since some of the information is confidential, the company is referred to as “ABC” plant throughout this chapter. “ABC” plant was established and was started its production before than 15 years ago. “ABC” supplied with the latest state-of-the-art technologies in the field of the tissue and toilet paper industry. All the products of the “ABC” plant go through a standard sequence of operations, but with different processing times depending on the batch being processed at the time. Table 1 represents the standard process plan from the moment of receiving the raw material until shipping it as final products to customers.

Table 1. Standard sequence of operations for tissue and toilet paper production in the considered plant

Seq.	Operation	Description
1	Receiving Raw Material	Unloading raw material from the loaded truck
2	Storing	Entering raw material into stores and inventory updating
3	Handling	Handling RM by Forklift from store to a conveyor belt
4	Pulping	Soft wood pulping
5	Pulping	Hard wood pulping
6	Mixing	Mixing soft wood pulp and hard wood pulp together
7	Curing	Paper preparation
8	Reeling and spool making	Paper is being rolled over to make the spool
9	Quality Check and weighting	Running all the quality tests
10	Handling	Moving spools to unwinding station
11	Unwinding	Unrolling of two or three spools to make one smaller shaft
12	Wrapping	Smaller rolls being wrapped as a unit
13	Handling	Rapped units to the transit area
14	Bar-Coding checking	
15	Handling	To warehouse
16	Handling	To the shipping dock, checked, bar-coded then loaded
17	Shipping	Loading product from dock to truck
18	Truck Departure	Released after weighing and filling check out document.

“ABC” plant is currently working in its full capacity; with no raw material shortages, no process line stoppages or operator shortages. But that does not mean that further improvement can not be done, or even that “ABC” plant is efficient enough to not try to increase its productivity, there is a saying "organizations keep running not to get ahead, just to not go backwards". In addition “ABC” plant looking towards continuous improvement or Kaizen. This chapter has been carried out to investigate the chances for continuous productivity increase and efficiency enhancement, as well as to introduce the lean thinking to the “ABC” plant.

The "ABC" plant produces 9 types of products; these are shown in table 2. For each product type 8 product grade can be produced, these grades are presented in table 3. The historical data of demand indicated that the highest demand was on the product F, 100 cm width, grade PR. Because product F, 100 cm width, grade PR (F_{100}^{PR}) has the highest demand then it will be considered as a primal product in this chapter, therefore the production process of F_{100}^{PR} will be analyzed by VMS and hence the objectives of this research will be achieved.

Table 2. Types of product that can be produced by "ABC" plant

Sequence	1	2	3	4	5	6	7	8	9
Product Type	F*	T	N	TO	W	M	J	P	C

*Highest demand was on the product F 100 Cm width grade PR.

Table 3. Grades that can be produced for any product type

Grade							
ST	PR	SP	BU	PU	CO	CB	LC

The process plan for producing the primal product F_{100}^{PR} is developed, time study also is conducted finally time standards computed and assigned for every operational step in the process plan; all of these have been presented in table 4.

Table 4. Process plan for producing 540 Kg of F_{100}^{PR} and corresponding time standards

Seq.	Operation	Standard Time (Minute)	
1	Receiving pulp/ forklift	0.5	1
2	Storing	0.5	
3	Handling H.W. pulp/ forklift/ 5-20 meter	1.5	3
4	Handling S.W. pulp/ forklift/ 5-20 meter	1.5	
5	Storing H.W. pulp in chest	5	26
6	Storing S.W. pulp in chest	3	
7	Mixing and dilution water	8	
8	Reeling and spool making (3515 kg/Spool)	8	
9	Quality check and weighting	2	
10	Handling of spools/ overhead crane	1	1
11	Waiting for second spool	9	9
12	Unwinding of paper shaft	9	9
13	Handling of paper shaft/ forklift	2	2
14	Waiting for rapping	3	3
15	Wrapping	4	4
16	Handling of semi-final product of 540 kg / forklift/ 5 meter	1	2
17	Bar-coding checking	1	
18	Handling of semi-final checked product/ forklift	2	2
19	Handling of customer order quantity/ forklift	1	3
20	Shipping customer order quantity	1	
21	Truck departure	1	

3. Developing the Current Value Stream Map

According to the process plan presented in table 4, current value stream map for producing F_{100}^{PR} has been developed and presented as depicted in Figure 1. All data for the current state map were collected. Data collection for the material flow started at the shipping department, and worked backward all the way to the suppliers, gathering snapshot data about processing times. Figure 1 shows the current state map that was constructed; the boxes in the map represent the process and the number inside the box is the processing time of the process. It should be noted that this data were collected whilst walking the shop floor and talking to the foreman and operators at each operational step. The processing times are all based on the average of historical data.

After collecting all the information and material flows, they are connected as indicated by arrows in the map, representing how each operational step receives its schedule from production planning. Total processing time (or value-added) is indicated at the bottom of the current state map, which is 66.0 minutes. This time is calculated by adding the processing time for each process in the value stream. The processing time for each process is the average, which was determined by using actual data from the company.

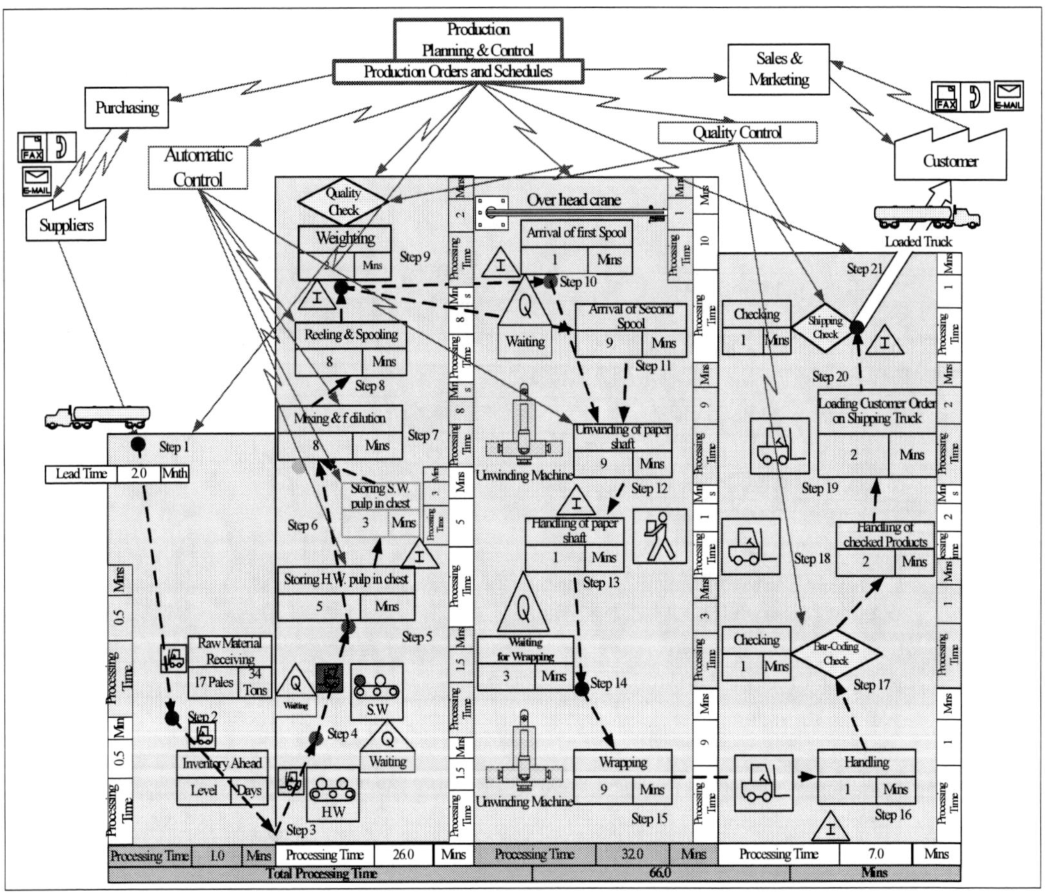

Figure 1. Current value stream map for producing 540 Kg of F100PR.

Tact time: refers to the time needed for producing one unit of production. Processing time of each work stage is translated to processing rates, results are depicted as shown in Figure 2 which indicates that the bottleneck stages are: step 7, step 8, step 11, step 12, and step 15. Processing rate for any bottleneck stage is equal to 0.1 roll per minute, therefore the tact time, which is the time needed for an output of one roll of 540 Kg weight is,

$$Tact\ Time = \frac{(1\ Roll)(1\ Mins)}{(0.1\ Roll)} = 10\text{Mins} \qquad (1)$$

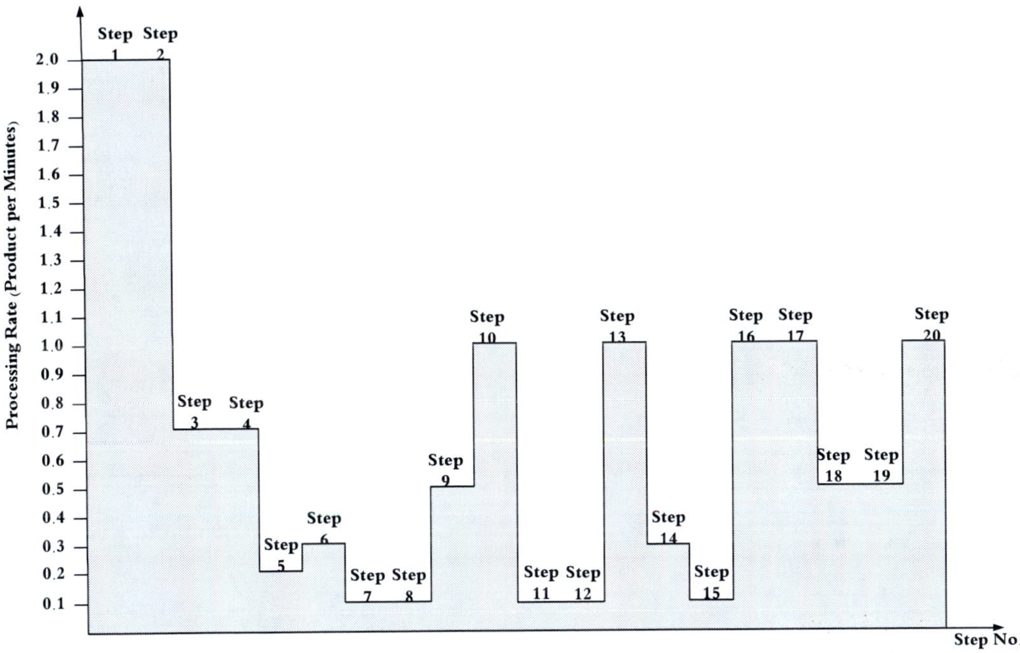

Figure 2. Processing rate for each work stage.

Queuing time: arrival rate to any step is considered as the processing rate of the previous step, arrival rate, processing rate, and queuing time for each step is derived and presented as shown in table 5. Note that equation (2) is used to compute the queuing time.

$$\text{Queuing time}(W_q) = \frac{\lambda}{\mu(\mu - \lambda)},\ \text{Where}\ \lambda < \mu \qquad (2)$$

Throughput time: sometimes referred to as flow time, refers to the time from when a job is released to the production shop until it is completed (Askin R.G. *et al.* 2002). During this time, the job is considered as apart of Work-In-Process (WIP). Therefore throughput time can be estimated as the summation of the total processing time and the total product queuing time as follows,

$$\text{Throughput time} = \text{Total processing time} + \text{Total queuing time} = 66.0 + 7.4 = 73.4\text{Mins} \qquad (3)$$

Average number of jobs in the production line (WIP): is the throughput time divided by the tact time, then

$$WIP = \frac{\textit{Throughput time}}{\text{Tact time}} = \frac{73.4}{10} = 8\ \textit{Product} \tag{4}$$

Table 5. Arrival and processing rate and their corresponding queuing time

Step	Arrival Rate	Processing Rate	Queuing Time (W_q) in Minutes	
	(λ)	(μ)	Product	Machine (delay)
1	2	2	Transient	Transient
2	2	2	Transient	Transient
3	2	0.7	-	2.3
4	0.7	0.7	Transient	Transient
5	0.7	0.2	-	7.1
6	0.2	0.3	4.5	-
7	0.5	0.1	-	10.4
8	0.1	0.1	Transient	Transient
9	0.1	0.5	0.7	-
10	0.5	1	1	-
11	1	0.1	-	11.6
12	0.1	0.1	Transient	Transient
13	0.1	1	0.1	-
14	1	0.3	-	4.5
15	0.3	0.1	-	13.5
16	0.1	1	0.1	-
17	1	1	Transient	Transient
18	1	0.5	-	4
19	0.5	0.5	Transient	Transient
20	0.5	1	1.0	-
Total Queuing Time (Minutes)			7.4	53.4

4. DEVELOPING THE FUTURE VALUE STREAM MAP

The process of defining the future state map, starts while developing the current state map, where target areas for improvement start to roll up. Looking at the current state map several things stick out: (a) Tact time, (b) Production policy, (c) throughput time, (d) Queuing time, and (e) WIP. These may be viewed as related issues since the more the inventory, the longer the queuing time and thus, the longer the lead-time. A new modified production process resulted from the raised question; can the arrival rate into a bottleneck stage be increased? Increasing the processing rate of stage 5, stage 6, stage 9, and stage 14 by doubled the number of severs in these stages lead to an increase in the arrival rate to the successive bottleneck stages 7, 11, and 15. The representative value stream map of this situation is presented in figure 3. The main objective now is to analyze this future map.

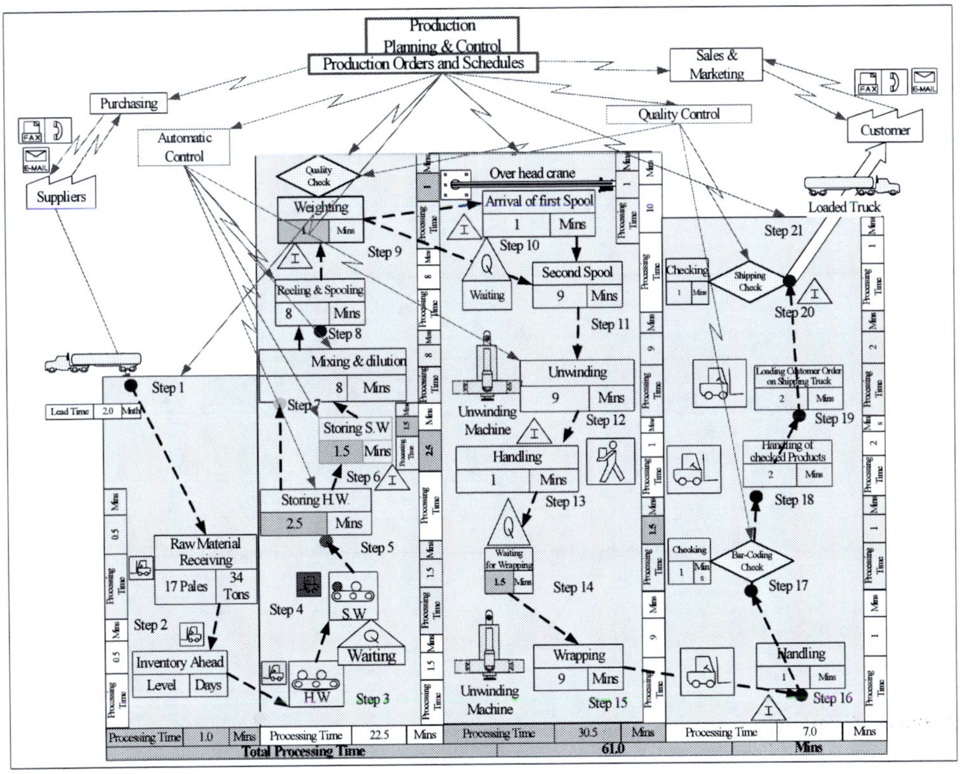

Figure 3. Future value stream map for producing 540 Kg of F_{100}^{PR}.

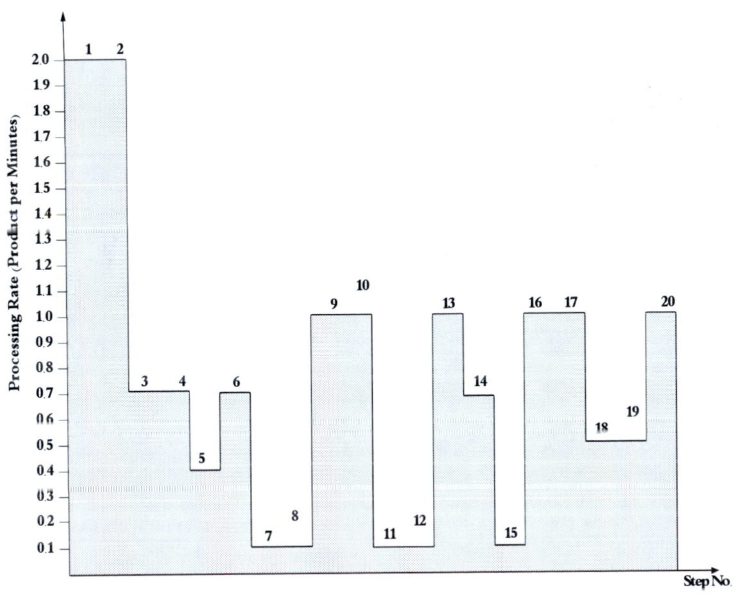

Figure 4. The new processing rate for each work stage.

Tact time. Figure 4 indicates that the bottleneck stages remain the same: step 7, step 8, step 11, step 12, and step 15. Also processing rate for these bottleneck stages remain 0.1 roll per minute, therefore the tact time, remains the same

$$Tact\ Time = \frac{(1\ Roll)(1\ Mins)}{(0.1\ Roll)} = 10\text{Mins} \tag{5}$$

Queuing time: new arrival rates, new processing rates, and new queuing time for each step is derived and presented as shown in table 6.

Throughput time: total processing time (or value-added) is indicated at the bottom of the current state map, which is 61.0 minutes

$$\text{Throughput time} = \text{Total processing time} + \text{Total queuing time} = 61.0 + 3.6 = 64.6\text{Mins} \tag{6}$$

Average number of jobs in the production line (WIP), under the new situation;

$$WIP = \frac{\textit{Throughput time}}{\text{Tact time}} = \frac{64.6}{10} = 7\ \textit{Product} \tag{7}$$

Table 6. New Arrival and processing rate and their corresponding queuing time

Step	Arrival Rate	Processing Rate	Queuing Time (W_q) Mins	
	(λ)	(μ)	Product	Machine (delay)
1	2.0	2.0	Transient	Transient
2	2.0	2.0	Transient	Transient
3	2.0	0.7	-	2.3
4	0.7	0.7	Transient	Transient
5	0.7	0.4	-	6.3
6	0.4	0.7	2.3	-
7	1.1	0.1	-	9.1
8	0.1	0.1	Transient	Transient
9	0.1	1.0	0.1	-
10	1.0	1.0	Transient	Transient
11	1.0	0.1	-	10.1
12	0.1	0.1	Transient	Transient
13	0.1	1.0	0.1	-
14	1.0	0.7	-	4.5
15	0.7	0.1	-	10.8
16	0.1	1.0	0.1	-
17	1.0	1.0	Transient	Transient
18	1.0	0.5	-	4.0
19	0.5	0.5	Transient	Transient
20	0.5	1.0	1.0	-
Total Queuing Time (Mins)			3.6	42.5

5. CONCLUDED REMARKS AND RECOMMENDATIONS

Detailed value stream mapping can be used to evaluate basic performance measures and analyze production system configurations. The availability of the information provided by the value stream mapping can facilitate and validate the decision to implement lean manufacturing and can also motivate organization during the actual implementation in order to obtain the desired results. It is concluded that value stream mapping is a powerful tool, which highlighted chances for process improvement. Value stream mapping helps identify wastes, bottlenecks, idling times, WIP inventory, and all the chances to improve and remedy a faulty wasteful situation into an improved more efficient one by elimination of non-added value activities.

Average number of jobs (WIP), throughput time, machine delay time, and product queuing time are used as measures for evaluating the performance of the process. Table 7 presents a summery for the values of these measures before and after process improvement.

Table 7. Measure value before and after process improvement

Performance Measure	Value		Improvement %
	Before	After	
Total processing time	66 Minutes	61 Minutes	7.6
Tact Time	10 Minutes	10 Minutes	0.0
Product queuing time	7.4 Minutes	3.6 Minutes	51.0
Machines Delay time	53.4 Minutes	42.5 Minutes	20.4
Throughput time	73.4 Minutes	64.6 Minutes	12.0
WIP	8 Unit	7 Unit	12.5

As a future research, a simulation model may conducted to contrast the current and the future scenarios in detail, in order to illustrate to managers potential benefits such as increased productivity as well as reduced flow time and work-in-process inventory. Finally it is worthy to mentioned that Microsoft (R) Visio (R) professional 2002 (10.0.525) is used as a mapping tool for this research.

REFERENCES

Abdulmalek, and F.A., Rajgopal, J., 2006.Aanalyzing the benefits of lean manufacturing and value stream mapping via simulation. A process sector case study. *International Journal Production Economics,* doi: 10.1016/j.ijpe.2006.09.009.

Rother, M., and Shook, J., 1999. *Learning to See: Value Stream Mapping to Add Value and Eliminate Muda.* The Lean Enterprise Institute, Inc., Brookline, MA.

McDonald, T., Van Aken, E.M., and Rentes, A.F., 2002. Utilizing simulation to enhance value stream mapping: a manufacturing case application. *International Journal of Logistics: Research and Applications* 5 (2), 213 232.

Tashtoush G., Gharaibeh B., and Massarweh W., 2001. Converting from mass to lean production, a future look for Jordanian industries, *1st International Industrial Engineering Conference*, Amman-Jordan, Sep. 23rd –27th.

Monden, Y., 1998. Toyota Production System— An Integrated Approach to Just-In-Time, third edition, Norcross, Georgia. Engineering and Management Press.

Feld, W.M., 2000. Lean Manufacturing: Tools, Techniques, and How To Use Them. The St. Lucie Press, London.

Nahmias, S., 2001. Production and Operations Analysis, fourth edition, New York McGraw Hill.

Vais A., Miron V., Pedersen M., and Folke J., 2006. "Lean and Green" at a Romanian secondary tissue paper and board mill—putting theory into practice. *Resources, Conservation and Recycling* 46 (2006) 44–74.

Askin R.G., and Goldberg J. B., 2002. *Design and analysis of lean production systems*, 1st edition, John Wiley and Sons.

In: Focus on Information, Intelligence and Knowledge
Editor: Rafiq Islam

ISBN 978-1-61209-979-8
2011 Nova Science Publishers, Inc.

Chapter 7

ADAPTIVE BUSINESS INTELLIGENCE FOR INFORMATION AND COMMUNICATION TECHNOLOGY MANAGEMENT

***Sueh Ing Su*[1] *and Raymond Chiong**[*2]

[1]School of Business & Enterprise
[2]School of Computing & Design
Swinburne University of Technology (Sarawak Campus)
Jalan Simpang Tiga, 93576 Kuching
Sarawak, Malaysia

ABSTRACT

With the rapid advancement of both business techniques and technologies in recent years, knowledge has become an important and strategic asset that determines the success or failure of an organisation. Analyses showed that a competitive advantage in business environment depends on the accessibility to adequate and reliable information in shortest time possible and the high selectivity in the creation and utilisation of information. An effective instrument to create, aggregate and share knowledge in an organisation has therefore become a key target of management. As the organisations today can no longer spend money on excessive infrastructure and technology that does not provide a quick impact on the business' bottom-line, Business Intelligence (BI) is becoming indis-pensable to an organistion's success in the emerging global economy. BI is currently one of the fastest developing directions in information and communication technology. In this paper, we introduce the concept of BI and address the importance of BI in revolutionising knowledge towards economics and business advancement. Ensuing sections then explore some specific BI applications, including discussion surrounding the technologies and frameworks which may facilitate BI in an e-business initiative. In the concluding remarks, the future of BI is drawn to wrap up the studies.

* Email: rchiong@swinburne.edu.my

1. Introduction

In today's business world, the availability of accurate information in a timely manner for management to make better business decision is important. Successful organisations that rise above others are normally those who can respond quickly and flexibly to market changes and opportunities. The key to this response is their effectiveness and efficiency in using data and information available to them. Information in the business world usually entails customer needs, customer decision-making processes, the competition conditions in the industry, and general economic, technological, and cultural trends. Most organisations collect vast amount of data from their business operations. To keep track of that information, a business would need to use a wide range of software programs and different database applications for various departments throughout their organisations. Using multiple software programs such as Customer Relationships Management (CRM), Enterprise Resource Planning (ERP), etc make it difficult to retrieve information in a timely manner and to perform analysis of data. An effective instrument to create, aggregate and share knowledge in an organisation has therefore become a key target of management. This results in the concept of Business Intelligence (BI) being introduced into the marketplace to enhance an organisation's response in making better business decision.

The objective of this paper is to discuss why the concept of BI has become increasingly important for organisations. We also present some of the key applications and technologies which facilitate BI in an e-Business initiative. Finally, the future of BI is addressed to envision the potential of its growth.

2. BI and Knowledge Management

BI is the process of using advanced technologies to gather, store and analyse information in the field of business. According to Pareek (2006), the term was first used and popularised by analyst Howard Dresner of the Gartner Group[12]. It typically describes the process of turning data into information and then into knowledge. The first probable reference to BI is said to be made in Sun Tzu's "Art of War", where he claimed that to succeed in war, full knowledge on one's strengths and weaknesses as well as the strengths and weaknesses of the enemy must be known. Applying this to the modern business world, BI becomes the art of wading and sieving through tons of data, and presenting the overloaded data as information that provides significant business value in improving the effectiveness of managerial decision making (Turban, King, Lee and Viehland, 2004). As such, BI is carried out not just for gaining sustainable competitive advantage, but it also has a valuable core competence in most instances.

Literature reviews have shown that knowledge can be managed successfully using many BI techniques. Knowledge Management (KM) is seen as the "helping hand of BI" (Haimila, 2001). According to Herschel and Jones (2005), KM is described as a "systematic process of finding, selecting, organising, distilling and presenting information in a way that improves an employee's comprehension in a specific area of interest". They argued that KM and BI are

[12] www.gartner.com

integrated as mutually critical components in the management of intellectual capital. Promoting the effective sharing and transfer of these intellectual assets is the centerpiece of KM (Call, 2005). Thus, BI must be seen as an integral part of a larger KM effort. The effectiveness of BI will help to promote and enhance knowledge to improve decision making and hence a firm's performance.

3. Growing Importance of BI

BI has provided a roadmap to deliver solutions for business analysis which includes data models, metadata and analytical applications. By having this roadmap, superior business value through improved return on investment (ROI), time value by enabling fast solution delivery, and technical value through open database enablement can be delivered. Time savings, single version of truth, better strategies, better tactics and decision are among the other top BI values.

According to a survey conducted by a market research firm, International Data Corporation (IDC) in 2005, organisations who have successfully integrated BI into their business process can achieve a significant ROI. IDC's Financial Impact of Business Analytics study interviewed over 40 companies in a wide variety of industries in North America and Europe, and the study found that a BI implementation generates a median five-year ROI of 112% with a mean payback of 1.6 years on average costs of $4.5 million. Among the organisations included in this study, 54% had an ROI of 101% or more (Eastwood, Vesset and Morris, 2005).

In a nutshell, the concept of BI has become increasingly important for organisations because of the following reasons:

a) Improve Competitive Response and Decision-Making Process

The competitive advantage in business depends on two factors: access to adequate and reliable information within a short period of time and high selectivity in the creation and utilisation of quality information. According to Porter (2001), competitive advantage would be long lasting only if a firm can do the things its competitors do, but do it better, or think differently from its competitors. To realise long lasting competitive advantage, an organisation needs to have rapid and continuous innovation and dynamic coupling of processes so that they cannot be easily duplicated. Moreover, firms also need to leverage on resources such as structural capital, human capital and relationship capital to achieve sustainable competitive advantage (Farhoomand, 2005). With BI systems connected to CRM, ERP, human resource and finance systems, information can be produced in a more accurate and timely manner. BI systems thus make up a complex solution that allows decision makers to create, aggregate and share knowledge in an organisation easily, along with greatly improved service quality for efficient decision making.

b) Easing Corporate Governance and Regulatory Compliance

The new rules in today's business mandate that organisations be able to show that their data is accurate. This means an organisation must be able to verify the lineage of data, starting at its source and tracking it through its various manipulations and aggregations, which in turn means having reliable metadata and data auditing that are consistent across the enterprise. In this respect, BI assists in ensuring accuracy of data across the enterprise. It allows reporting and query results to be consistent when they are produced. This is crucial for corporate governance and regulatory compliance when auditing do not find discrepancies in the reports produced.

c) Customer Relationship Management

CRM system can be part of BI concept which provides an organisation with the ability to categorise or segment its existing customer base and prospects. This may be done based on the products or services that the client purchases, demographic information for consumer clients, industry sector or company size for corporate clients and so on (Robson, 2005). CRM allows the organisation to better understand the trend or buying pattern of its existing customers and the segmentation of customer base would assist the sales and marketing to sell products or services which their customers want. This enhances the relationship between the organisation and its customers. With good relationship, customers' loyalty can then be achieved where future sales can be generated. With timely information available for executives and managers to make better and faster decision, the organisation will be able to provide quality service to its customers and respond effectively to changing business condition.

Nevertheless, IDC annual end-user survey continuum 2005 revealed that more than 50% of the respondents in the Asia-Pacific region said they have yet to invest in any BI software. This shows that many organisations in the region have some way to go before they attain optimal business performance management. IDC research indicates that obtaining management buy-in through education and training, and the availability of user-friendly and industry-specific BI solutions are the factors in determining the adoption of BI in enterprises.

4. Four Dimensions of BI Systems

BI supports a range of business applications to acquire up-to-date and consolidated information. Some main tasks of BI systems include intelligent exploration, integration, aggregation and a multidimensional analysis of data originating from various information resources.

According to Olszak and Ziemba (2003), there are four basic dimensions which require serious consideration in the design and implementation of BI systems, as shown in Figure 1 below:

BUSINESS Methods and techniques of management	FUNCTIONALITY Functional range
TECHNOLOGY Technological methods and tools	ORGANISATION Methodologies of implementation and utilisation

Figure 1. Four basic dimensions of BI systems.

4.1. Business Aspect of BI

BI systems provide an integrated analysis and evaluation of an organisation by means of utilisation of financial and non-financial indicators. One of the concepts to be applied is the Balanced Scorecard (BSC). BSC facilitates a multidimensional overview of an organisation. It highlights intellectual resources management as an important and integrating factor in the realisation of an organisation's strategy through four perspectives: financial, customer, internal processes, and development and growth.

4.2. Functional Aspect of BI

BI may be utilised mainly for strategic planning in realisation of the organisation's strategy, mission, goals and tasks; improving relations with customers to meet their needs; analysing profitability of products and services; analysing internal processes and operational efficiency of an organisation; controlling and management accounting by means of analysis of actual costs and financial flows.

4.3. Organisational Aspect of BI

Building BI systems require setting business goals, selection criteria of data warehouse implementation, and tools for supporting both the process of multi-criteria analysis of data as well as data exploration. Besides that, training users to develop their skills of identifying, modelling, verifying, coding and sharing knowledge is another important issue for proper implementation of BI technology.

4.4. Technological Aspect of BI

The technological aspect primarily involves methods of knowledge creation, sources of knowledge and tools. Naturally, the Internet, extranet and intranet have facilitated knowledge creation. Sources of knowledge could be acquired from the BI reference process that covers three main process steps in (1) data integration, (2) data storage and (3) data usage, through tools such as data warehouse, data marts, reporting, online analytical processing (OLAP) and data mining.

5. BI PROVIDERS

As aforementioned, the concept of BI has become increasingly important for organisations to remain competitive in the marketplace. Based on the research work of Schlegel, Hostmann, Bitterer, and Burton (2006), we can categorise the BI providers into a few categories, namely the leaders, challengers, visionaries and niche players (see Figure 2). The two well-known market leaders in BI today are Cognos[13] and Business Objects[14].

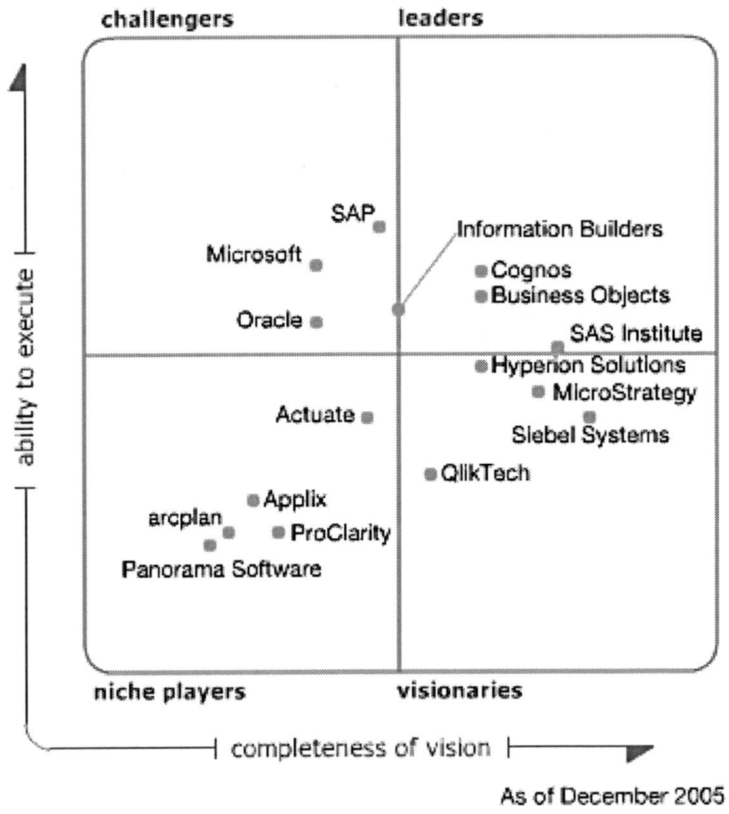

Figure 2. Categories of BI providers.

13 http://www.cognos.com
14 http://www.businessobjects.com

Cognos

Cognos is a well-known BI leader that has introduced new reporting software called Cognos ReportNet, which is built on a Web services-based architecture. It takes a single product approach with Cognos 8 for all reporting requirements serving all users, with simplified deployment, maintenance, and integration, drag-and-drop ease of use, and multilingual reporting. The open data strategy of Cognos 8 ensures the access of data from all sources and structures, including all relational databases, widely deployed ERP systems, enterprise data warehouses and marts, OLAP sources, modern data sources such as XML, JDBC and LDAP, satellite sources, mainframe sources, content management data and portal standards. Butler Research remarked that Cognos has created an architecture that delivers robust and scalable reporting solutions, which can act as a future platform for an organisation's entire BI strategy. The ROI Cognos delivers to its customers is proven by independent industry watchers, and backed up by customers around the world.

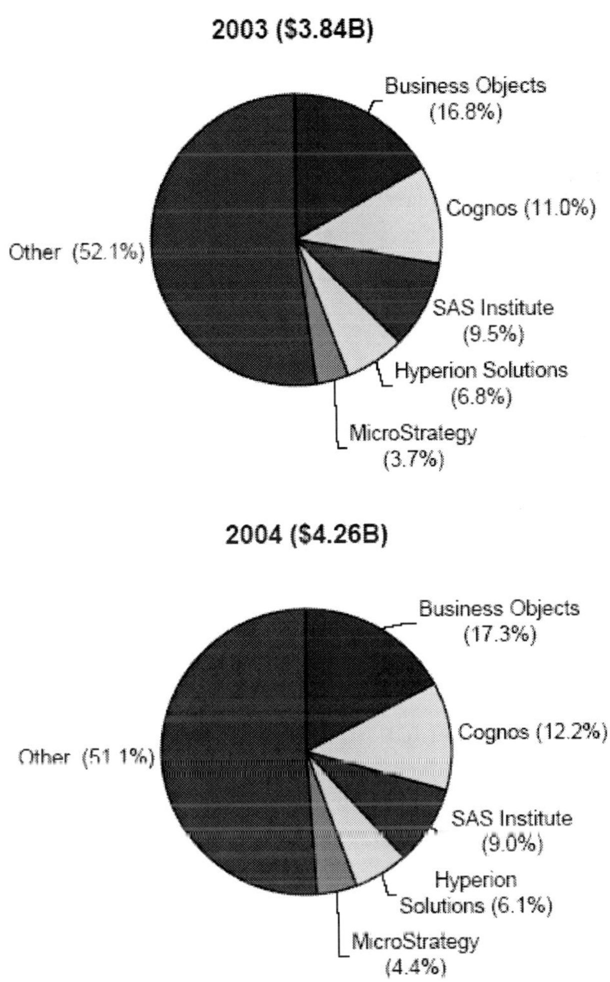

Figure 3. Worldwide BI tools market share by leading vendors in 2003 and 2004.

Business Objects

Business Objects is another world's leading BI software company. About 82% of Fortune 500 companies choose Business Objects as it allows users to track performance, understand business drivers, and manage the business. Business Objects offers a single platform to address all BI needs and improve end-user efficiency, as well as administrative services for security and auditing. The company's solutions leverage the information that is stored in an array of corporate databases, ERP and CRM systems. It also reduces duplication in the back-end infrastructure, and lowers total cost of ownership by consolidating the number of BI tools and technologies in use. Business Objects' combination of Web Intelligence, OLAP Intelligence and Desktop Intelligence within the Business Objects XI platform makes for a strong, integrated offering that includes products for both analytic and enterprise reporting.

Apart from Cognos and Business Objects, almost all the other BI providers also offer some comprehensive sets of BI functionalities that encompass OLAP, ad-hoc query, end-user reporting, enterprise reporting, and some types of analytic dashboard or even scorecarding application. Figure 3 shows the market share of the leading BI providers in 2003 and 2004.

6. Applications of BI

Nowadays, information sharing and communication are normally done via the Internet or some web-based applications, especially in an e-business environment. BI is thus applicable to various industries such as communications, education, financial services, government and public sector, consumer product goods, healthcare, manufacturing, pharmaceuticals, retail, technology, etc for the purposes of improving productivity and enhancing decision-making process. The following subsections discuss some examples of different industries where BI has been applied to.

6.1. Banking and Financial Services

According to Business Objects, BI has been applied in the banking and financial services industry to perform customer risk analysis and customer valuation on issues such as current profitability, lifetime potential of a customer based on costs, revenues and future predictive behaviour, etc. BI has the ability to do profiling by identifying the like-minded customers, thus enabling segmentation. It also has the ability to create, refine and target campaigns to win profitable new customers. With these capabilities, banking and financial institutions are able to effectively serve their customers and gain deeper customer insight. In addition, BI automates compliance with reporting standards and regulations which are crucial to these institutions to improve control and productivity.

6.2. Manufacturing

Cognos in their website mentioned that BI helps to improve sales performance, increase profitability, improve operational effectiveness and optimise the supply chain. According to Cognos, BI has been applied in the manufacturing industry to analyse cost components and drivers, in order to reduce the cost of goods sold. Meanwhile, BI is also able to gain visibility into demand and sales trends in order to minimise inventory investment. With BI systems, manufacturing companies can identify and analyse excess, obsolete or slow-moving inventory that can be scrapped or repurposed. The companies can also receive timely notification of events such as late supplier delivery, changes in customer demand or production stoppages. With the help of BI, these companies are able to respond quickly to changing markets and company sensitivities.

6.3. Pharmaceuticals and Life Sciences

Besides the manufacturing companies, BI has been applied in the pharmaceuticals and life sciences services industry too. BI systems help to identify products that yield better results with a clinical trial research process that segments, tracks, analyses and shares test results. The systems can also analyse and report on clinical performance data to identify best practices and decide whether to continue, pursue or terminate a project. Business Objects complimented BI's capability in maximising the supply chain by sharing production information with suppliers, tracking product quality, optimising stock replenishment, and monitoring vendor performance. BI optimises campaign creation and implementation by analysing the effectiveness of marketing strategies, and increases sales productivity by understanding the sales activities, territory, and representative performances. Cognos further attributed the ability of BI to improve operational, sales and marketing performance and compliance with regulatory requirements.

6.4. Retail

The world's leading retailers have turned to Business Objects' BI to improve purchasing, forecasting, and distribution management. BI systems help these retailers to optimise product profitability and increase effectiveness of marketing campaigns, identify and segment customers to enable retention strategies, integrate management and financial reporting for improved performance, reduce and control operational costs through optimised store performance and effective customer service. According to Cognos, organisations are able to improve store performance by improving visibility and accountability for top controllable expenses like labour and cost of goods sold, by communicating sales and margin information across the chain with enterprise-scale reporting, through identification of high or low performing divisions, stores, channels, products, and staff, by optimising staffing levels through headcount planning and workforce analytics, and monitoring turnover, customer satisfaction, returns, sales trends, and employee utilisation through retail scorecards. In addition, BI is able to help in increasing profit margins by aligning corporate and store operations around critical revenue and profitable targets, providing sales and item movement information to operations, marketing, and merchandising, improving customer satisfaction by identifying trends and responding to customer buying needs and behaviour, and optimising profitability by planning and adjusting resources accordingly.

6.5. Medical Services

Hospitals need to analyse and report critical business data accurately and quickly by generating reports that reveal vital information about processes, efficiency levels and revenue as well as expenditures to optimise patient care. BI systems can automatically initiate the generation and printing of standard letters such as patient reminder letters, re-invitation letters for monitoring purposes and follow-up letters to general physicians to obtain patient feedback. These letters or reports can improve not only the fluctuation in waiting times, but also monitor the quality of care.

According to Hsu, Lee, Liu and Ling (2000), Singapore has introduced a regular screening program for diabetic patients. Patient information, clinical symptoms, eye-disease diagnosis and treatments are captured in a database. The availability of historical data leads to the application of data mining techniques to discover interesting patterns. The objective is to find rules that can be used by medical doctors to improve daily work. As a result, doctors are able to gain a much better understanding on how diabetes progresses over time and how different treatments affect its progress. The challenge in this according to Apte (2003) is to continue on developing more sophisticated techniques that can assist users in analysing the discovered knowledge more easily and quickly.

7. Technologies of BI

In developing a BI system, BI tools, data warehouse management tools, ETL (Extract, Transform and Load) and data integration tools are some of the essential elements.

The main BI activities include query, reporting, OLAP, statistical analysis, forecasting, data mining and decision support. Ensuing subsections will describe data warehousing, data mining and OLAP.

Figure 4 below shows an example of BI solution architecture (Eastwood, Vesset and Morris, 2005).

7.1. Data Warehousing

In the past, data warehouse used to store aggregated and historical data in a homogeneous database, where the physical storage of data was in a single, centralised data pool (Baumgartner, Frolich, Gottlob, Harz, Herzog and Lehmann, 2005). However, this approach is no longer suitable because data warehouses nowadays contain different business models which have their own update requirements that change over time. An important characteristic of the modern data warehouse is therefore the integration of heterogeneous, distributed, internal and external data.

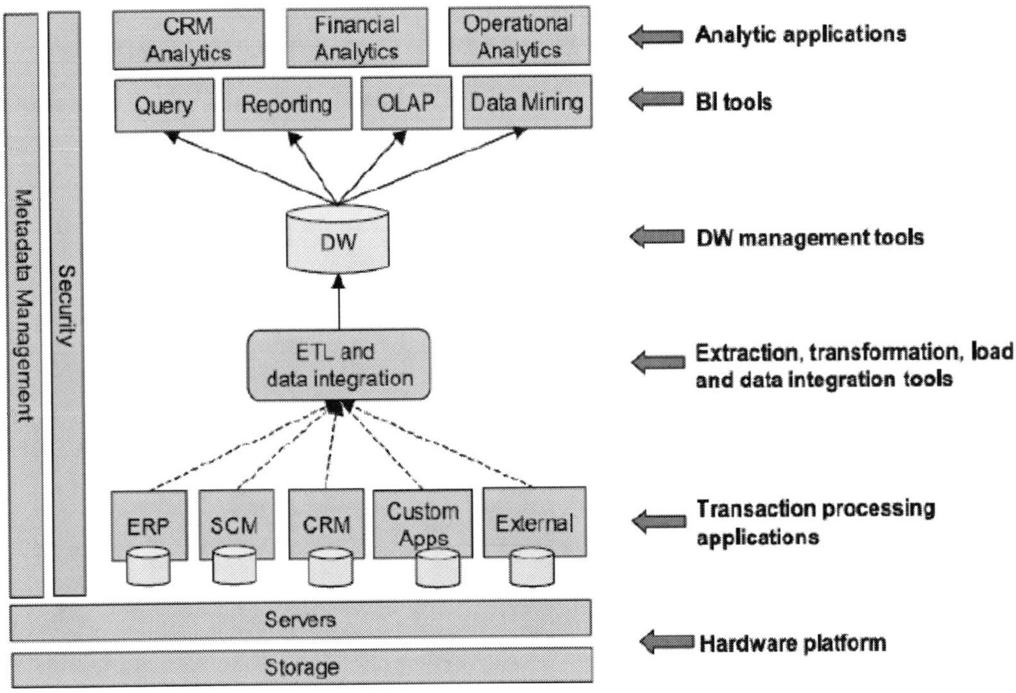

Figure 4. BI Solution Architecture.

The ETL tool, as shown in Figure 4, is used to extract, transform, load and integrate all the data from operational databases into the data warehouse. Data warehouse thus contains consolidated data from several operational databases over potentially long periods of time. The BI tools will then query and analyse the information from data warehouse and present the information to users based on the definitions that have been setup. Information can be presented in the form of reports, graphs, inquiries, figures, etc, and are normally accessible via web applications. Data warehousing thus opens up new possibilities in terms of decision support systems.

7.2. Data Mining

Data mining has been referred to as the core of knowledge discovery process (see Figure 5). It centers on automated discovery of new business information and relationships in data. It is the process of analysing and discovering actionable insights or patterns from large volume of data and generating useful information. It is being increasingly used and embedded in vertical solutions for BI and decision support.

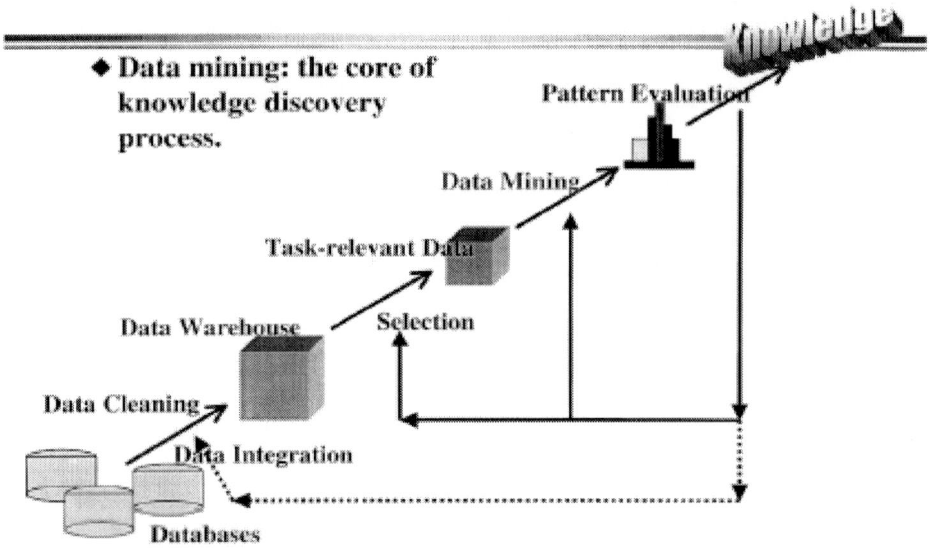

Figure 5. Data mining: a knowledge discovery process (Han and Kamber, 2001).

According to IBM Research Division[15], typical BI applications using data mining include risk analysis and management, targeted marketing, customer retention, and fraud detection. Data mining software uses technologies such as neural networks, rule induction, and clustering to discover relationships in data and make predictions. To support decision making, data in a data warehouse has to be well-organised in order to fulfill different end-user requirements, such as the predefined reporting for occasional users, ad-hoc data analysis for knowledge workers, or data mining for data analysts (Baumgartner, Frolich, Gottlob, Harz, Herzog and Lehmann, 2005). A statement from IBM Research speculated that data mining will soon join Mathematical Programming and Optimisation as a key scientific technology for building decision support systems, and the use of data mining should eventually become as easy and pervasive as working with databases and spreadsheets.

7.3. OLAP

According to Turban, King, Lee and Viehland (2004), OLAP accommodates queries that would otherwise have been computationally impossible on a relational database management system. OLAP technology is characterised by multidimensionality, and the concept represents one of its key features. Generally, OLAP involves many data items in complex relationships. Its objectives are to analyse these relationships and look for patterns, trends, exceptions, and to answer queries. It is necessary to note that the whole process of OLAP must be carried out online with rapid response time to provide just-in-time information for effective decision making. OLAP operations typically include rollup (increasing the level of aggregation) and drill-down (decreasing the level of aggregation or increasing detail) along with one or more dimension hierarchies, slice and dice (selection and projection), and pivot (re-orienting the multidimensional view of data) (Chaudhuri and Dayal, 1997). For example, it allows users to

[15] http://www.almaden.ibm.com/laborday/index.html

analyse data such as revenue from any dimension such as region or product line anytime of the year. The ability to present data in different perspectives involves complex calculations between data elements, thus enables users to pursue an analytical process without being thwarted by the system.

8. The Integration of BI with KM

It is crucial to have the right information at the right time that suits the right circumstance. As such, there should be an interaction effect between KM activities and BI efforts. KM technologies help to process routine or non-routine and structured or unstructured information so as to enhance search capabilities, organise and extract value from various information sources. With the valuable processed information stored in data warehouse, BI efforts will further serve decision-making objectives of the users. However, one should not underestimate the influence of culture and people in the organisation. Resistance to change, to learn new technologies, and users' inability to determine how to take action based on the results of BI analysis are among the challenges. Therefore, fostering a culture of sharing and exchanging of knowledge in the workforce is crucial to enable successful KM and BI approaches. Effective leadership is necessary to influence the organisational culture in order to achieve superior knowledge integration within people for better decision-making process. In year 2002, Sunassee and Sewry proposed a framework which contains three main interlinked components, as shown in Figure 6. They proposed that an organisation needs to achieve a balance between these three subsystems in order to achieve a successful KM effort. KM allows users to utilise all the knowledge collected by the organisation and provides new insights for decision-making process.

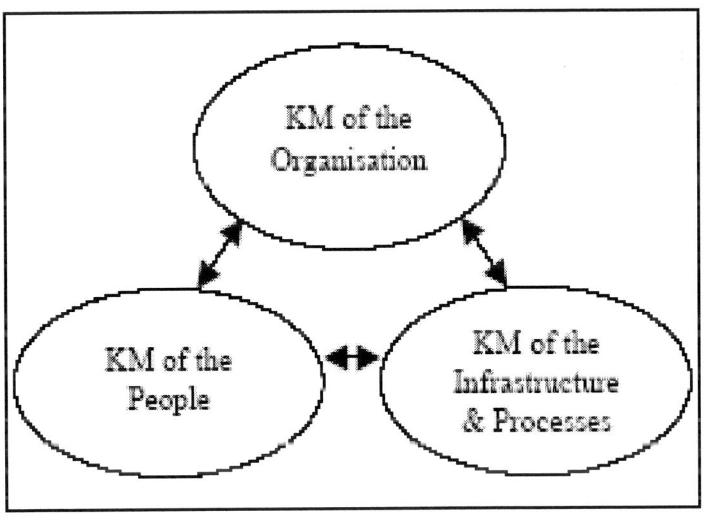

Figure 6. Proposed Knowledge Management Framework (Sunassee and Sewry, 2002).

9. THE FUTURE OF BI

Based on a study conducted by The Data Warehousing Institute[16] in 2002, US businesses lose more than $600 billion a year to data quality problems. Much of the blame can be attributed to a lack of proper method for data integration. As such, BI as an effective instrument for providing highest quality information on which successful decision is built upon has a tremendously bright future.

Table 1 shows an estimated market growth for BI till 2009. In 2004, the BI market grew 11% to reach $4.27 billion in worldwide license and maintenance revenue. This growth rate was believed to be double to that achieved in 2003. In 2009, it is expected that the market will hit $6.63 billion.

The main concern now is knowing how to apply the technology to its greatest effect and for maximum benefit. If implemented properly, a reliable decision network based on BI will drive employee performance to the next level as the employees know they can make the right decision confidently. With empowered employees and BI offerings tailored to particular verticals, organisations will be armed with the right tools (more intuitive and visual) for smarter decision planning, thus open up new revenue streams and outpace competitors.

Table 1. Estimated BI market growth from 2003 to 2009

Worldwide Business Intelligence Tools, 2003–2009 ($M)

	2003	2004	2005	2006	2007	2008	2009	2004–2009 CAGR (%)
BI tools	3,842.8	4,270.2	4,661.9	5,092.2	5,558.7	6,068.8	6,635.4	9.2
Growth (%)		11.1	9.2	9.2	9.2	9.2	9.3	

As Figure 7 has shown, IDC predicts that in the next few years, BI's popularity will increase in a faster rate than during the past 15 years. BI will be more widely adopted in the coming years as the costs of deploying large data warehouses drop, and as BI becomes more helpful and user friendly.

A common complaint about BI is that it takes too long to see the results. Operational and efficiency benefits of BI often offer payback on a strategic level. Nevertheless, according to IDC, nearly half of the companies that buy into BI gradually see the payback within a year (Eastwood, Vesset and Morris, 2005). Direct benefits such as cost reduction and revenue growth appear quickly in reports and are easy to track, though it tends to take years to see the indirect benefits of BI such as risk reduction, cost avoidance and competitive gain (Strassmann, 1990).

16 http://www.dw-institute.com

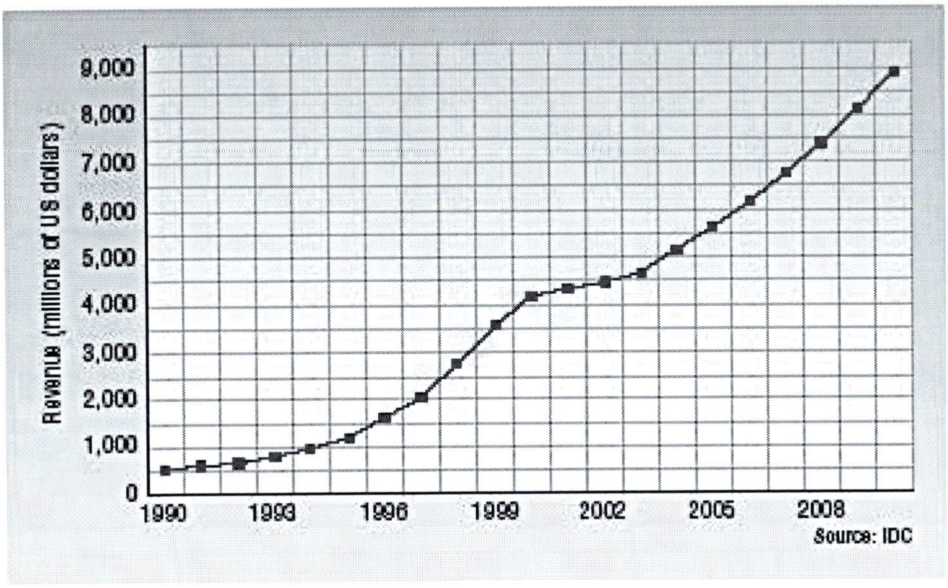

Figure 7. Market research firm IDC prediction of BI's popularity (Lawton, 2006).

Security is another major concern on the implementation of BI as BI relies very much on the Internet for distributing information. Legislation can be enacted by the authorities where security of data is enhanced by using encryption technology for data storage and access. In the outlook, it is anticipated that the BI applications will evolve in more intelligent and innovative ways in handling structured and unstructured data, and they will soon be part of knowledge and performance management to be incorporated within business processes. BI technology will shift to more agile, distributed intelligence, and become more dynamic in the responses to changing market demands. The future will also see more integration of systems and tools that focus on specialised areas of BI.

Concluding Remarks

In this paper, we have seen that organisations who successfully integrate BI into their processes can achieve significant return on their investments. BI can improve an organisation's performance and productivity by consolidating and producing information in a timely manner for making faster and better decision. The concept of BI is thus important for organisations to improve competitive response and decision-making process, ease corporate governance and regulatory compliance, and maintain customer relationship.

BI can be applied across various industries. The analyses, such as customer profiling, customer risk analysis, analysis of cost component to reduce costs of goods sold, identification of demand and sales trends to manage inventory effectively, and so on, are to assist the organisation in providing better services to the customers, optimising productivity and assisting management in decision-making process. In an e-business environment, BI can assist in personalising websites, services or product offerings based on customers' predictive behaviour and risk analysis. Customers' demand and sales trend statistical analysis and forecasting assist in inventory and suppliers management. Inventory and suppliers information between the organisation and various suppliers can be made available on the Internet via web-based B2B application.

The technologies that are available for BI are query, reporting, data mining, OLAP and data warehousing. With the availability of accurate and timely information, organisations would be able to respond effectively to changing business condition and gain competitive edge in the marketplace. Nowadays, an integrated BI system that provides quality information and meets the end-users requirements is crucial for sustaining competitive advantage. Proper applications and technologies are needed for successful implementation of BI. However, competitors may also be able to source the same solutions to stay in competition. Organisations must therefore continue to find new ways and be innovative in turning information into useful knowledge.

REFERENCES

Apte, C. (2003). *Data Mining Analytics for Business Intelligence and Decision Support*. White Paper, IBM.

Baumgartner, R., Frolich, O., Gottlob, G., Harz, P., Herzog, M., and Lehmann, P. (2005). Web Data Extraction for Business Intelligence: the Lixto Approach. In *Proceedings of Datenbanksysteme in Business, Technologie und Web (BTW 2005)*, Karlsruhe, Germany.

Call, D. (2005). Knowledge Management – not Rocket Science. *Journal of Knowledge Management*, 9(2), 19-30.

Chaudhuri, S., and Dayal, U. (1997). An Overview of Data Warehousing and OLAP Technology. *ACM SIGMOD Record, 26*(1), 65-74.

Eastwood, M., Vesset, D., and Morris, H. D. (2005). *HP: Delivering Value in Business Intelligence*. White Paper #05C4359. Retrieved November 15, 2006, from http://cxosc.leadix.com/sites/dataexplosion/files/HP_BusIntelligence_print.pdf, IDC.

Farhoomand, A. (2005). *Managing e-Business Transformation: A Global Perspective*. China: Palgrave MacMillan.

Haimila, S. (2001). KM in Practice: the Helping Hand of BI. *KMWorld*, 10(10).

Han, J. and Kamber, M. (2001). *Data Mining: Concepts and Techniques*. San Francisco, CA: Morgan Kaufmann.

Herschel, R. T., and Jones, N. E. (2005). Knowledge Management and Business Intelligence: the Importance of Integration. *Journal of Knowledge Management*, 9(4), 45-55.

Hsu, W., Lee, M. L., Liu, B., and Ling, T. W. (2000). Knowledge Discovery and Data Mining. In *Proceedings of the Sixth ACM SIGKDD International Conference*. New York: ACM Press, 430-436.

Lawton, G. (2006) Making Business Intelligence More Useful. *Computer*, IEEE Computer Society, 39(9), 14-16.

Olszak, C. M., and Ziemba, E. (2003). Business Intelligence as a Key to Management of an Enterprise. In *Proceedings of Informing Science and IT Education*. Santa Rosa: The Informing Science Institute.

Pareek, D. (2006). *Business Intelligence for Telecommunications*. New York: CRC Press.

Porter, M. E. (2001). Strategy and the Internet. *Harvard Business Review*, 79(3), 63-78.

Robson, A. (2005). *CRM: The Concept and the Technology*. White Paper, Perfect Image Ltd.

Schlegel, K., Hostmann, B, Bitterer, A., and Burton, B. (2006). *Magic Quadrant for Business Intelligence Platforms, 1Q06*. Technical Report G00136660, Gartner, Inc.

Strassmann, P. (1990). *The Business Value of Computers*. New Canaan: The Information Economics Press.

Sunassee, N. N., and Sewry, D. A. (2002). A Theoretical Framework for Knowledge Management Implementation. In *Proceedings of the Annual Conference of the South African Institute of Computer Scientists and Information Technologists (SAICSIT 2002)*, Port Elizabeth, South Africa, 235-45.

Turban, E., King, D., Lee, J., and Viehland, D. (2004). *Electronic Commerce 2004: A Managerial Perspective*. Upper Saddle River, New Jersey: Prentice Hall.

In: Focus on Information, Intelligence and Knowledge
Editor: Rafiq Islam
ISBN 978-1-61209-979-8
2011 Nova Science Publishers, Inc.

Chapter 8

Adaptive System for Tower Structure

T. Rengaraja[*] and P. Devadas Manoharan[**]
[*]R M K College of Engineering and Technology, Puduvoyal-601 206, Tamil Nadu, India
[**]Anna University, Chennai- 600 025, Tamil Nadu, India

Abstract

Truss and tower systems are normally designed for specified loads by using respective codes. In certain cases, they may be subjected to loads more than the design values. Thus a need arises to protect these systems when the specified loads are exceeded by certain margins. This paper focuses on providing in-built smartness in a three-dimensional, two-panel tower system to handle both force and deformation when unanticipated loads up to 55 percent increase over a short duration act on these systems. Analysis is made for loads at the rate of 1.25, 1.35, 1.45 and 1.55 times the design load of the tower. The paper explains suitable control forces that are to be generated and how the system under combined action of unanticipated and control forces are balanced to keep the structural system integral during short duration of unanticipated loads. This has been highlighted with an example.

Keywords: Smart Structure, Adaptive Structure, Unanticipated Loads, Seismic Loads, Tower Systems, Sensors, Actuators, Active Control.

Introduction

Structures built by men are constantly under threat by natural disasters like earthquake, hurricane, rainstorm, thunderbolt, flooding and landslides. Added to this Structural degradation like material aging, corrosion and environmental changes lead to structural deterioration. Natural disasters are difficult to forecast accurately. Structural degradation can not be avoided. Human expects future engineering should sense and respond to internal and

[*] E-mail: rajatr90@gmail.com

external environmental changes at optimum conditions on time. It is needed to have neural systems, which can sense or forecast hidden troubles and make the structures to adapt to changes and survive under such eventuality. Such systems, are called 'intelligent structures and systems'.

The protection of civil engineering structures, including their material contents and human occupants, is a world-wide priority. Such protection may range from reliable operation and comfort, on the one hand, to survivability on the other. Examples of such structures include buildings, offshore rigs, towers, roads, bridges, and pipelines. In like manner, events which cause the need for such protective measures include earthquakes, winds, and waves.

The common feature of the different proposed approaches is the modification of the dynamic interaction between the structure and the dynamic loads. The focus of the modification is to minimize the damage and vibrations throughout the structure. The result is that well-designed controlled systems display enhanced safety and occupant comfort.

Various means of controlling structural vibrations produced by earthquake or wind have been investigated by the structural engineering community. These means include modifying rigidities, masses, damping, or shape through the provision of countering forces. A structure that is designed solely on the basis of strength requirements does not necessarily ensure that the building will respond dynamically in such a way that the comfort and safety of the occupants is maintained.

BASIC PRINCIPLES

Active control concerns with the design of systems that are able to act so as to force a system to behave according to some prescribed specifications. The behavior of the system is described by means of a dynamic system characterized by state variables. The relationship between the system and the external environment is described in terms of inputs variables, which represent external actions on the system, and output variables, which are measurable response of the system. Two basic controller arrangements are open-loop and closed-loop. In open-loop control, the controller manipulates the system by trying to drive output to a specified behavior without using information regarding the output. This type of control suffers by not allowing it to compensate for errors due to perturbations in either the input, from unknown or unmeasured disturbances, or system perturbations. In contrast, closed-loop control uses information about the system response as feedback information. The basic elements found in closed-loop control system are shown in Figure 1.

Studies into active control of civil engineering studies have flourished since its introduction to the field by Yao (1972). B.F. Spencer and Sain (1997); Suharidjo et al. 1992; as per Luciana R. Barroso (1999). B.F. Spencer et al. (1994) have used H_2 and H_∞ control to their research. Sliding mode control was adopted by Yang et al. (1995). Saturation control was adopted by Chase and Smith (1996). B.F. Spencer et al. (1992); R.V. Field et al. (1996); R.V. Field and Bergman (1997) was used reliability-based control Fuzzy control was developed by Nagarajaiah (1994). Venini and Wen (1994); Ghaboussi and Joghataie 1995 used neural control and Agrawal and an Yang 1996 used nonlinear control . Modeling issues in control of civil engineering structures were discussed by Smith and Schemmann (1996);

Dyke et al. (1995). B.F. Spencer et al. (1997) developed benchmark studies A good reference to the current state-of-the-art in active control can be found in Housner et al. (1997).

Common actuator systems found in civil engineering applications include the active mass driver (AMD) and the active tendon brace (ATB) systems. The AMD actuation system is composed of a small, relative to the structure, additional mass, that is activated by servo-hydraulic actuators which connects the mass to the structure. The system is typically located near the top of the structure so as to be most effective at controlling the first mode of the structure. The ATB actuation system uses a tendon/pulley system to transmit the force generated by the hydraulic actuator to the structure.

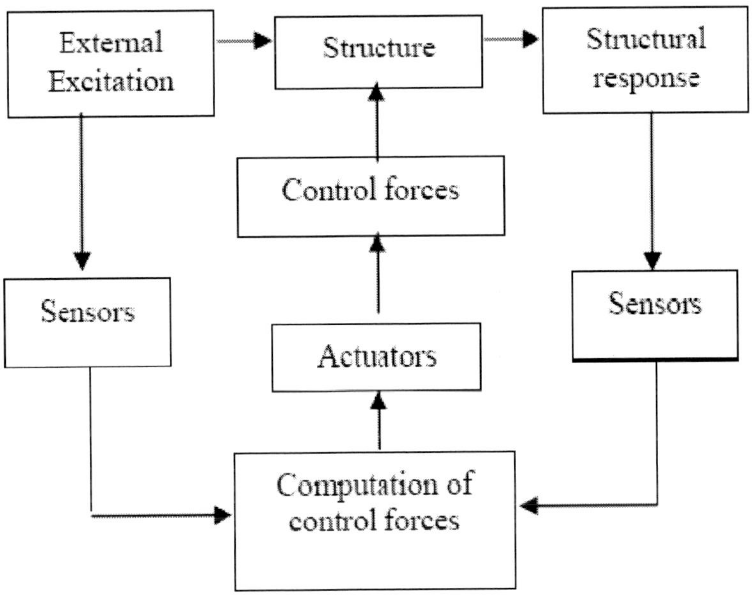

Figure 1. Active control feedback loop.

An adaptive structure is defined as a structural system whose geometrical configuration and inherent structural characteristics such as damping, can be changed in order to adapt to environmental changes or to meet the mission requirements (F.Yang and R. Sedaghati, 2005).

Power and communication virtually control the various activities of the society. The lines of power and communication become life line systems of the modern age. The hardware aspect of these life line systems namely steel towers, need to be properly designed so that failure is prevented under any circumstance. In the power and communication systems, the unanticipated loads are mostly in the forms of cyclone and earth quake. The design codes and material strengths are so formulated that probability of failure for unanticipated loads is kept within certain limits. In spite of this every year in our country, failures of transmission line towers and micro wave towers are quite common. Hence, instead of putting a premium on the design, it is preferable to look up for an alternative approach, through which a structure can withstand unanticipated loads without failure and return to its normal state under normal loads. This concept is called SMART or adaptive structural system. At some point it may no longer be prudent to rely entirely on the strength of the structure and its ability to dissipate energy to withstand these extreme loads. Active control strategies for structural systems have been developed to minimize the effects of these loads [T.T. Soong, 1990].

FEATURES OF ADAPTIVE STRUCTURAL SYSTEMS

Normally structures are designed and constructed based on criteria stipulated in design codes and left unattended till their design life. In modern parlance, this is called passive design, [Senoul Utku, 1998] mainly because the system does not know its performance and is incapable of reacting to any unanticipated situations. Historically this has been the only design methodology available and practiced by structural engineers. With advances in sensor, actuator, and microprocessor technologies, it has, now become feasible to monitor the performance of a system in terms of its deflections or stresses. Based on this information, it is feasible to actively control some of the responses of passively designed structure in real time to help in reacting to unusual situations. An adaptive structure is an engineering structure whose response to excitations can be sensed and if in excess, can be controlled in real time by either the insertion of internal deformations or inducement of force-type loading through appropriately placed actuators that are part of the structure.

When a structure at its reference state is subjected to a load, it assumes another state. The difference between these two states is called the response of the structure. They are generally displacement, stress and strain. In adaptive structures, one or more response components are monitored, i.e., measured continuously. The devices that make the measurements are sensors. When the measures of monitored response quantities exceed, they are reduced by subjecting the structure to an additional set of excitations, which may minimally alter the existing stress or internal force. The device that creates and applies these differential excitations on the structure is actuator. The actuators are devices that compensate deviation from the nominal value in number of controlled response components. If these components become unacceptably large, they can be compensated by a control loading that would create opposing forces to the measured displacements.[Shen, S, 2002].

If an actuator induces force-type loading at the nodes then it is called a force inducing actuator. The measurement of the disturbances, and the computation of control to compensate them, may be depicted through a feedback loop as shown in Figure 1.

In this study the control mechanism is illustrated through an example of a tower system. Tower is chosen because of the wide application in areas of power and communication. The focus of the study is to introduce smartness in such systems, so that the buckling can be overcome by external actuation eliminating lower buckling modes, [Rao G.V. Singh G, 2001] to ensure that the failure of the system does not take place even during critical times.

Truss System Chosen for the Study

To demonstrate the introduction of smartness, a 32-member tower under lateral load is chosen. Grouping of the members into design groups namely leg members, horizontal bracing members and diagonal bracing members was done from the design point of view. The geometry and load details are given in Table 1, 2 and Figure.2.

Table 1. Structural details of the tower

Description of Members	Members Chosen	Sectional Area Sq. mm.
Bottom Panel Leg	ISA 65 × 65 × 6	744
Top panel leg	ISA 40 × 40 × 6	447
Diagonal Bracings	ISA 45 × 45 × 6	507
Horizontal Bracings	ISA 40 × 40 × 6	447

Table 2. Design load details

Sl.No.	Node	Design load in N		
		Fx	Fy	Fz
1	3	50000	0.00	0.00
2	9	50000	0.00	0.00

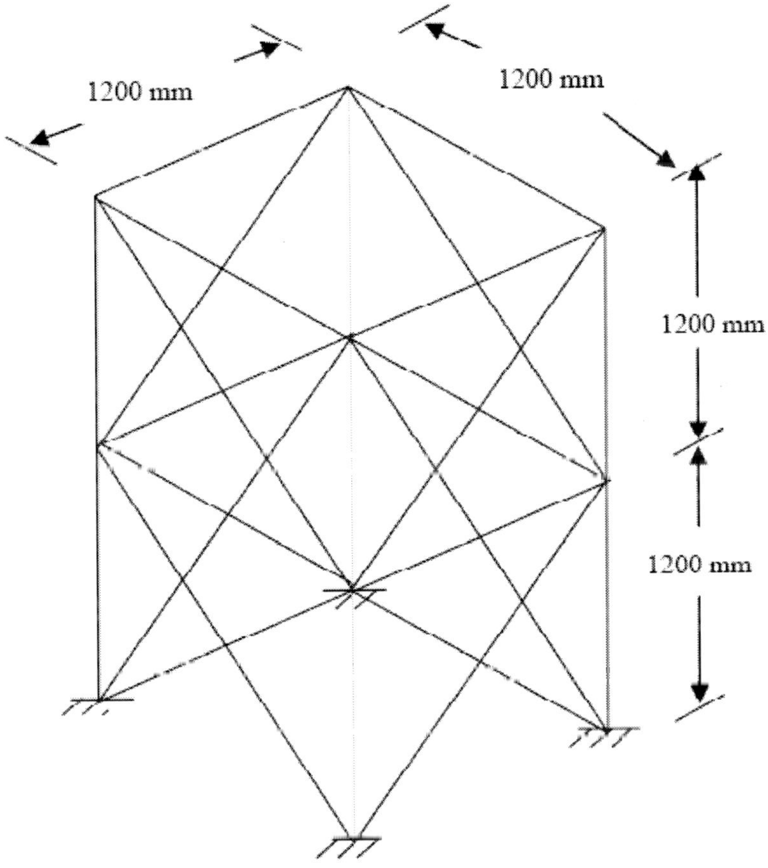

Figure 2. Tower chosen for study.

METHODOLOGY

Unanticipated Loads

Loads on structural systems are normally classified into service or working and ultimate or extreme. These are very well enunciated in various codes of practice. Once the design is done, it is the philosophy of this concept that structures will perform within the limits prescribed both for material and function. But a different scenario is becoming more prominent and this relates to short term and high amplitude dynamic loads causing responses much higher than the designed load and occurring for a very short duration. This may be because of cyclones, earthquake, man-made disasters like bombs etc. It is not worth designing for this situation as it is for a short period and may not occur frequently. Instead if the system is allowed to have a mechanism by which it can sense, be aware and take action to counter this extra load, the system is saved with marginal extra cost. In this paper, an attempt is made to counter the unanticipated load occurring for a short period, by incorporating smartness.

Method of Introducing Smartness

The actuators are used to insert deformation into tower elements. Figure.3 shows the block diagram of the sensors and control system provided in one face of the tower chosen. Sensors monitor the response of the towers continuously and pass the information to the PC with control board. Based on the predetermined control algorithms, the controller has been instructed to actuate the actuators, fixed in the leg members of the towers. Actuators develop required strains to counteract the excess response developed due to the unexpected excitations. This process will continue, till the unexpected load, which exists beyond the design load.

The study is carried out by increasing the design load in four stages.

Stage I : 25 % increase in design load
Stage II : 35 % increase in design load
Stage III: 45 % increase in design load
Stage IV:55 % increase in design load

The strain developed due to the design load in the leg members are assumed as e1, e2, e3, e4, e5, e6, e7 and e8. e1, e2, e3 and e4 in the left side leg members are tensile strain and e5, e6, e7 and e8 in the right side leg members are compressive strain. These values should not be exceeded to maintain the structural safety. When the design load is increased, these strain values increases to $e^{1}5$, $e^{1}6$, $e^{1}7$ and $e^{1}8$ respectively resulting in excess compressive values strain of ($e^{1}5$-e5), ($e^{1}6$-e6), ($e^{1}7$-e7) and ($e^{1}8$-e8). These additional strains are developed in the compression leg members. To compensate the excess strain, external force is applied through actuators, which develop counter strain in these members as well as tension leg members. To estimate the strain required to be developed in the leg members by the actuators, four set of equal strains, say e is applied in the four pairs of columns. Each pair consists of compressive strain –e in the tension leg members and tensile strain e in the compression leg members. Due to the first pair of applied strain, the strain developed in the compression leg members are e51, e61, e71 and e81. Similarly strain developed in the compression leg members due to the other three pairs of strain, can be determined.

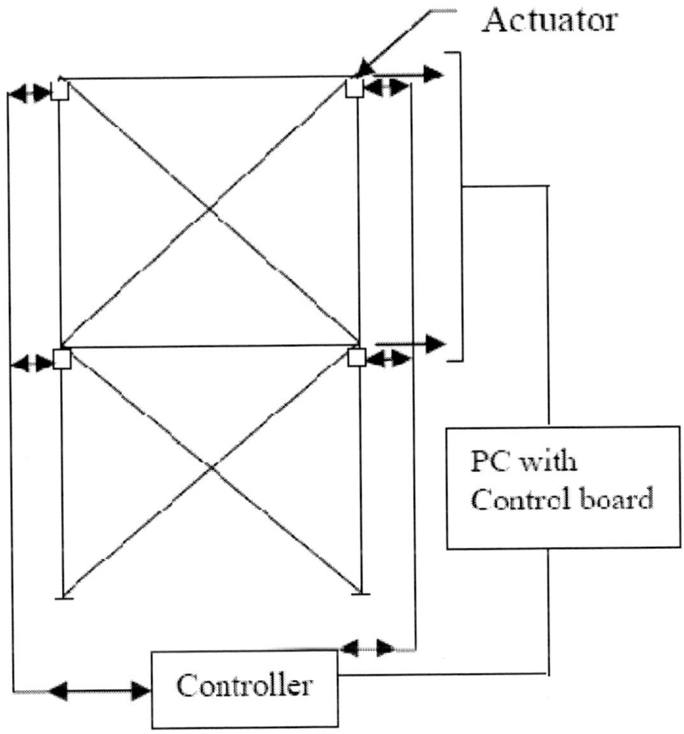

Figure 3. Block diagram of control system.

Let A, B, C and D be the values to be multiplied with the strain value e to obtain the actual strain to be applied in each of the four pairs of leg members. The value of A, B, C and D can be determined from the following equations.

$$\begin{bmatrix} e51 & e52 & e53 & e54 \\ e61 & e62 & e63 & e64 \\ e71 & e72 & e73 & e74 \\ e81 & e82 & e83 & e84 \end{bmatrix} \begin{Bmatrix} A \\ B \\ C \\ D \end{Bmatrix} = \begin{bmatrix} e^{1}5-e5 \\ e^{1}6-e6 \\ e^{1}7-e7 \\ e^{1}8-e8 \end{bmatrix}$$

A * e, B *e, C * e and D * e respectively are the strain required in each of the four pairs to balance the excess load.

Results and Discussion

Stage I: 25% Increase in Design Load

An increase in the load by 25% more than the design value produce forces higher than the design values in nine members. Based on the calculations as mentioned in this paper, four pairs of strain calculated using the above equation were applied in the leg members. 0.00019117 and -0.00019117 strain in the two pairs of bottom panel leg members and 0.00006407 and -0.00006407 strain in the two pairs of top panel leg members. These responses by the actuators balance the excess force developed in the leg members and partly balance the other members also. Figure 4 shows the status of the forces in various members, after and before the balance force in effect.

Dynamic analysis is also carried out by taking the 25%more than the design load as rectangular pulse load. The duration of each pulse is 0.01 times the natural period. The natural frequency of the structure is 0.33. The duration of each pulse is 0.03 seconds. Figure 5 shows, how effective the control strategy is in controlling the maximum deflection of the tower.

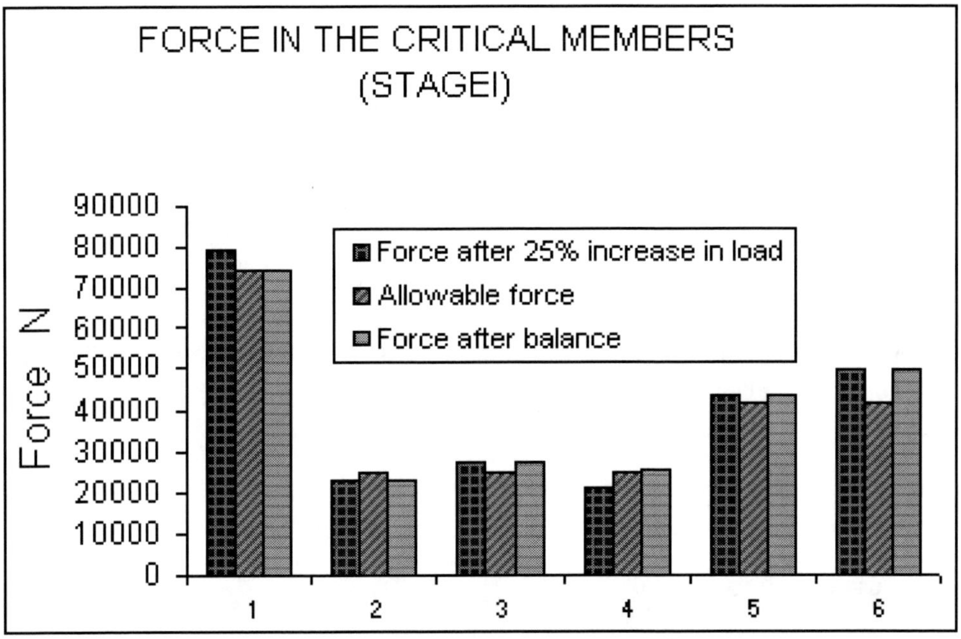

1.bottom compression leg member, 2.top compression leg member, 3.top horizontal bracing member,4. bottom horizontal bracing member, 5.bottom diagonal bracing member, 6.top diagonal bracing member.

Figure 4. Force in the critical members (Stage I).

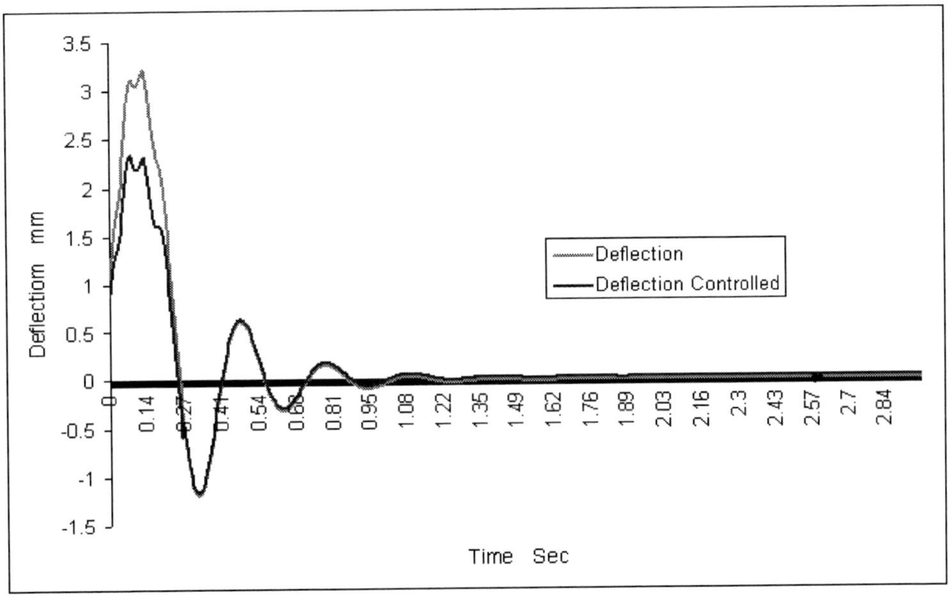

Figure 5. Deflection 25% excess load.

Stage II: 35% Increase in Design Load

To balance the 35% excess load, 0.00044001 and - 0.00044001 strain in the two pairs of bottom panel leg members, 0.00006407 and -0.00006407 strain in the two pairs of top panel leg members were introduced through the actuators. The forces and deflection in the critical members before and after the application of the actuation are shown in Figure 6 and Figure 7.

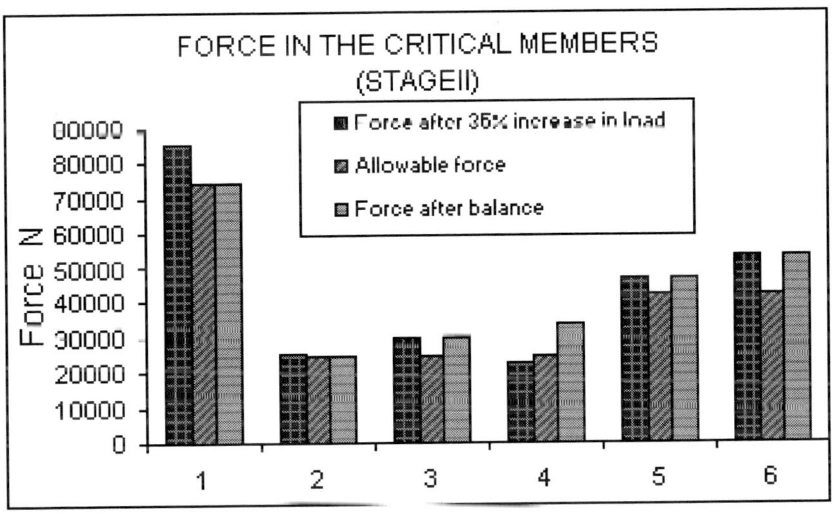

1.bottom compression leg member, 2.top compression leg member, 3.top horizontal bracing member,4. bottom horizontal bracing member, 5.bottom diagonal bracing member, 6.top diagonal bracing member.

Figure 6. Force in the critical members (Stage II).

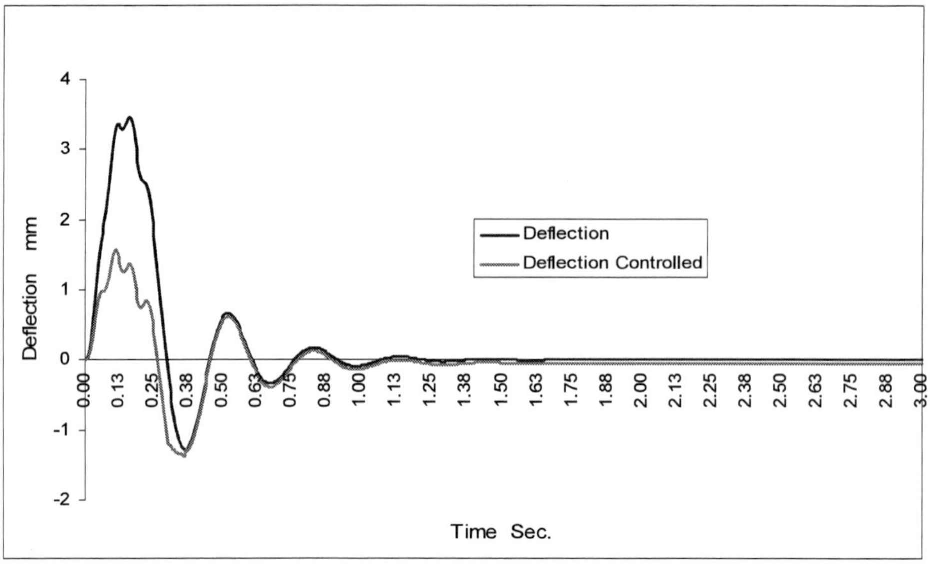

Figure 7. Deflection 35% excess load.

Stage III: 45% Increase in Design Load

To balance the 45% excess load, 0.00071099 and - 0.00071099 strain in the two pairs of bottom panel leg members, 0.00018843 and -0.00018843 strain in the two pairs of top panel leg members were introduced through the actuators. The forces and deflection in the critical members before and after the application of the actuation are shown in Figure 8and Figure.9.

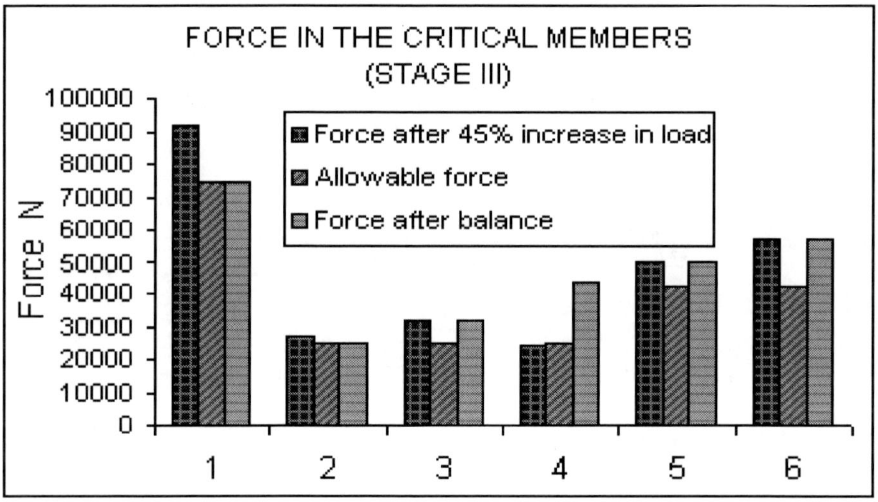

1.bottom compression leg member, 2.top compression leg member, 3.top horizontal bracing member,4. bottom horizontal bracing member, 5.bottom diagonal bracing member, 6.top diagonal bracing member.

Figure 8. Force in the critical members (Stage III).

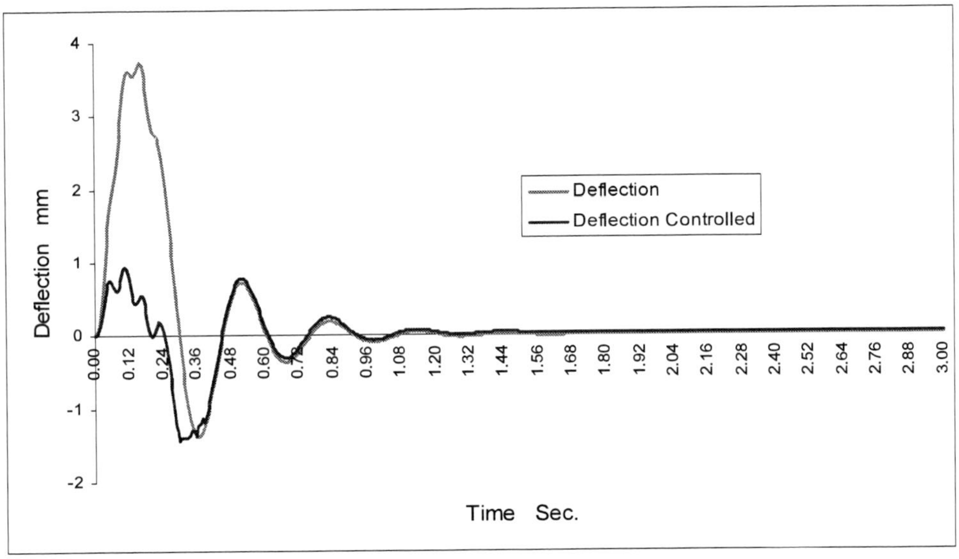

Figure 9. Deflection 45% excess load.

Stage IV: 55% Increase In Design Load

To balance the 55% excess load, 0.00097598 and - 0.00097598 strain in the two pairs of bottom panel leg members, 0.00031281 and -0.00031281 strain in the two pairs of top panel leg members were introduced through the actuators. The forces and deflection in the critical members before and after the application of the actuation are shown in Figure 10 and Figure 11.

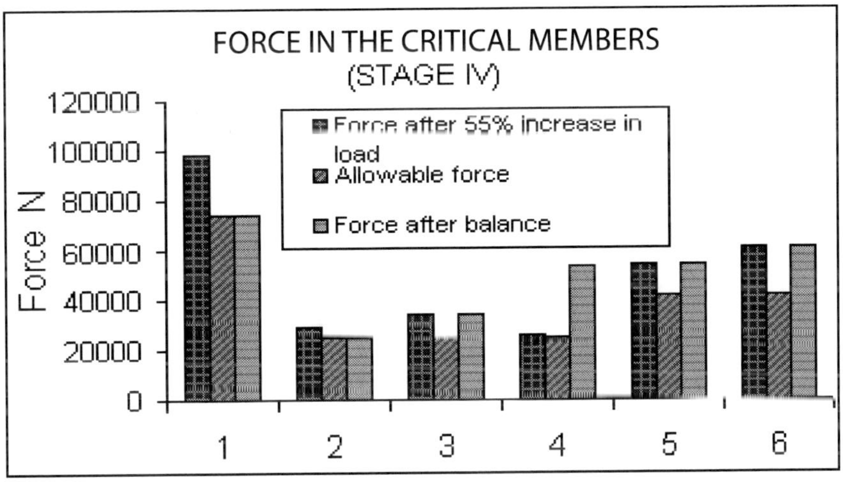

1.bottom compression leg member, 2.top compression leg member, 3.top horizontal bracing member,4. bottom horizontal bracing member, 5.bottom diagonal bracing member, 6.top diagonal bracing member.

Figure 10. Force in the critical members (Stage IV).

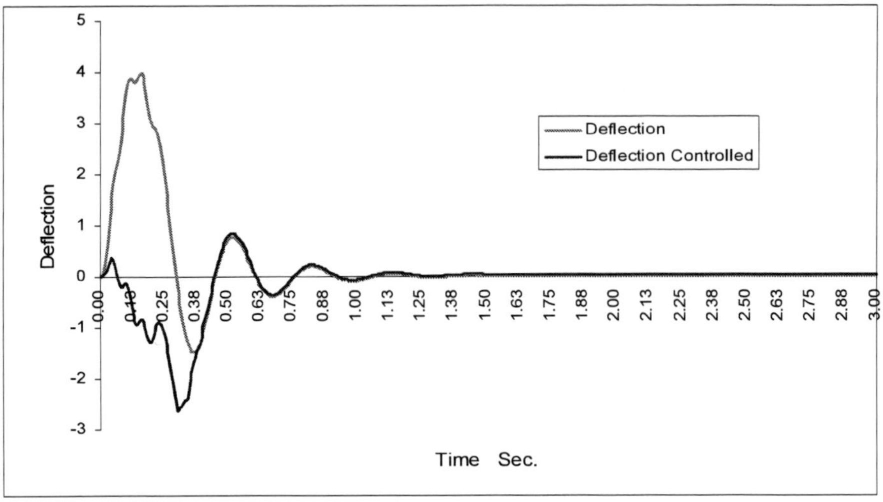

Figure 11. Deflection 55% excess load.

Discussion

From the results it can be clearly seen that the strain produced by the actuators are able to successfully balance the excess force in the leg members. However during the process the force in the bracings exceeds their design values marginally. These excess forces in the bracings are not so predominant and it is worth making the structure smart by introducing strain inducing actuators. The bracings can be redesigned for the changed force or the bracing pattern itself can be changed to withstand the excess load.

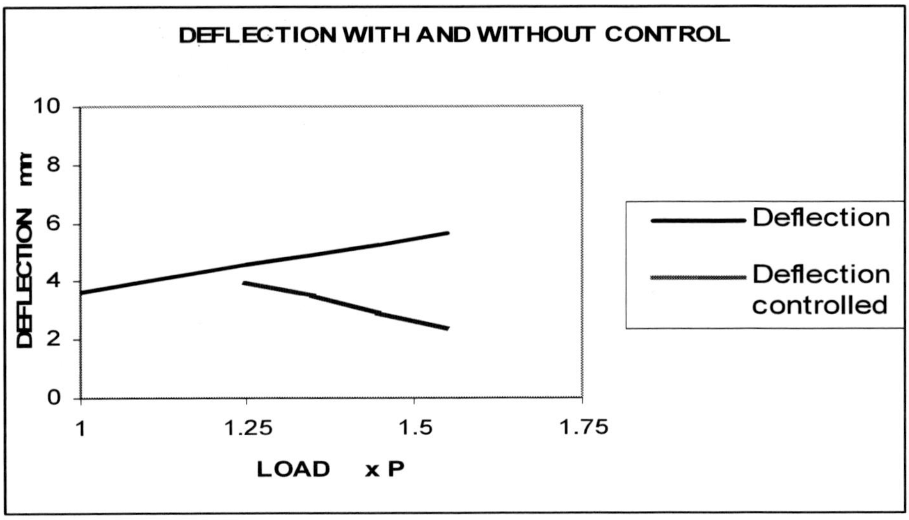

Figure 12. Maximum deflection of the tower at various stages of over load before and after control.

Figure 12 shows the maximum deflection of the tower at various stages of over load before and after control. It clearly indicates the effectiveness of the control through reduction in deflection.

Conclusion

- Certain smartness or adaptive character can be built in to a structure through introducing sensors, actuators and control algorithms as a part of the structural system, such that the structure can resist to certain extent of unanticipated overload.
- The example presented illustrates the possibility and effectiveness of this concept.
- The reduction in compressive strength due to buckling can also be overcome by external actuation eliminating lower buckling modes using this concept.
- The smartness that can be in-built into the system has certain limitations and requires alteration in the design of the bracings.
- The same actuators can be effectively utilized to provide control in either of the two directions.
- The results presented above are encouraging and show that this simple control system can be effective for tower structure.

References

Agrawal, A.K. and Yang, J.N. (1996). Optimal Polynomial Control for Seismically Excited Non-linear and Hysteretic Structures. *Earthquake Engineering and Structural Dynamics, 25(11)*, pp.1211-1230.

Chase, J. Geoffrey and Allison Smith, H. (1196). Robust Control Considering Actuator Saturation – I: Theory, II: Applications. *Journal of Engineering Mechanics, 122(10)*, pp.976-993.

Dyke, S.J., Jr., Spencer, B.F, Quast, P. and Sain, M.K. (1995). The Role of Control Structure Interaction in Protective System Design. *Journal of Engineering Mechanics, 121(2)*, pp.322-338.

Field, R.V, Jr. and Bergman, L.A. (1997). Reliablity-Based Covariance Control Design. In *Proceedings of the 1997 American Control Conference, Volume WA01*, Albuquenrque, New Mexico, pp.11-15.

Field, R.V. Jr., Voulgaris, P.G. and Bergman, L.A (1996). Probabilistic Stability Robustness of Structural System, *Journal of Engineering Mechanics, 122(10)*, pp.1012-1021.

Housner, G.W., Bergman, L.a. Caughey, T.K., Chassiakos, A.G, Claus, R.O. Masri, S.f, Skelton, R.E, Soong,T.T, Spencer, B.F, and Yao, J.T.P. (1997). "Structural Control: Past, Present, and Future. "*Journalof Engineering Mechanics,* 123(9), pp.897-971.

Lee, J. D., Shen, S., Manzari, M. T. and Shen, Y.L. (2000), Structural control algorithms in earthquake resistant design, *Journal of Earthquake Engineering, 4(1)*, pp.67-96.

Luciana R. Barroso (1999), Performance evaluation of vibration controlled steel structures under seismic loading, Ph.D. Thesis.

Luco, J.E. (1998), A Simple Model for Structural control including Soil-Structure Interaction Effects, *Journal of Earthquake Engineering and Structural Dynamics*, pp.225-242.

Nagarajaiah, S. (1994), Fuzzy Controller for Structures with Hybrid Isolation System, In *Proceedings of the 1st World Conference on Structural Control, Volume TA2*, Los Angeles, California, pp.67-76.

Rao G.V. and Singh G., *Smart Structure Concept for the buckling load enhancement of columns*, Smart Material and Structure Journal (2001).

Sato, T. and Toki, K. (1992), Predictive Control of Seismic Response of Structure Taking into account The Soil-structure Interaction, Proceedings of the Ist European conference on Smart Structures and Materials, Glasgow, Scotland, pp.245- 250.

Sedaghati, R. (2005), 'Optimal Placement of active bars in Smart Structures, Proceedings of IEEE *International Conference on Mechatronics and Automation*', Niagara Falls, Canada.

Senoul Utku (1998), 'Theory of Adaptive Structures: Incorporating Intelligence into Engineering Products, CRC Press.

Shen, S. (2002), Active Control of Planar Framed Structures In Earthquake Resistant Design, Doctor of science dissertation, George Washington University, Washington, C., USA.

Sivakumar, P., Samuel Knight G.M. and Rajaraman A. (1999), Smart Towers for Life-Line systems in Power and Communications, *International Conference on Smart Materials, Structures and Systems*, Bangalore, pp.695-700.

Smith, H.A. and Schemman, A.G. (1996), Modeling Issues Associated with Vibration Control of Cable-Stayed Bridges subjected to Multiple Support Excitation, In *Proceedings of the 2nd International Workshop on Structural Control,* Hong Kong, pp.509-520.

Soong, T.T (1990), Active Structural Control: Theory and Practice, Longman Scientific and Technical.

Spencer B.F, Jr., , Suhardjo J,and Sain M.K (1994). "Frequency Domain Optimal Control Strategies for Aseismic Protection". *Journal of Engineering Mechanics* 120(1), pp.135-159.

Spencer B. F, Jr., , and Sain M.K. (1997). "Controlling Buildings: A New Frontier in Feedback". *IEEE Control Systems Magazine* 17(6), pp.19-35.

Suhardjo, J., Jr Spencer, N.F., and Kareem A. (1992), Frequency Domain Optimal Control of Wing Excited Buildings, *Journal of Engineering Mechanics, 16*, pp.1177-1188.

Venini, P. and Wen, Y.K. (1994), 'Hybrid Vibratiion Control of MDOF Hysteretic Structure with Neural Networks', In *Proceedings of the 1st World Conference on Structural Control,* Volume TA3, Los Angeles, California, pp.53-62.

Yang, F. and Yang, J.N., Wu, J.C. and Agrawal A.K. (1995), Sliding Mode Control for Seismically Excited Linear Structures, *Journal of Engineering Mechanics, 121(12),* pp.1386-1390.

Yao, J.T.P. (1972), Concept of Structural Control, *ASCE Journal of the Structural Division, 98*, pp.1567-1874.

In: Focus on Information, Intelligence and Knowledge
Editor: Rafiq Islam

ISBN 978-1-61209-979-8
2011 Nova Science Publishers, Inc.

Chapter 9

ADAPTIVE FORECASTING OF INDUCED TRAFFIC VOLUME BASED ON INDUCED ECONOMIC GROWTH

Xin Miao**, *Bao Xi** and *Hui-min Zou**
*School of Management, Harbin Institute of Technology, Harbin, 150001, China
**Business School, Nankai University, Tianjin, 300071, China

ABSTRACT

The objective of this chapter is to study the quantitative forecasting method of induced traffic volume in Chinese situation. With sufficient consideration of the mechanism and evolution of induced traffic, the relation between induced traffic volume and induced economic growth is discovered. Considering their close relation, adaptive forecasting of induced traffic volume in keeping with induced economic growth is feasible. Accordingly, the modified spatial economic approachable model is adopted to bring forward the forecasting method. The technical route is expounded step by step. Empirical study is provided by preliminary test of the method on the Ha-Shuang freeway. The test result shows the applicability of the proposed method.

Keywords: Induced traffic volume, economic growth, forecasting, infrastructure, freeway, spatial economics, mathematical model, empirical study.

1. INTRODUCTION

A common response to traffic congestion is to build new roads, while a common claim in transport planning is that more road capacity generates additional car travel. Some people

* Corresponding author: E-mail: miaoxin@yahoo.cn

argue that these new roads do not reduce congestion but instead induce additional traffic (Uchiyama et al., 2006). Most researchers believe that this additional traffic reflects an increase in the consumer surplus of the travelers; others see these additional trips and miles as wasteful. The induced traffic demand from increased infrastructure supply over time typically coincides with an increase in population and a higher per capita income in the study area. To isolate the effects of the different variables is a complicated task. There is also a danger of oversimplification in answering this complex problem. This occurs quite frequently during discussions of traffic growth (Fröhlich and Zürich, 2003).

The induced traffic usually indicates new traffic volume brought by the construction or rebuild of highways that promote the development of regional economy (Goodwin 1996; Mokhtarian et al 2002). "Induced" is a term implying that a particular condition is indirectly caused by another condition. In the case of traffic volumes, the term arose from the phenomenon that improvements to a highway, especially capacity improvements, seemed to result in more traffic choosing to use the road than would be the case if the highway were not improved. To an economist, this is an example of demand elasticity (Lee et al., 1999). The induced traffic volume has significant meaning in economic evaluation of projects, selection of road technical standards and transportation management of new opening roads, and therefore, draws the attentions of researchers to the relevant research areas.

Bonsall (1996) addressed the feasibility of measuring induced traffic by means of appropriately designed surveys. The problems which confront any attempt to measure induced traffic are described and discussed. However, even with considerable expenditure, adequate precision may be difficult to achieve and attribution of cause impossible to make (Bonsall, 1996).

Mathematical models of transportation systems have played a prominent role in transportation planning throughout the world since the 1960s. These models are hypotheses of how people use transport systems. They provide a means of extrapolating the necessarily limited empirical evidence of how road users respond to changes in the road system. Mathematical models can tell us two things: (a) the scale of the various elements of induced traffic, and (b) the implications for the economic benefits of road schemes (Coombe 1996).

As for Chinese especial situation that the railway, the freeway and the low-grade highway are parallel coexist in most areas, the general forecasting methods for induced traffic volume are not well suit for Chinese instance. Domestic scholars have done some researches on induced traffic volume forecasting, for example, the have-without contrast method, the microcosmic forecasting model and the macroscopic forecasting model (Wang et al., 2002; Wang and Shen, 2003). But the above methods are either experiential or insufficient in the mechanism analysis of induced traffic volume. Therefore, quantitative forecasting methods need to be developed based on penetrated consideration of the mechanism of induced traffic.

2. Mechanism and Evolution of Induced Traffic Volume

2.1. Constitutive Mechanism of Induced Traffic Volume

The structure of road network has been modified by the construction of freeways. Since original traffic balance has been broken, traffic volume is to be redistributed in the whole network. In this process, the productive effect of freeways not only leads to the shift of the original traffic volume, but also simulates the generation of new traffic volume that form induced traffic volume.

The induced traffic volume constitutes of potential traffic volume and developing traffic volume. The former indicate the potential passenger and freight volume that are restrained by the inconvenience, the expensive transportation expense or the long distance before the construction of freeway. The construction of freeways stimulates regional social and economic developments in terms of a mass of industrial parks, business networks and clustered towns in forming a belt-shape or corridor-liked industrial zone. Investment in new or up-graded roads both raises the level and alters the pattern of accessibility over the whole area served by the road system. Vehicle-users will perceive the opportunities that this increased accessibility offers and respond in various ways, most of which can lead to more rather than less travel on the system. To the extent that travel increases overall, it can be said to have been induced by the road-improvement. Conversely, congestion as it spreads on the network will deter some travel and can be said to have a 'traffic suppression' effect (Hills, 1996).

2.2. Evolutional Process of Induced Traffic Volume

The evolution of induced traffic volume usually experiences three stages, namely, gathering and germination stage, fast increasing stage and gradually steady stage (Zhou and Wang, 2002).

With the construction of freeways, the land property along the freeways begin to change companied by the emergence of induced traffic volume, while the volume is small and mainly manifest itself by the gathering and germination process. After the freeways come into operation, with the enhancement of land exploitation and further adjustment of industrial structure along the corridor-liked industrial zone, the induced traffic volume increases rapidly and steps into the fast increasing stage. When the land exploitation along the corridor-liked industrial zone comes to certain degree and the exploitation rate becomes gradually slow, then the third stage is coming.

3. Adaptive Forecasting of Induced Traffic Volume

3.1. Technical Route

The forecasting method presented in this chapter suits for the evaluation of a newly built freeway. The main procedures are listed as follows:

- Compartmentalize traffic influencing zone and traffic sub-zones;
- Determine the distributing matrix of the initial induced traffic volume;
- Economic forecasting for traffic sub-zones;
- Establish economic forecasting model to forecasting the generation and distribution of induced traffic volume in the future characteristic year;

- Forecasting assignment of induced traffic volume.

The compartmentalization of traffic influencing zone and traffic sub-zones should comply with engineering feasibility study. In the above process, establishing the forecasting model is the kernel part of this chapter.

3.2. Determination of Initial Induced Traffic Volume

The initial total induced traffic volume in the direct influencing zone can be calculated by formula (1) (Zhou and Wang, 2002).

$$Q_u = Q_n + Q_o - Q_t - Q_o' \qquad (1)$$

Q_u the initial total induced traffic volume in the direct influencing zone

Q_n the observational freeway traffic volume in the direct influencing zone

Q_o the observed traffic volume on original roads in the direct influencing zone

Q_t the shifting traffic volume from other transportation manners

Q_o' the forecasted trend traffic volume when there is no highway

Based on the current OD (Origin-Destination) table and the economic investigation results that come from the feasibility study report, we can distribute initial induced traffic volume to each traffic sub-zone in the direct influencing zone and thus obtain the distribution matrix of initial induce traffic volume. This matrix is the foundation of the forecasting.

3.3. Forecasting of Induced Economic Growth

There is close relationship between induced traffic volume and economic growth. In traffic zones, the economic growth caused by freeway projects is called induced economic growth. In this chapter, the spatial economic approachable model (Wang et al., 2001) is applied to forecasting the induced economic growth. The model constitutes of time distance formula, economic potential determining formula and economic induced effect formula.

1. Time distance formula

$$ACC_i = \sum_{j=1}^{n} p_i \cdot e^{-\delta \cdot D_{ij}} \qquad (2)$$

ACC_i the economic approachable coefficient of traffic sub-zone i in the characteristic year

p_i the population in traffic sub-zone i in the characteristic year

D_{ij} the time distance between traffic sub-zone i and traffic sub-zone j in the characteristic year

The economic approachable coefficient can be differentiated as $ACC_i(W)$ and $ACC_i(Y)$:

$ACC_i(W)$ economic approachable coefficient when there is no freeway project
$ACC_i(Y)$ economic approachable coefficient when there is a freeway project

The difference between them is the different time distance D_{ij}.

2. Economic Potential Determining Formula

$$POT_i = e^{\alpha} p_i^{\beta} ACC_i^{\eta} \quad (3)$$

POT_i economic growth coefficient of traffic sub-zone i in the characteristic year
α, β, η parameters of the formula

Economic growth coefficient can also be differentiated as $POT_i(W)$ and $POT_i(Y)$:

$POT_i(W)$ economic growth coefficient when there is no freeway project
$POT_i(Y)$ economic growth coefficient when there is a freeway project

3. Economic Induced Effect Formula

$$\theta_i = \frac{POT_i(Y)}{POT_i(W)} \quad (4)$$

θ_i coefficient of induced economic growth in traffic sub-zone i in the characteristic year

$$u_i = (\theta_i - 1) \cdot u_i^* \quad (5)$$

u_i^* economic accrual index of traffic sub-zone i in the characteristic year
u_i induced economic growth in traffic sub-zone i in the characteristic year

3.4 The Forecasting Model of Induced Traffic Volume

3.4.1. Basic Assumptions

The basic assumption is that the induced traffic distribution between two traffic sub-zones is directly proportional to the induced traffic generation, induced traffic attraction and induced economic growth in each sub-zone. This assumptive relation can be expressed by formula (6).

$$q_{ij} = K_{ij} \left(P_i u_i^{\gamma}\right) \cdot \left(A_j u_j^{\gamma}\right) \quad (6)$$

q_{ij} the induced traffic distribution from traffic sub-zone i to j

u_i the induced economic growth in traffic sub-zone i

u_j the induced economic growth in traffic sub-zone j

P_i the generation of induced traffic volume in sub-zone i

A_j the attraction of induced traffic volume in sub-zone j

γ pending parameter

k_{ij} regulative coefficient that reflects influence from other factors

3.4.2. Parameter Estimation

Based on the initial induced traffic distribution matrix shown in table 1 and the initial induced economic growth in each traffic sub-zone, further investigation can be done about induced traffic distribution and economic growth in typical sub-zones to determine the parameter γ. Once the γ is determined, the k_{ij} can be calculated by formula (7).

Table 1. Distribution matrix of initial induced traffic volume

	1	…	j	…	n	P_i^0	u_i^0
1	0	…	q_{1j}^0	…	q_{1n}^0	P_1^0	u_1^0
⋮	⋮	⋱	⋮	⋱	⋮	⋮	⋮
i	q_{i1}^0	…	q_{ij}^0	…	q_{in}^0	P_i^0	u_i^0
⋮	⋮	⋱	⋮	⋱	⋮	⋮	⋮
n	q_{n1}^0	…	q_{nj}^0	…	0	P_n^0	u_n^0
A_j^0	A_1^0	…	A_j^0	…	A_n^0	/	/
u_j^0	u_1^0	…	u_j^0	…	u_n^0	/	/

$$\begin{cases} r_{ij} = \dfrac{q_{ij}^0}{\left[P_i^0 \left(u_i^0\right)^\gamma\right] \cdot \left[A_j^0 \left(u_j^0\right)^\gamma\right]} \\ Y_{ij} = q_{ij}^0 / P_i^0 \\ k_{ij} = \dfrac{1 - Y_{ij}}{1 - Y_{ij} r_{ij}} r_{ij} \end{cases} \tag{7}$$

P_i^0 the generation of induced traffic in traffic sub-zone i in the basic year

A_j^0 the attraction of induced traffic in traffic sub-zone j in the basic year

u_i^0 induced economic growth in traffic sub-zone i in the basic year

u_j^0 induced economic growth in traffic sub-zone j in the basic year

q_{ij}^0 induced traffic volume distribution from i to j in the basic year

3.4.3. Forecasting of Traffic Generation and Distribution

Firstly, the induced economic growth in each sub-zone in the future characteristic year should be forecasted by formula (2), (3) and (4), then the generation and distribution forecasting may be carried out. The kernel will be the calculation of the generation and attraction by applying inverse matrix.

P_i, the generation of induced traffic in traffic sub-zone i in future characteristic year can be denoted as formula (8):

$$\begin{aligned} P_i &= \sum_j q_{ij} = \sum_j k_{ij} \left(P_i \cdot u_i^{\gamma}\right)\left(A_j \cdot u_j^{\gamma}\right) \\ &= P_i u_i^{\gamma} \sum_j \left(k_{ij} u_j^{\gamma} A_j\right) \end{aligned} \tag{8}$$

Formula (8) can be translated into linear equation group and corresponding matrix, as is shown in formula (9) and (10)

$$\begin{cases} u_1^{\gamma}\left(k_{11}u_1^{\gamma}A_1 + k_{12}u_2^{\gamma}A_2 + \cdots + k_{1n}u_n^{\gamma}A_n\right) = 1 \\ u_2^{\gamma}\left(k_{21}u_1^{\gamma}A_1 + k_{22}u_2^{\gamma}A_2 + \cdots + k_{2n}u_n^{\gamma}A_n\right) = 1 \\ \vdots \\ u_n^{\gamma}\left(k_{n1}u_1^{\gamma}A_1 + k_{n2}u_2^{\gamma}A_2 + \cdots + k_{nn}u_n^{\gamma}A_n\right) = 1 \end{cases} \tag{9}$$

$$\begin{bmatrix} k_{11}u_1^{\gamma} & k_{12}u_2^{\gamma} & \cdots & k_{1n}u_n^{\gamma} \\ k_{21}u_1^{\gamma} & k_{22}u_2^{\gamma} & \cdots & k_{2n}u_n^{\gamma} \\ \vdots & \vdots & \ddots & \vdots \\ k_{n1}u_1^{\gamma} & k_{n2}u_2^{\gamma} & \cdots & k_{nn}u_n^{\gamma} \end{bmatrix} \times \begin{bmatrix} A_1 \\ A_2 \\ \vdots \\ A_n \end{bmatrix} = \begin{bmatrix} 1/u_1^{\gamma} \\ 1/u_2^{\gamma} \\ \vdots \\ 1/u_n^{\gamma} \end{bmatrix} \tag{10}$$

Through formula (10), we know that if we obtain the inverse matrix of the coefficient matrix, the attraction of induced traffic volume in each sub-zone in the future characteristic year can be predicted. According to the above process, the generation of induced traffic volume can be predicted by the same way, too. After obtaining the attraction and generation, the induced traffic volume distribution can be predicted by formula (6).

3.4.4. Forecasting of Traffic Assignment

The forecasting of induced traffic assignment may use the usual methods such as capacity restrained iterative equilibrium method (Yu and Yan, 2001) or the multi-path probability equilibrium method (Li and Peng, 2001), which need not to be depicted here.

4. Empirical Study

The method raised in this chapter has been applied to forecasting the induced traffic volume brought by Ha-Shuang (Harbin-Shuangcheng) freeway. The Ha-Shuang freeway is one of the critical highway construction projects in the national "ninth five-year planning". It is the highest invested freeway in Heilongjiang province, and the total investment is 36 hundred million Yuan. It arises from Dongfeng town in the city of Harbin, traversing Shuangcheng city and terminates in the border between Heilongjiang province and Jilin province. It has four-lane carriageway and its total length is 100.94 kilometers.

The initial distribution matrix of induced traffic volume in the year 2001 is shown in table 2, the parameters ($\gamma = 0.43$) and the regulative coefficient are shown in table 3.

Table 2. Induced traffic volume distribution of Ha-Shuang freeway in 2001

OD	Harbin	Shuangcheng	Jilin	Generation
Harbin	0	50	161	211
Shuangcheng	48	0	41	89
Jilin	155	37	0	192
Attraction	203	87	202	492

Table 3. Regulative coefficient

k_{ij}	Harbin	Shuangcheng	Jilin
Harbin	0	0.148	0.397
Shuangcheng	0.14	0	0.081
Jilin	0.349	0.068	0

Based on the forecasting method presented in this chapter, the generation and distribution of induced traffic volume in the year 2005 are calculated and shown in table 4. The actual distribution matrix in the same year is shown in table 5.

Table 4. Forecasting results of induced traffic volume distribution in 2005

OD	Harbin	Shuangcheng	Jilin	Generation
Harbin	0	124	241	365
Shuangcheng	100	0	60	160
Jilin	306	83	0	389
Attraction	406	207	301	914

Table 5. Actual distribution matrix of induced traffic volume in 2005

OD	Harbin	Shuangcheng	Jilin	Generation
Harbin	0	117	252	369
Shuangcheng	94	0	65	157
Jilin	290	77	0	367
Attraction	382	194	317	893

The error and relative error can be calculated by formula (11) and (12).

$$\varepsilon(i) = x(i) - \hat{x}(i) \tag{11}$$

$x(i)$ the actual value

$\hat{x}(i)$ the forecasting value

$\varepsilon(i)$ the forecasting error

$$e(i) = \frac{\varepsilon(i)}{x(i)} \times 100\% \tag{12}$$

$e(i)$ the relative forecasting error

Through calculating the relative forecasting error, we can evaluate the forecasting results and further validate the forecasting model. The relative forecasting error matrix is shown in table 6.

Table 6. Relative forecasting error matrix

	Harbin	Shuangcheng	Jilin	Generation
Harbin	0	-5.98%	4.37%	1.08%
Shuangcheng	-6.38%	0	7.69%	1.91%
Jilin	-5.52%	-7.79%	0	-5.99%
Attraction	-6.28%	-6.70%	5.05%	-2.35%

The smaller the absolute value of the relative forecasting error the better the forecasting results. From table 6, we can see that the maximum absolute value of the relative forecasting error is less than 8%. Considering many stochastic and uncertain factors may take place and thus influence the forecasting results, this error range in induced traffic volume forecasting can be considered as acceptable.

From table 6, we can also see that the relative error has plus value and minus value in the distribution forecasting, this phenomenon shows that the forecasting model is stable and will not led to the inflation of total error.

Conclusion

In this chapter, through analysis of the mechanism and evolution of induced traffic volume, the quantitative forecasting model based on induced economic growth is put forward. Our preliminary test results from application in the forecasting of Ha-Shuang freeway have revealed the applicability of the proposed method. Considering the foundation of the model is closely related to the mechanism of induced traffic, the forecasting method presented in this chapter is relatively rational.

Despite the potential advantages mentioned above, further tests as well as modifications may be needed to verify the robustness of the proposed method, and its applicability to diverse cases.

ACKNOWLEDGMENTS

This chapter is supported by National Center of Technology, Policy and Management at Harbin Institute of Technology.

REFERENCES

Bonsall, P. (1996). Can induced traffic be measured by surveys?. *Transportation*, Vol. 23, no. 1, pp. 17-34.

Coombe, D. (1996). Induced traffic: What do transportation models tell us?. *Transportation*, Vol. 23, No. 1, pp. 83-101.

Fröhlich, P., and Zürich, E. T. H. (2003). Induced Traffic: Review of the explanatory models. *3rd Swiss Transport Research Conference*, Monte Verità / Ascona, March 19-21, Session Microsimulation, pp. 1-16.

Goodwin, P. B. (1996). Empirical evidence on induced traffic. *Transportation*, Vol. 23, No.1, pp. 35-54.

Hills, P. J. (1996). What is induced traffic?. *Transportation*, Vol. 23, No. 1, pp. 5-16.

Lee, D. B. J., Klein, L. A. and Camus, G. (1999). Induced traffic and induced demand. *Transportation Research Record*, No. 1659, pp. 68-75.

Li, J. and Peng, G. X. (2001). A study on estimating OD trips from observed link flows-based on the method of iterative procedure to calculate OD on the multi-path assignment model. *Journal of traffic and transportation engineering*, Vol.1, No.2, pp. 78-82.

Mokhtarian, P. L., Samaniego, F. J., Shumway, R. H. and Willits, N. H. (2002). Revisiting the notion of induced traffic through a matched-pairs study. *Transportation*, Vol. 29, No. 2, pp. 193-220.

Uchiyama, T., Maruyama, T. and Harata, N. (2006). Validating the influence of induced traffic using the combined travel model and person trip data. *Proceedings of the Fifth International Conference on Traffic and Transportation Studies, ICTTS 2006*, Xi'an, China, pp. 625-634.

Wang, B., Dong, Z. H. and Cao, Z. J. (2002). Study on induced-increasing traffic theory. *Journal of Heibei university of technology*, Vol. 31, no. 6, pp. 102-106.

Wang, H. and Shen, S. J. (2003). Induced traffic and forecasting model. *Journal of highway and transportation research and development*, Vol. 20, No. 3, pp. 147-150.

Wang, W. F., Lin, H. F. and Chen, X. H. (2001). Two major factors in volume forecasting of freeway. *China Journal of Highway and Transport*, Vol. 14, Sup. pp. 77-80.

Yu, L. J. and Yan, B. J. (2001). The relation between two types of traffic assignment models algorithm and it's application. *Journal of Xi'an highway university*, Vol. 21, No.2, pp. 74-76.

Zhou, W. and Wang, Y. (2002). Analysis for induced traffic volume on expressway. *Journal of Chang'an University*, Vol. 22, No.1, pp. 49-52.

In: Focus on Information, Intelligence and Knowledge
Editor: Rafiq Islam
ISBN 978-1-61209-979-8
2011 Nova Science Publishers, Inc.

Chapter 10

EFFICIENCY OF MACHINE LEARNING TECHNIQUES IN BIOMEDICAL DATA

S. Kotsiantis*[*1] and I. Tsagaraki[2]
[1]Department of Computer Science and Technology
University of Peloponnese, Greece
[2]Institute of Biology, NCSR "Demokritos", Greece

ABSTRACT

Machine learning and statistical techniques applied to gene expression data have been used to address the questions of distinguishing tumor morphology, predicting post treatment outcome, and finding molecular markers for disease. Today the classification of different morphologies, lineages and cell histologies can be performed successfully in many instances. The performance in predicting treatment outcome or drug response has been more limited but some of the results are quite promising. The scope of the research reported here is to investigate the efficiency of machine learning techniques in a number of different biomedical problems. To this end, a number of experiments have been conducted using representative learning algorithms. The decision of which particular method to choose is a complicated problem. A good alternative to choosing only one method is to create a hybrid forecasting system incorporating a number of possible solution methods as components (an ensemble of classifiers). For this purpose, we have implemented a hybrid decision support system that combines the representative algorithms using a stacking variant methodology and achieves better performance than any examined simple and ensemble method.

1. INTRODUCTION

During the last century biology has changed dramatically, from a rather qualitative science that was based on observations of whole organisms to a more quantitative science that is now based on measurements at the molecular level. These measurements resulted in a

* E-mail: sotos@ math.upatras.gr

tremendous amount of biological data of different types. This made it difficult or even impossible for biologists and biochemists to analyze the data by hand. As a result, computational means are needed to automate the process of finding patterns within the data and to derive general laws from the data, in short, to discover new knowledge within known data. This process, sometimes termed knowledge discovery in databases (KDD) or just knowledge discovery (KD), became very popular in computational biology, not only because the computational power as well as the size of biological databases increased rapidly over the last few decades, but also because the various data mining techniques (the most important step in KD) were improved and new methods were invented. However, the availability of numerous different methods makes the process not simpler. Rather, the opposite is true.

Sommer and Golz [22] compare two comprehensive classification algorithms, support vector machines (SVM) and several variants of learning vector quantization (LVQ) in two biomedical datasets. The paper [12] presents Bayesian network-based analysis of infertility patient data, which have been collected from the IVF clinic in a general hospital for two years. Through learning Bayesian networks from the clinical data, they identify the significant factors and their dependence relationships in determining the pregnancy of an infertility patient they classify the patient data into two classes (pregnant and not-pregnant) using the learned Bayesian network classifiers.

Searching for best prediction methods is still in progress. This study provides a critical analysis of most commonly used learning methods. Thus, we use a representative algorithm for each one of the most common Machine Learning (ML) techniques so as to investigate the efficiency of ML techniques in a number of different biomedical problems.

The decision of which particular learning algorithm to choose for the specific problem, is a complicated problem. A good alternative to choosing only one method is to create a hybrid forecasting system incorporating a number of possible solution methods as components (an ensemble of classifiers). For this purpose, we have implemented a hybrid decision support system that combines the representative algorithms using a stacking variant methodology and achieves better performance than any simple method.

The following section describes the data preparation process and the data sets used in our study. Some elementary Machine Learning definitions and a more detailed description of the used techniques and algorithms are given in section 3. It is hoped that the brief introduction to methodological details of these models would be of great use to those with recent interest in this field. Section 4 presents the proposed technique and the experimental results for the representative compared algorithms. Finally, section 5 discusses the conclusions and some future research directions.

2. Data Preparation and Data Description

The data preparation step deals with imperfect data. Parts of the data may be missing, because it can not be measured in principle or was not collected during the measurements. One effect of this may be that one or more types of instances are over- or under-represented. To avoid such a misrepresentation, the instances must be selected in a balanced way or must be weighted with a numeric value—rare types of instances get a higher weight than those which are frequent. Stratified sampling is applicable when the class values are not uniformly distributed in the training sets. Instances of the minority class(es) are selected with a greater frequency in order to even out the distribution.

Another effect may be that some instances have missing values, which is attributes without a specific value. This may be a problem if the subsequent methods can not deal with missing values. Analogically with the case, the expert has to choose from a number of methods for handling missing data:

- *Method of Ignoring Instances with Unknown Feature Values:* This method is the simplest: just ignore the instances, which have at least one unknown feature value.
- *Most Common Feature Value:* The value of the feature that occurs most often is selected to be the value for all the unknown values of the feature.
- *Concept Most Common Feature Value:* This time the value of the feature, which occurs the most common within the same class is selected to be the value for all the unknown values of the feature.
- *Mean substitution:* Substitute a feature's mean value computed from available cases to fill in missing data values on the remaining cases. A smarter solution than using the "general" feature mean is to use the feature mean for all samples belonging to the same class to fill in the missing value
- *Regression or classification methods:* Develop a regression or classification model based on complete case data for a given feature, treating it as the outcome and using all other relevant features as predictors.
- *Hot deck imputation:* Identify the most similar case to the case with a missing value and substitute the most similar case's Y value for the missing case's Y value.
- *Method of Treating Missing Feature Values as Special Values:* treating "unknown" itself as a new value for the features that contain missing values.

In the case of biological sequences (RNA, DNA and amino acid sequences) the crucial point is often how to derive a set of numerical or nominal attributes—a so-called feature set—for a given sequence, since most standard data mining methods do not operate on strings. Of course, there are specific solutions to this problem, e.g. one may use the InterProScan software [28] to transform an amino acid sequence into a set of numeric attributes (a vector counting the occurrence of several patterns within a sequence).

Even if an appropriate transformation is found it is not certain that all features contribute equally to the overall performance of the data mining algorithm applied in the next step. Some features can make the results even worse than without and too many features may slow down the process. Generally, features are characterized as:

- Relevant: These are features have an influence on the output and their role can not be assumed by the rest
- Irrelevant: Irrelevant features are defined as those features not having any influence on the output, and whose values are generated at random for each example.
- Redundant: A redundancy exists whenever a feature can take the role of another (perhaps the simplest way to model redundancy).

As a result, in some cases it may be necessary to select only a subset of the possible features. This is called feature selection and also helps to avoid the overfitting of an algorithm. Overfitting refers to the case where a model describes the known data nearly perfectly but fails on future data because the model's parameters are fitted too strongly to the known data. Attempts to overcome this problem have been made using feature generation (construction) and feature selection. [29]

The choice between feature selection and feature construction depends on the application domain and the specific training data, which are available. Feature selection leads to savings in measurements cost since some of the features are discarded and the selected features retain their original physical interpretation. In addition, the retained features may be important for understanding the physical process that generates the patterns. On the other hand, transformed features generated by feature construction may provide a better discriminative ability than the best subset of given features, but these new features may not have a clear physical meaning.

A source of biological data are gene expression experiments, in which dozens or even hundreds of different genes are observed with regard to their expression intensity under different circumstances, e. g. in different environments or in different cell phases. In general, such an experiment is repeated multiple times resulting in a matrix, where the rows represent the measured genes and the columns the expression levels under different conditions, or vice versa. To find subsets of co-regulated genes, i.e. genes that are up or down-regulated (indicated by lower or higher intensity) at the same time, data mining methods are commonly used.

For our study, we used the following datasets from the bibliography:

- DLBCL-Stanford [3]: Distinct types of diffuse large B-cell lymphoma (DLBCL) using gene expression data. There are 47 samples, 24 of them are from "germinal centre B-like" group while 23 are "activated B-like" group. Each sample is described by 4026 genes.
- CentralNervousSystem-outcome [17]: Patients outcome prediction for central nervous system embryonal tumor. Survivors are patients who are alive after treatment whiles the failures are those who succumbed to their disease. The data set contains 60 patient samples, 21 are survivors and 39 are failures. There are 7129 genes in the dataset.
- ColonTumor [2]: Contains 62 samples collected from colon-cancer patients. Among them, 40 tumor biopsies are from tumors and 22 normal biopsies are from healthy parts of the colons of the same patients. Two thousand out of around 6500 genes were selected based on the confidence in the measured expression levels.
- DLBCLOutcome and DLBCLTumor [21]: There are two kinds of classifications about diffuse large b-cell lymphoma (DLBCL) addressed in these data. First one is DLBCL versus Follicular Lymphoma (FL) morphology. This set of data contains 58 DLBCL samples and 19 FL samples. The second problem is to predict the patient outcome of DLBCL. Among 58 DLBCL patient samples, 32 of them are from cured patients while 26 of them are from patients with fatal or refractory disease. The expression profile contains 6817 genes.
- Lung-Michigan [5]: 86 primary lung adenocarcinomas samples and 10 non-neoplastic lung samples are included. Each sample is described by 7129 genes.
- Lungcancer-ontario [25]: Gene expression data on tumor specimens from a total of 39 NSCLC samples. Among these samples, 24 patients had experienced relapse of their tumor either locally or as a distant metastasis. The remaining 15 patients are

disease-free based on both clinical and radiological testing. The processed data are described by 2880 genes.

- Prostate_outcome [20]: Prediction of clinical outcome. 21 patients were evaluable with respect to recurrence following surgery with 8 patients having relapsed and 13 patients having remained relapse free ("non-relapse") for at least 4 years.

3. MACHINE LEARNING ISSUES

Predicting unknown attribute values from a given set of known attribute values is a special case of data mining called prescriptive mining or, considering the process to solve this problem, supervised machine learning. A large number of techniques have been developed based on Artificial Intelligence (Logic-based techniques, Perceptron-based techniques) and Statistics (Bayesian Networks, Instance-based techniques).

A recent overview of existing work in decision trees is provided by [14]. Decision trees are trees that classify examples by sorting them based on attribute values. Each node in a decision tree represents an attribute in an example to be classified, and each branch represents a value that the node can take. Examples are classified starting at the root node and sorting them based on their attribute values. The attribute that best divides the training data would be the root node of the tree. The same process is then repeated on each partition of the divided data, creating sub trees until the training data sets are divided into subsets of the same class.

In rule induction systems, a decision rule is defined as a sequence of Boolean clauses linked by logical AND operators that together imply membership in a particular class [9]. The general goal is to construct the smallest rule-set that is consistent with the training data. A large number of learned rules is usually a sign that the learning algorithm tries to "remember" the training set, instead of discovering the assumptions that govern it. During classification, the left hand sides of the rules are applied sequentially until one of them evaluates to true, and then the implied class label from the right hand side of the rule is offered as the class prediction.

Artificial Neural Networks (ANNs) are another method of inductive learning based on computational models of biological neurons and networks of neurons as found in the central nervous system of humans [13]. A multi layer neural network consists of large number of units (neurons) joined together in a pattern of connections. Units in a net are usually segregated into three classes: input units, which receive information to be processed, output units where the results of the processing are found, and units in between called hidden units. Classification with a neural network takes place in two distinct phases. First, the network is trained on a set of paired data to determine the input-output mapping. The weights of the connections between neurons are then fixed and the network is used to determine the classifications of a new set of data.

An excellent book about the Bayesian networks is provided by [10]. A Bayesian network is a graphical model for probabilistic relationships among a set of attributes. The Bayesian network structure S is a directed acyclic graph (DAG) and the nodes in S are in one-to-one correspondence with the attributes. The arcs represent casual influences among the variables while the *lack* of possible arcs in S encodes conditional independencies. Moreover, an attribute (node) is conditionally independent of its non-descendants given its parents. Using a suitable training method, one can induce the structure of the Bayesian Network from a given training set. The classifier based on this network and on the given set of attributes $X_1, X_2, \ldots,$ X_n, returns the label c that maximizes the posterior probability $p(c \mid X_1, X_2, \ldots, X_n)$.

Instance-based learning algorithms belong in the category of lazy-learning algorithms [1], as they defer in the induction or generalization process until classification is performed. One of the most straightforward instance-based learning algorithms is the nearest neighbour algorithm. K-Nearest Neighbour (kNN) is based on the principal that the examples within a data set will generally exist in close proximity with other examples that have similar properties. If the examples are tagged with a classification label, then the value of the label of an unclassified example can be determined by observing the class of its nearest neighbours.

The SVM technique revolves around the notion of a 'margin', either side of a hyperplane that separates two data classes. Maximizing the margin, and thereby creating the largest possible distance between the separating hyperplane and the instances on either side of it, is proven to reduce an upper bound on the expected generalization error [6]. Nevertheless, most real-world problems involve non-separable data for which no hyperplane exists that successfully separates the positive from negative instances in the training set. One solution to the inseparability problem is to map the data into a higher-dimensional space and define a separating hyperplane there. This higher-dimensional space is called the feature space, as opposed to the input space occupied by the training instances. Generally, with an appropriately chosen feature space of sufficient dimensionality, any consistent training set can be made separable.

4. Proposed Technique and Experimental Results

For the purpose of the present study, a representative algorithm for each described machine learning technique was selected. The most commonly used C4.5 algorithm [16] was the representative of the decision trees in our study. At each level in the partitioning process a statistical property known as *information gain* is used by C4.5 algorithm to determine which attribute best divides the training examples. Naive Bayes algorithm was the representative of the Bayesian networks [8]. It is a simple learning that captures the assumption that every attribute is independent from the rest of the attributes, given the state of the class attribute. We also used the 3NN algorithm, with Euclidean distance as distance metric as instance based learner [1]. The RIPPER [7] algorithm was the representative of the rule-learning techniques because it is one of the most usually used methods that produce classification rules. RIPPER forms rules through a process of repeated growing and pruning. The grow heuristic used in RIPPER is the information gain function. A well-known learning algorithm for building a neural network - RBF algorithm [13] - was the representative of the ANNs. Finally, the Sequential Minimal Optimization (or SMO) algorithm was the representative of the SVMs in our study because it is one of the fastest methods to train SVMs [15].

The most commonly used measure of success in classification problems is the classification accuracy, which is defined as the ratio of the number of the correct classifications to the number of cases examined. A standard procedure for calculating classification accuracy is the so called 10-cross validation. It must be mentioned that we used the free available source code by [26] for our experiments. In Table 1, the classification accuracy of each algorithm for all the testing datasets is presented.

Table 1. Comparing learning algorithms

Datasets	C4.5	RIPPER	NB	3NN	SMO	RBF
centralNervousSystem	58.33	58.33	61.67	56.67	68.33	71.67
colonTumor	82.26	79.03	53.23	75.81	85.48	79.03
DLBCLOutcome	53.45	58.62	43.10	39.66	55.17	41.38
DLBCL-Stanford	78.72	76.60	97.87	76.60	97.87	85.11
DLBCLTumor	72.73	76.60	80.52	93.51	96.10	85.71
lungcancer-ontario	84.62	56.41	69.23	56.41	79.49	61.54
lung-Michigan	98.96	95.83	100	98.96	100	95.83
prostate_outcome	33.33	47.62	33.33	52.38	47.62	42.86

No single learning algorithm can perform well and uniformly outperform other algorithms over all data sets.

Recently in the area of machine learning and data mining the concept of combining classifiers is proposed as a new direction for the improvement of the performance of individual classifiers. These classifiers could be based on a variety of classification methodologies, and could achieve different rate of correctly classified individuals. The goal of classification result integration algorithms is to generate more certain, precise and accurate system results. Witten [26] provides an accessible and informal reasoning, from statistical, computational and representational viewpoints, of why ensembles can improve results.

An *ensemble of classifiers* is a set of classifiers whose individual decisions are combined in some way to classify new instances (Figure 1). Techniques using ensemble of classifiers consist of two phases: classifier generation and classifier combination. It is important to generate a set of classifiers that are diverse in the sense that they make errors in different ways.

Numerous methods have been suggested for the creation of ensemble of classifiers. Mechanisms that are used to build ensemble of classifiers include: i) Using different subset of training data with a single learning method, ii) Using different training parameters with a single training method (e.g. using different initial weights for each neural network in an ensemble), iii) Using different learning methods.

The most usual used method is to use a variety of learning algorithms on all of the training data and combine their predictions according to a voting scheme.

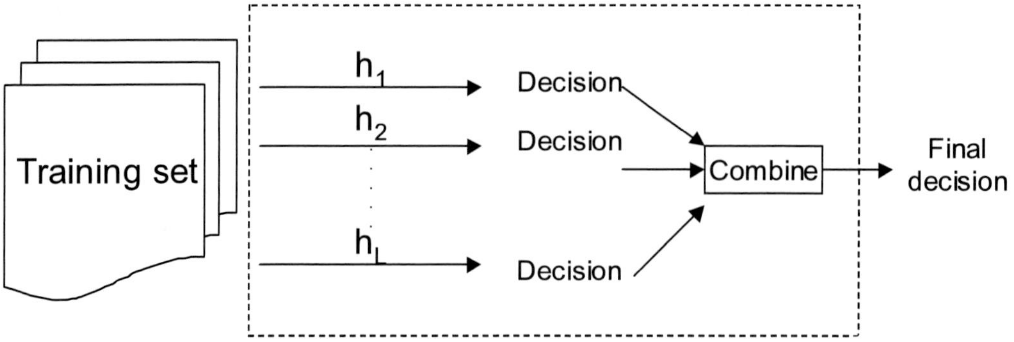

Figure 1. An ensemble of classifiers

This technique attempts to achieve diversity in the errors of the learned models by using different learning algorithms, which vary in their method of search and/or representation. The intuition is that the models generated using different learning biases are more likely to make errors in different ways.

In this study, we combined the previous tested algorithms using a stacking variant methodology. Its basic idea may be derived as a generalization of voting as follows. Let us consider the voting step as a separate classification problem, whose input is the vector of the responses of the base classifiers. Simple voting uses a predetermined algorithm for this, namely to count the number of predictions for each class in the input and to predict the most frequently predicted class. Stacking replaces this with a trainable classifier [27]. This is possible, since for the training set, we have both the predictions of the base learners and the true class. The matrix containing the predictions of the base learners as predictors and the true class for each training case will be called the meta-data set. The classifier trained on this matrix will be called the meta-classifier or the classifier at the meta-level. While stacking [23] uses all class probabilities for all models, our method uses only the class probabilities associated with the true class [18]. The dimensionality of the meta-data set is reduced by a factor equal to the number of classes, which leads to faster learning. Concerning the choice of the algorithm for learning at the meta-level, we have explored the use of model trees instead of MLR [18], [23] since model trees naturally extend MLR to construct piecewise linear approximations. Model trees have the same structure as decision trees, with one difference: they employ a linear regression function at each leaf node to make a prediction. The most well known model tree inducer - M5′ [24] – is used by our system. In the following, we briefly describe the learning algorithms that are used as base learners.

Subsequently, we compare the proposed stacking methodology (Stacking′) (Table 2) with:

- The methodology of selecting the best classifier according to 10-cross validation (BestCV) [26].
- Grading methodology using the instance based classifier IBk with ten nearest neighbors as the meta level classifier [19]. In grading, the meta-level classifier predicts whether the base-level classifier is to be trusted (i.e., whether its prediction will be correct). The base-level attributes are used also as meta-level attributes, while the meta-level class values are + (correct) and − (incorrect). Only the base-level

classifiers that are predicted to be correct are taken and their predictions combined by summing up the probability distributions predicted.

- Simple Voting methodology using the same base classifiers [11].
- Stacking methodology that constructs the meta-data set by adding the entire predicted class probability distribution instead of only the most likely class using MLR as meta-level classifier [23].

In the following Table, win (v) indicates that the specific method (in the column) performed statistically better than the proposed stacking method. It must be mentioned that the resulting differences between algorithms were assumed statistically significant when $p<0.05$ is not strict enough. Loss(*) indicates that the stacking process performed statistically better than the specific method (in the column). In all the other cases, there is no significant statistical difference between the results. In the following Table, we also present the average accuracy of all tested data set for each ensemble.

Table 2. Accuracy of ensembles in our datasets

Datasets	Stacking′	Voting	BestCV	Grading	Stacking
centralNervousSystem	68.33	71.67v	68.33	63.33*	63.33*
colonTumor	83.81	83.81	84.05	83.81	80.71*
DLBCLOutcome	52.00	44.00*	54.33	52.00	43.33*
DLBCL-Stanford	97.50	97.50	100v	93.50*	91.50*
DLBCLTumor	96.07	89.64*	90.71*	92.14*	94.82
lungcancer-ontario	78.33	78.33	78.33	70.83*	63.33*
lung-Michigan	100	100	100	100	100
prostate_outcome	66.67	41.67*	41.67*	41.67*	41.67*
Average	80.34	75.83	73.92	74.66	72.34
W-D-L		1/5/3	0/6/2	0/4/4	0/2/6

In the last raw of the Table 2 one can see the aggregated results. The proposed ensemble (Stacking′) is significantly more accurate than voting in 3 out of the 8 data sets, while it is significantly less accurate in one data set. Similarly, the proposed ensemble is significantly more accurate than stacking using MLR procedure in 6 out of the 8 data sets, while it is significantly less accurate in none data set. Moreover, the proposed ensemble is significantly more accurate than BestCV procedure in 2 out of the 8 data sets, while it is significantly less accurate in none data set. Finally, the proposed ensemble is significantly more accurate than Grading in 4 out of the 8 data sets, while it is significantly less accurate in none data set.

The average relative accuracy improvement of the proposed stacking methodology is at least 6% in relation to the remaining methods. As a conclusion, our approach performs better than selecting the best classifier from the ensemble by cross validation and other examined ensemble methods.

Conclusion and Future Work

Mining biological data involves many steps which are common to all knowledge discovery processes. However, there are also tasks which are special to biological problems:

- Biologists use their own data formats and quite often the data format is simply an ad hoc implementation.
- Sequences are a special case because most machine learning algorithms operate on numeric or nominal values and not on strings.
- The default implementations of known algorithms do not always perform well on biological data sets. Therefore, it may be necessary to incorporate biological background knowledge into a classification algorithm or to modify an algorithm in another way to achieve reasonable results.

The concept of combining learning algorithms is proposed as a new direction for the improvement of the performance of individual classifiers. The goal of classification result integration algorithms is to generate more certain, precise and accurate system results.

The stacked generalization architecture for classifier combination has still many open questions. For example, there are currently no strict rules saying which base classifiers should be used and what features of the training set should be used to train the combining classifier.

In a future work, we will use a feature selection pre-process before the usage of the stacking. Feature subset selection is the process of identifying and removing as much irrelevant and redundant features as possible. This will reduce the dimensionality of the data enabling the proposed ensemble to operate faster and maybe more effectively.

References

[1] Aha, D. (1997), *Lazy Learning*, Dordrecht: Kluwer Academic Publishers.

[2] Alon, U. et al. (1999), "Broad Patterns of Gene Expression Revealed by Clustering Analysis of Tumor and Normal Colon Tissues Probed by Oligonucleotide Arrays", *PNAS*, 96:6745-6750.

[3] Ash A. Alizadeh, et al. (2000), "Distinct types of diffuse large B-cell lymphoma identified by gene expression profiling". *Nature*, 403:503-511

[4] Bauer, E. and Kohavi, R. (1999). An empirical comparison of voting classification algorithms: Bagging, boosting, and variants. *Machine Learning*, 36, 105–139.

[5] Beer, D. et al. (2002), "Gene-expression Profiles Predict Survival of Patients with Lung Adenocarcinoma". *Nature Medicine*, 8(8):816-823.

[6] Burges, C. (1998). A tutorial on support vector machines for pattern recognition. *Data Mining and Knowledge Discovery*. 2(2):1-47.

[7] Cohen, W. (1995), "Fast Effective Rule Induction", *Proceeding of International Conference on Machine Learning* 1995, pp. 115-123.

[8] Domingos, P. and Pazzani, M. (1997), "On the optimality of the simple Bayesian classifier under zero-one loss". *Machine Learning*, Vol. 29, pp. 103-130.

[9] Furnkranz, J. (1999), "Separate-and-Conquer Rule Learning", *Artificial Intelligence Review*, Vol. 13, pp. 3-54.
[10] Jensen, F. (1996), *An Introduction to Bayesian Networks*, Springer.
[11] Ji, C., Ma, S., 1997. Combinations of weak classifiers. *IEEE Transaction on Neural Networks* 8, 32–42.
[12] Kim, In-Cheol, Jung, Yong-Gyu, Using Bayesian Networks to Analyze Medical Data, *Lecture Notes in Computer Science*, Volume 2734/2003, pp. 317-327.
[13] Mitchell, T. (1997), *Machine Learning*, McGraw Hill.
[14] Murthy, S. (1998), "Automatic Construction of Decision Trees from Data: A Multi-Disciplinary Survey", *Data Mining and Knowledge Discovery*, Vol. 2, pp. 345–389.
[15] Platt, J. (1999), Using sparseness and analytic QP to speed training of support vector machines. In M. S. Kearns, S. A. Solla, and D. A. Cohn (Eds.), *Advances in neural information processing systems* 11. MA: MIT Press.
[16] Quinlan, J. R. (1993), C4.5: *Programs for machine learning*, Morgan Kaufmann, San Francisco.
[17] Scott L. Pomeroy, et al. (2002), "Prediction of Central Nervous System Embryonal Tumour Outcome Based on Gene Expression", *Nature*, 415:436-442, January 2002.
[18] Seewald, A.K, 2002. How to Make Stacking Better and Faster While Also Taking Care of an Unknown Weakness, in Sammut C., Hoffmann A. (eds.), *Proceedings of the Nineteenth International Conference on Machine Learning (ICML 2002)*, Morgan Kaufmann Publishers, pp.554-561.
[19] Seewald, A. K., Furnkranz, J., 2001. An evaluation of grading classifiers. In *Advances in Intelligent Data Analysis*: *Proceedings of the Fourth International Symposium (IDA-01),* pages 221–232, Berlin, Springer.
[20] Singh, D. et al. (2002), "Gene Expression Correlates of Clinical Prostate Cancer Behavior". *Cancer Cell*, 1:203-209.
[21] Shipp, Margaret A., (2002), et al. "Diffuse Large B-cell Lymphoma Outcome Prediction by Gene-expression Profiling and Supervised Machine Learning". *Nature* Medicine, 8(1):68-74.
[22] Sommer, D., Golz, M., Comparison of Validation Methods for Learning Vector Quantization and for Support Vector Machines on Two Biomedical Data Sets, From *Data and Information Analysis to Knowledge Engineering Proceedings of the 29th Annual Conference of the Gesellschaft für Klassifikation e.V. University of Magdeburg,* March 9–11, 2005, pp. 150-157.
[23] Ting, K., and Witten, I., 1999. Issues in Stacked Generalization, *Artificial Intelligence Research* 10, 271-289, Morgan Kaufmann.
[24] Wang, Y., Witten, I., 1997, Induction of model trees for predicting continuous classes, In *Proc. of the Poster Papers of the European Conference on ML*, Prague, 128–137.
[25] Wigle, D. et al. (2002), "Molecular Profiling of Non-Small Cell Lung Cancer and Correlation with Disease-free Survival". *Cancer Research*, 62:3005-3008.
[26] Witten, I. and Frank E. (2000), Data Mining: *Practical Machine Learning Tools and Techniques with Java Implementations*, Morgan Kaufmann, San Mateo.
[27] Wolpert, D. (1992). "Stacked Generalization". *Neural Networks*, 5(2):241–260.
[28] Zdobnov, E. and Apweiler, R. (2001), InterProScanan integration platform for the signature recognition methods in InterPro. *Bioinformatics*, 17(9):847–848.

[29] Zeng, F., Yap C.H.R. and Wong, L. (2002). Using Feature Generation and Feature Selection for Accurate Prediction of Translation Initiation Sites. *Genome Informatics*, 13, 192-200.

In: Focus on Information, Intelligence and Knowledge
Editor: Rafiq Islam

ISBN 978-1-61209-979-8
2011 Nova Science Publishers, Inc.

Chapter 11

'SAHAM': SHARED ADAPTIVE HYPERMEDIA APPLICATION MODEL

Arpita Aggarwal*[*] and Naveen Kumar*[**]

[*]Department of Computer Science,
PGDAV College, University of Delhi, Delhi – 65, India
[**]Department of Computer Science,
University of Delhi, Delhi – 7, India

ABSTRACT

Many reference models have been suggested for the adaptive hypermedia systems but they lack the capability to share their information about the user with other AH systems. In the present paper, we propose a reference model where AH systems have two user models – one integral to the storage layer and having domain dependent characteristics and another 'shared user model' having domain independent attributes and characteristics. Similarly, AH systems would have two adaptation models – one integral having adaptation rules for domain dependent data and another 'shared adaptation model' having domain independent adaptation rules. The shared user model and shared adaptation models are part of the 'shared module' which is shared among different AH systems. 'Shared Module' provides the benefit of initializing the AH systems with more mature and accurate user model, saving of time and effort in creating individual user models for different AH systems and enforcing some sort of standardization and synergistic behaviour among them. Apart from the 'Shared Module', the paper also suggest the division of runtime layer into input acquisition component and presentation component, also the division of the presentation specification interface into inference engine and adaptation engine in individual AH systems.

[*] Email: arpita_aggarwal@hotmail.com

1. Introduction

Adaptive Hypermedia (AH) systems have the mechanism for observing the environment and the user behaviour, storing them and then drawing non-trivial inferences from them. Based on these derived inferences, adaptation decisions are taken dynamically by the system and the next information is presented in a personalized format to the user. Such dynamic adaptations by the hypermedia systems give the users freedom from the problems such as 'cognitive overload' and 'confusion' created by the availability of multiple hyperlinks to choose from. The presented information is tailor-made for users taking into account their goals, history and preferences.

AH systems have been applied in various application areas such as education (ELM-ART (Brusilovsky, Schwarz and Weber, 1996), AHA! (de Bra and Calvi, 1998)), On-line information system (Hypadapter (Hohl, Böcker and Gunzenhäuser 1998), MetaDoc (Boyle and Encarnation, 1994)), On-line help systems (WING-MIT (Kim, 1995)), Information retrieval systems (HYPERFLEX (Kaplan, Fenwick and Chen, 1993)), Intelligent interfaces, Natural language systems (Whalster and Kobsa, 1986), and Adaptive recommender systems (Ardissino and Goy, 1999). Although these systems have been in existence for sometime now, they are all standalone systems, with little or no co-operation of any kind with each other. There are no standards to describe their functioning or the development process which results in systems being very different from each other in terms of their components like user model, adaptation model, domain model or user interface. This makes sharing of any kind of information impossible between two or more AH systems.

A reference model of an AH system promotes the development of standards and common concepts and terminology among the people working on the system by providing the division of conceptual components of the system based on their functionality and the dataflow between them. Without going into the implementation details, a reference model attempts to capture, both formally and informally, the important abstractions found in existing and future AH systems.

Literature provides us with many examples of reference models for AH systems. Dexter Hypertext Model (Halasz and Schwartz, 1994) is one of the important models that serves as the basis for a number of AH system's reference models such as Adaptive Hypermedia Application Model (Wu, de Bra and Houben, 1999), and Munich reference model (Koch, 2000). Goldsmiths Adaptive Hypermedia Model (Ohene-Djan, 2000) is a formal model, which differentiates presentation from adaptation. Another reference model for AH systems, based on the state-machine view of applications, is Generic Adaptivity Model (Vrieze, Bommel and Weide, 2004). Current reference models for AH systems do not share user information among different AH systems. With the growing trend towards the AH systems in the current times, it would be very useful to be able to share as much knowledge about the user as possible.

In this paper, we propose a reference model for AH systems – *SAHAM – Shared Adaptive Hypermedia Application Model* where the use of a 'Shared Module' among AH systems is advocated. This 'Shared Module (SM)' consists of a 'Shared User Model (SUM)' containing the domain-independent characteristics of user profile and a 'Shared Adaptation Model (SAM)' containing the adaptation rules for this shared user model. This way, all AH systems will have two user models – one with the domain-dependent characteristics and is an integral part of the AH system, and another with the domain-independent characteristics and is a part of the 'shared module'. Similarly, AH systems will also have two adaptation models – an integral one with domain-dependent adaptation rules and one with domain-independent adaptation rules and is part of 'shared module'.

This 'Shared Module (SM)' can be used for the initialization of a new AH system from an earlier, more mature, more accurate one. Any updations made in the shared module are reflected in all the systems using it. Another benefit is the saving in time and effort of developers of AH systems with the use of SM.

In the next section, we give a brief overview of some of the existing reference models followed by their weaknesses in section 3. Detailed description of the reference model – SAHAM is given in section 4. Section 5 specifies the advantages of SAHAM and section 6 gives the conclusion.

2. Various Reference Models

2.1. Dexter Hypertext Reference Model

Proposed by Halasz and Schwartz (1994), Dexter Hypertext Reference Model was the result of two small workshops on hypertext. It was built with the intension of providing a standard for comparing the characteristics and functionality of various hypertext systems and for developing standards for interoperability and interchange among them.

The Dexter model is not suitable for AH systems, but it forms the basis for many reference models of AH systems, including SAHAM. This model divides the hypertext system into three layers – the *runtime layer*, the *storage layer* and the *within-component layer*, connected by the interfaces *presentation specification* and *anchoring*. Figure 1 shows the different layers of Dexter model. The main focus of the model is on the storage layer, which describes the basic node/link network structure, i.e. the mechanisms by which the links and nodes are related to each other. Nodes are the data-containing "components" which are interconnected by "links". The within-component layer focuses on the contents and structure within the components of the hypertext network but due to the large range of possible contents in a component (e.g. text, image, audio, and video), it has not been elaborated upon. Similarly, the runtime layer containing the description of the presentation of nodes and links, user interaction and the dynamics of the application has been provided bare bone mechanism due to broad range of possible tools for accessing, viewing and manipulating networks.

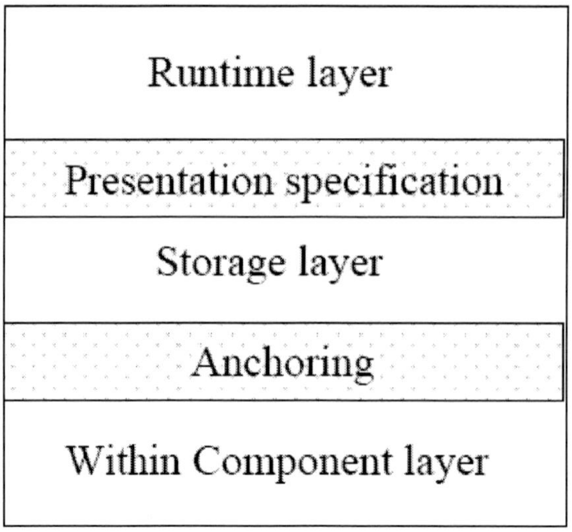

Figure 1. Layers of Dexter model.

The important aspect of the Dexter model are the two interfaces: (i) *Presentation specification*: present between storage layer and runtime layer, it provides mechanism for encoding the information about a component's presentation for the user, and (ii) *Anchoring*: present between storage layer and within component layer, it provides the mechanism for referring locations or items within the content of individual components.

Although the Dexter model is one of the most important milestones of hypermedia development, it focuses mainly on the storage layer describing the network of nodes and links. The model does not make clear distinction between atomic and composite components and lacks support for open hypertext systems (Lowe and Hall, 1999) or large-scale hypertext systems (Leggett and Schnase, 1994) because of its assumption that components are aware of each other and are tightly bound. Moreover, the details of the content of general containers of data in within-component layer and the user interaction in runtime layer have not been covered. The treatment of storage layer and within component layer as essentially passive data structures and non-addressing of the issues of tailoring of interaction process between user and system itself makes it unsuitable for AH systems.

2.2. Adaptive Hypermedia Application Model (AHAM)

AHAM (Wu, de Bra and Houben, 1999) is an extension of the Dexter reference model to provide the adaptation functionality to hypermedia systems. Similar to the Dexter model, this model is also divided into the same three layers i.e. *runtime layer, storage layer and within-component layer*. Further, it advocates clear separation of storage layer into: (a) domain model (b) user model (c) adaptation model. Figure 2 shows the layers of AHAM.

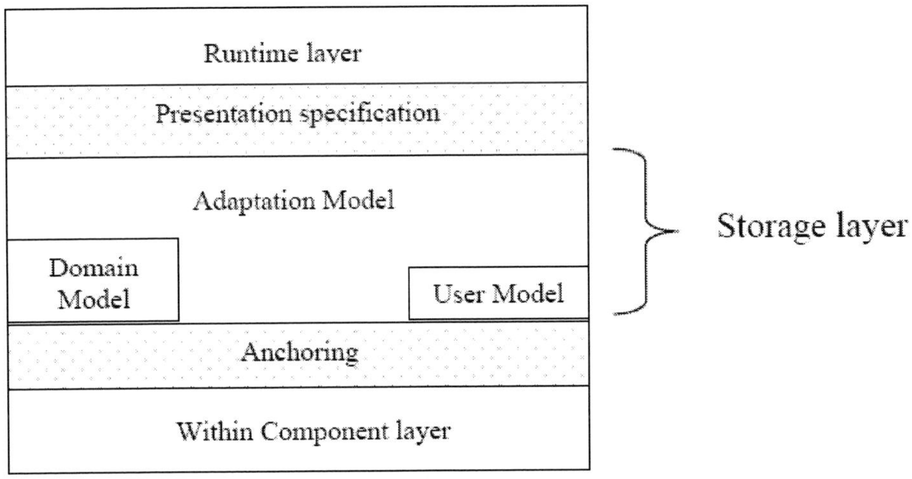

Figure 2. Layers of AHAM.

- *Domain Model (DM):* DM is used to describe how the information in the hypermedia is structured. It uses 'concept component' for an abstract representation of information, both at the conceptual as well as data fragment and page level, by means of atomic and composite nodes, arranged in a directed acyclic graph. Links among concept components are represented by 'concept relationships'.
- *User Model (UM)*: Overlaid on top of DM, UM is used for storing stable personal characteristics of the user such as interests, goals, background, aspects of user's environment such as place, time, used equipment, and the relation between the user and the concepts in the DM such as 'level of knowledge' about each concept.
- *Adaptation Model (AM)*: It consists of the adaptation rules for updating the user model and for the adaptation done by using information from DM and UM.

As in the Dexter model, the raw data is stored in the '*within-component layer*' while '*runtime layer*' is responsible for the user interaction. The storage layer uses '*anchoring*' and the '*presentation specification*' as interfaces to the within-component layer and the runtime layer respectively.

AHAM strongly advocates a clear separation of the storage layer into domain model, user model and the adaptation model, explicitly stating their roles and relationships with each other. This makes it possible to write different adaptation rules for the same information domain. Fragments, concepts and concept relationships in DM support advanced adaptation and different attributes of UM allow user-centred as well as system-centred control over the adaptation.

AHAM has been designed to operate on predefined data and not the dynamically generated data at run time. This restriction has the advantage of allowing identification of non-terminating rules at design time itself with full knowledge of adaptation rules but also limits the ability of AHAM systems to offer personalization actions to users such as creating annotations on hypertext objects such as pages or chunks (Wu, Kort, de Bra, 2001).

2.3. Munich Reference Model

Another model based on Dexter reference model, developed independently around the same time as AHAM, was the Munich reference model (Koch, 2000). In essence, this model is very similar to AHAM and extends Dexter's storage layer to have a domain model, user model and adaptation model as part of it. The other layers and the interfaces have been retained as in Dexter model. Figure 3 shows the Munich reference model.

Munich reference model provides a formal, object-oriented approach written using the graphical specification of UML (1999) and the constraints specified in OCL (Warmer and Kleppe, 1999). This model forms the basis of UML-based web engineering (UWE) (Koch and Kraus, 2002), an extension to UML, supporting web-application design and development.

The main difference between AHAM and Munich reference model is that AHAM addresses pedagogical applications, performing adaptations using adaptation rule language based on the user model represented by tables and the domain model. On the other hand, Munich reference model uses object-oriented specification written in UML and OCL.

2.4. Goldsmiths Adaptive Hypermedia Model (GAHM)

Developed at Goldsmiths College, University of London, Goldsmiths Adaptive Hypermedia Model (GAHM) (Ohene-Djan, 2000) is a formal model for personalisable, AH systems.

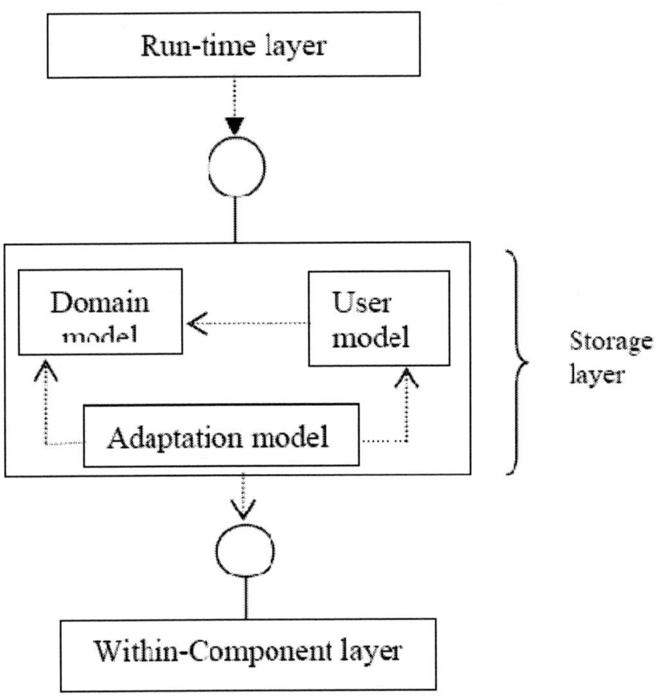

Figure 3. Munich Reference Model (Koch 2000).

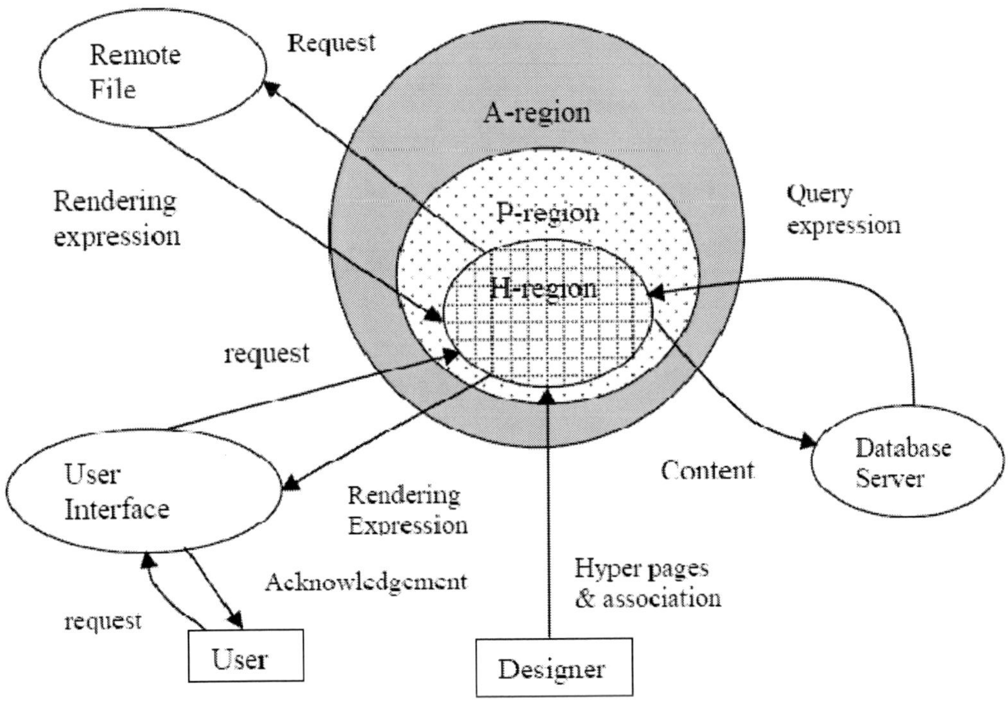

Figure 4. Abstract representation of GAHM ((Ohene-Djan, 2000).

The model has three layers, starting with core hyperlink based functionality referred to as the H-region that models the non-personalisable hypermedia based interaction. This core is extended by P-region's abstract model of personalisability having functions that model the user-initiated tailoring of hypermedia content. A-region's system-initiated tailoring or adaptivity of hypermedia content using active rules as the decision-making component in turn extends this region. Figure 4 gives an abstract view of the GAHM model.

GAHM isolates the scope of personalization and adaptivity (PandA) and formally induces set of PandA actions needed while other models treat this issue of PandA compounded together with no distinction between them. GAHM yields one complete picture of the subject of personalization, PandA actions and the formal consequences of issuing these actions but its functional nature might make the use of object-oriented methodology more tedious for the system developers. Moreover, the tightly coupled hyperpage structure of model's personalization language can make it unsuitable for some settings.

2.5. Generic Adaptivity Model (GAM)

With the aim to provide a generic user model which is not limited to hypermedia applications, Vrieze et al (Vrieze, Bommel and Weide, 2004) proposed a state-machine based model, Generic Adaptivity Model (GAM), and used the concepts of push and pull adaptations (Vrieze, Bommel, Weide and Klok, 2005) for making the user models. This model consists of adaptation engine, interface model, domain model, adaptation model and user model.

Compared to other models, GAM is more low-level and does not provide hypermedia specific constructs by itself. It uses overlay user model and provides an explicit interface model allowing it to offer question system that stores intermediate values and properties at the right level in the user model. GAM requires addition of high-level, specification models on top of itself.

3. Weaknesses of Current Reference Models

Current reference models for AH systems do not share the user information between two or more AH systems. It becomes particularly significant when the user model needs to be initialised since it should know as many user characteristics as possible to provide an effective adaptation. This is difficult to achieve with the regular techniques of initialization such as using default values, by explicit questioning of user or by testing the user to gain insight into his current level of knowledge. Initialization using default values (user stereotypes) do not offer enough information in some cases like student model because they do not permit the necessary fine-grained analysis. Unwillingness of users to answer some questions during explicit questioning and testing of his level might result in inadequate personalization.

Other problems with the current reference models are of redundancy and related issues because of storing separate UM in individual AH systems. Developers spend more time and effort in developing individual UMs and no standards are followed for developing them.

In our view, the solution to these problems is to use a single UM that is most accurate and personalized in all AH systems. This can be achieved by using a user model in cooperation with another UM of an older AH system. The problem arises when both the AH systems do not co-operate in the task of modelling user characteristics – especially the domain-dependent ones. Therefore, we suggest having two user models in all AH systems- one UM with domain dependent characteristics – and another 'shared user model (SUM)' – with domain independent characteristics. The characteristics of SUM can be used by other AH systems for initialization of their presentations such as presentation preferences or the desired language. This can be achieved by having a 'shared module' among AH systems consisting of a 'shared user model (SUM)' and a 'shared adaptation model (SAM)'. In the next section we give a detailed description of our proposed model – SAHAM.

4. Shared Adaptive Hypermedia Application Model - SAHAM

The proposed model, SAHAM, is based on AHAM developed by Wu et al (Wu, de Bra and Houben, 1999), which in turn, is based on the Dexter model (Halasz and Schwartz, 1994). AHAM is a generic model that is simple, clear and has an interface with fixed set of events available within a web, has a strong adaptation rule language and does not need specification of any more high-level models on top of itself. Therefore, it proves to be a good basis for our model.

SAHAM divides the AH system into the same three layers as in AHAM – *Runtime layer, Storage layer* and the *Within-component layer*, connected by the interfaces – *Presentation specification* and *Anchoring*. Figure 5 shows the detailed view of the proposed reference model of an individual AH system without the 'Shared Module'. Figure 6 shows the AH systems with the shared module. Description of each layer and interface along with its components is as follows:

- *Runtime layer*: This layer provides the user interface – taking inputs from the user, system and environment and presenting the personalized and adapted information to the user. Therefore it has two specific components - input acquisition component and presentation component.
  - *Input acquisition component*: This deals with the various ways of obtaining the input by the AH system – from the user (keystrokes, mouse clicks etc), and from the environment (device, network, bandwidth, etc). Kobsa et al (Kobsa, Koenemann, and Pohl, 2001) have stated a number of ways in which inputs can be taken by the application.
  - *Presentation component*: This component presents the final hyperpage to the user that has been prepared based on the adaptation decisions taken by the adaptation engine.
- *Presentation Specification Interface*: This acts as the interface between runtime layer and storage layer. Two main components of this interface are the inference engine and the adaptation engine.
  - *Inference Engine (IE)*: Inference engine takes the input from the input acquisition module of runtime layer, extracts or infers meaning and useful information from that input and bifurcates the information to pass it on to the storage layer's UM or the shared module's SUM for further processing.
  - *Adaptation Engine (AE)*: This module actually performs the adaptation with the adaptation rule language using the inputs from the storage layer and the shared module and passes it to the presentation component of the runtime layer. It executes the rules specified within AM.
- *Storage Layer*: Similar to AHAM, it consists of domain model, adaptation model, and user model.
  - *Domain Model (DM)*: It consists of the information or the knowledge of the domain of the AH system.
  - *Adaptation Model (AM)*: It consists of adaptation rules that are specific for the domain related attributes only.
  - *User Model (UM)*: The user model in this layer can be implicit or explicit but contains data related to the domain only i.e. *domain dependent data* only.
- *Anchoring*: It is the mechanism for addressing locations or items within the contents of an individual component.
- *Within–component layer*: It specifies the actual implementation of the contents and the structures inside the nodes of the individual components. Since the nodes can contain a range of possible contents and structures, implementation of this layer has been left to the AH system developer's choice.
- *Shared Module(SM)*: Shared Module acts as the bridge between two or more AH systems. It consists of a 'shared user model' and a 'shared adaptation model' containing user characteristics and adaptation rules respectively, which are independent of the domain of the AH systems.

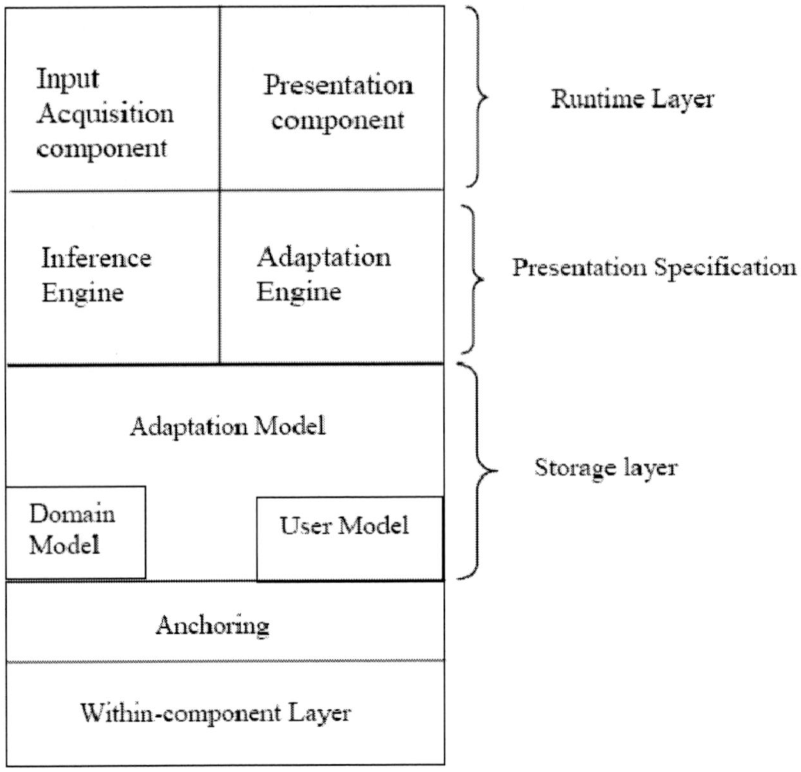

Figure 5. The SAHAM layers.

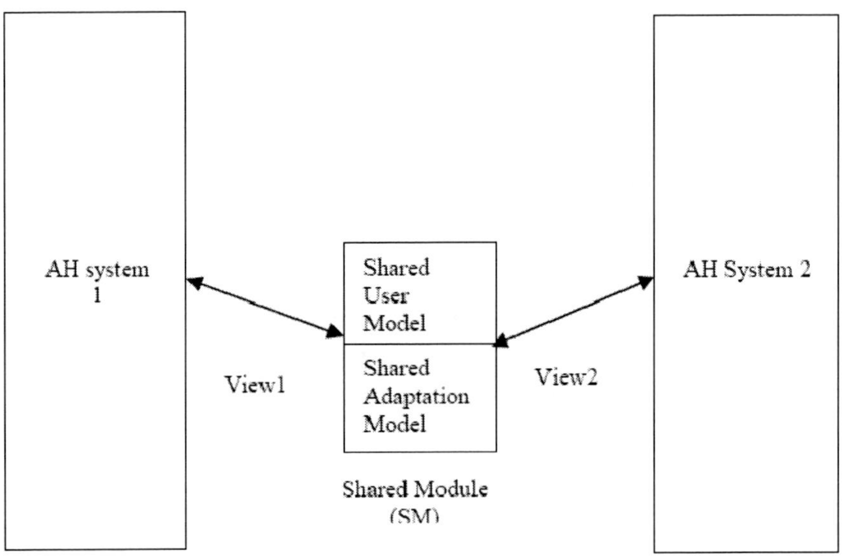

Figure 6. Proposed Reference Model SAHAM.

Kobsa and Wahlster (1989) have stressed that a UM is a knowledge source which could be separated from the rest of the knowledge. But in practice it has been proven to be difficult to find domain problems sharing a UM (Koch, 2000).

In our view, it is difficult to share the UM because it contains several domain dependent attributes which might not match with the other AH system's domain dependent attributes. There are many domain independent attributes also in the UM which could be shared. Koch (2000) distinguishes three types of user models based on their contents: *Domain-knowledge model*: containing the knowledge which the system assumes that the user has about the domain. *Domain-independent-knowledge model*: containing general knowledge about the user that is not at all domain specific, such as areas of interest of user or his background. *Psychological or cognitive model*: containing data about the user that is not domain-specific but psychological in nature such as user's preferences, abilities or personality traits. These three different types of user models form the basis for our 'shared module' concept. We propose that every AH system should have two separate user models, instead of having only one:

(1) Domain-dependent user model (2) Shared domain-independent user model.

- *Domain-dependent User Model (UM)*: This UM is an integral part of the storage layer and contains all the attributes which denote the system's assumption of the user's knowledge of the features specific to the domain of the application. For example, what pages the user has read, which ones are ready to be read?
- *Shared Domain-independent User Model (SUM)*: SUM is external to the system, having explicit representation and containing attributes that are much needed for the adaptation but are not domain specific attributes such as user's background, his area of interest and attributes related to his psychology such as his personality traits, his abilities and disabilities, his preferred learning style (holistic or serialized, exploratory or directed), his thinking style (whether inductive or deductive). Since these attributes are not related to any feature of domain, they can be shared by various AH systems for adapting the ultimate presentation for the user according to his cognitive skill. These attributes are assigned a value that usually belongs to a range of valid values such as Boolean (true for known or false for unknown), discrete (e.g. 1 for low, 2 for middle or 3 for high), or a combination of values (e.g. novice, intermediate or expert).

SUM could be implemented as a table with each user having a unique identity and attributes being represented as combination of attribute-value pair in the table. Table 1 shows an example of an instance of SUM.

Definition of SUM: The shared user model (SUM) is a set of <uid, uinfo> pairs, where uid is the unique identifier of user and uinfo is the attribute-value pair stored in SUM.

Table 1. Example of Shared user model instance

uid	Goal	Images	examples	Preferred media type	language	Possible disability
U1	searching	With	without	video	english	colorblind
U2	learning	With	with	audio	english	none
U3	exercising	without	with	text	french	none

To have an effective use of the two user models, we propose to have two separate adaptation models also –

- *Adaptation Model (AM)*: An integral part of the storage layer, the adaptation model contains the adaptation rules that are used for managing the domain dependent user model (UM).
- *Shareable Adaptation Model (SAM)*: Like SUM, this again, is external to the storage layer and contains adaptation rules to manage and use the data stored in shareable domain-independent user model (SUM) for the adaptation purpose. SAM differs from AM in the sense that it supports different 'views' of SUM made available to each individual AH system since one AH system might need a few components of SUM while other system might need some other components.

SAM defines how the data is read and updated in the SUM by the authorized AH systems. Authorization is needed so that every external system is not able to read and/or update the SUM. For this we define two functions in SAM – GetValue and SetValue functions.

GetValue is the function that could be used to access the required value from the SUM.

GetValue (uid, auth, uinfo)

where *uid* is the unique user identifier, *auth* is a system-dependent authorization and *uinfo* is the attribute-value pair whose value is required.

SetValue is the function that allows the AH systems to update the SUM.

SetValue (uid, auth, attr, value)

where *uid* is the unique user identifier, *auth* is the system-dependent authorization, *attr* is the attribute whose value is to be updated and *value* is the new value for this attribute.

The AH systems calling the *SetValue* function can acquire the new value of the attribute to be changed manually or (semi)automatically depending on the features provided in the system. On getting the value of the required attributes by using *GetValue* function, adaptation engine of the presentation specification interface can manipulate the presentation accordingly before delivering it to the user.

Ideally, SM should be independent of implementation of AH systems which are using it. Figure 6 shows our proposed reference model - SAHAM. The Shared Module containing SUM and SAM is being shared between two AH systems 1 and 2. Each AH system communicates with the 'Shared module (SM)', as shown in figure, for providing the required 'view' of adaptations related to general, domain-independent characteristics of the user.

4.1. Implementation of Shared Module

There could be a number of ways in which 'shared module' is implemented in an application independent way. Use of the web service technology is one of them where the User Model Web Server (WS) could be used to store the SM and provide functionality to AH system to obtain information stored in the SUM and for updating it. Kuruc (2005) and Kay et al (Kay, Kummerfeld and Lauder, 2002) have used this technology where for ensuring explicit semantics definition of the information stored in UM server, ontology based representations like RDF(S) or other higher level ontology languages like OWL were used. The systems could communicate with UM WS by SOAP (simple object access protocol) protocol over http (Hypertext transfer protocol) protocol using SSL (Secure Socket layer).

Orwant (1995) built a user model server with user models in Lisp-like languages. Paiva et al (Machado, Martins, Paiva, 1999) built a server that could support multiple teaching agents. Ohene-Djan et al (Ohene-Djan, Gorle, Bailey, Wills, Davis, 2003) gave a general, open architecture for hypermedia using client technology as the core of AH functionality loosely coupled to one or more user interface servers and database servers.

Any of the above approaches could be used for the implementation but the important aspect is that the SUM should be independent of the AH systems, having explicitly defined semantics of the information stored and having SAM capable of managing the SUM independent of the implementation of AH systems that are using it and allowing user to observe and modify information stored in SUM.

5. Advantages of SAHAM

By using Shared Module (SM), an AH system can be initialized by a more accurate user model, thus saving the time spent on going through the intermediate stages for arriving at a certain accuracy that has been achieved by another AH system. Moreover, any future updations done on the SUM are visible to all the systems at once. Sharing promotes low redundancy and, therefore, few redundancies related problems of the user models.

Another benefit of SM is in terms of the space utilized for storing the SUM and SAM. Since AH systems are sharing SM, it can be stored at one location thus saving the memory space used in storing it individually in all AH systems. A secondary benefit is that developers of the AH systems can use an existing SM instead of spending time and effort in developing it since the SM can act as the black box for the developers with the specified set of interface instructions. This promotes the independent development and maintenance of the SM. Fink (1998) has also specified some of these advantages of central maintenance of user modelling activities.

In a nutshell, the advantages of having a shared module are:

- The current, more accurate user model is available to all the AH systems.
- Initialization of AH system with current UM.
- Availability of similar stereotypes for different AH systems.
- Less space is required for storing SM then storing individual UMs.
- It results in low redundancy and therefore, minimum problems related to redundancy, in individual user model content.
- Saves developer's time and effort in developing individual SUM and SAM for individual AH systems.
- SM can act as the black box for the developers of AH systems.
- The development and maintenance of the shared module can be performed independently of the AH system.
- Last but not the least; it promotes use of standards for developing and using SM among developers and also promotes synergistic effect among AH systems.

CONCLUSION

In this paper, we proposed a reference model, SAHAM, for AH systems where a 'shared module (SM)' consisting of shared user model (SUM) and shared adaptation model (SAM) was introduced. It was built upon AHAM and extends it by subdividing user model into two parts – domain-dependent user model – which is part of the storage layer and the domain-independent user model – which is outside the AH system and is shared among two or more AH systems. The adaptation model has also been divided into two parts – adaptation model with domain-dependent adaptation rules and is part of the storage layer. The second is shared adaptation model – having domain-independent adaptation rules and rules for viewing and managing the shared user model.

SAHAM also proposes the runtime layer to have two sub-components – *input acquisition component* – for taking inputs from user, environment and system and *presentation component* – for finally presenting the adapted hyperpage to the user. Similarly, the presentation specification interface also has two subcomponents – *inference engine* – to extract useful information from all the inputs, and *adaptation engine* – to take adaptation decisions based on the UM, SUM, AM, SAM and DM. The storage layer is same as in AHAM except the fact that the UM and AM here have just the domain dependent attributes and rules.

The main benefit of this reference model is that a mature, more accurate user model of one AH system can be utilized to initialize another AH system. This results in providing same stereotypes in different AH systems, lower redundancy, lesser system development time and effort and use of standards for developing AH systems.

Our future work will involve developing a system based on SAHAM having a user profile of domain independent attributes stored in SUM and developing adaptation rules and strategies to act on it. Details regarding the contents of SUM, authorization checks and updation of SUM are being worked upon. Since there can be problems of different AH systems using different metadata or different interpretations of the data stored in SUM, we need to standardize the data and its format for storing in SUM. Our future work will focus on this issue where we plan to implement the 'shared module' using RDF and web service technology.

REFERENCES

Ardissino, L. and Goy, A., (1999) 'Tailoring the interaction with users in electronics shops', in Kay, J. (Ed.), *User Modelling: Proceedings of the Seventh International Conference*, UM99, Vienna, New York: Springer, pp 35-44.

Benyon, D. R. and Murray, D. M., (1993) 'Adaptive Systems: From Intelligent Tutoring to Autonomous Agents', *Knowledge-Based Systems*, 6(4), 197-219.

Boyle, C., Encarnation, A.O., (1994) 'Metadoc: An Adaptive Hypertext Reading System', User *modeling and User-Adapted Interaction*, Vol. 4, no. 1, pp 1-19.

Brusilovsky, P., E. Schwarz, and G. Weber, (1996) 'ELM-ART: An Intelligent Tutoring System on World Wide Web', *Third International Conference on Intelligent Tutoring Systems*, ITS-96, Montreal, Canada.

de Bra, P. and Calvi, L., (1998) 'AHA! An Open Adaptive Hypermedia Architecture', *The New Review of Hypermedia and Multimedia*, 4, pp 115-139.

Fink J. (1998). Implikationen aus dem Datenbank- und Transaktionsmanagement für anwendungs-orientierte Serversysteme zur Benutzermodellierung. ABIS-98: Workshop on Adaptivitiy and User Modeling in Interactive Software Systems, *FORWISS Report*, 7-15.

Halasz, F. and Schwartz M., (1994) 'The Dexter Hypertext Reference Model', *Communications of the ACM*, 37(2), Grønbæk K. and Trigg R. (Eds.), pp 30-39.

Hohl, H., Böcker, H-D, Gunzenhäuser, R., (1998) 'Hypadapter: An Adaptive Hypertext System for Exploratory Learning and Programming', P. Brusilovsky et al. (eds.), *Adaptive Hypertext and Hypermedia*, Kluwer Academic Publishers, pp 117-142.

Jameson, A. (2001) '*Systems That Adapt to their Users: An Integrative Perspective*', Saarbrücken: Saarland University, Germany.

Kaplan, C., Fenwick, J., and Chen J., (1993) 'Adaptive Hypertext Navigation Based on User Goals and Context', *User Modelling and User-Adapted Interaction*, 3, pp 193-220.

Kay, J., Kummerfeld, B., and Lauder, P., (2002) 'Personis: A Server for User Models', P. De Bra, P. Brusilovsky, and R. Conejo (Eds.): *Adaptive Hypermedia 2002*, Springer-Verlag, Berlin, Heidelberg, LNCS2347, pp 203–212.

Kim, D. W., (1995) 'WING-MIT: Das auf einer multimedialen und intelligenten Benutzerschnittstelle basierenade tutorielle Hilfesystem für das Werkstoffinformationssystem WING–M2', Workshop Adaptivität und Benutzermodellierung in interaktiven Softwaresystemen (ABIS 95), München.

Kobsa, A., Koenemann, J., and Pohl, W., (2001) 'Personalised Hypermedia Presentation Techniques for Improving Online Customer Relationships', *The Knowledge Engineering Review*, Vol. 16:2, pp 111-155.

Kobsa, A., and Wahlster, W., (1989) '*User Models in Dialog Systems*', Springer Verlag.

Koch, Nora, (2000) 'Software Engineering for Adaptive Hypermedia Systems: Reference Model, Modeling Techniques and Development Process', PhD Thesis, Ludwig-Maximilians-University of Munich, Germany.

Koch, N. and Kraus, A., (2002) 'The Expressive Power of UML-based Web Engineering', D. Schwabe, Pastor, O., Rossi, G. and Olsina, L. (Eds), *Proceedings of the Second International Workshop on web-oriented Software Technology* (IWWOST'02).

Kuruc, J., (2005) 'Sharing a User Model between Several Adaptive Hypermedia Applications', M. Bieliková (Ed.), *IIT.SRC 2005*, pp 249-256.

Lowe, D. and Hall, W., (1999) '*Hypermedia and the Web: an Engineering Approach*', Wiley, ISBN 0-471-98312-8.

Leggett, J. J., and Schnase, J. L., (1994) 'Viewing Dexter with Open Eyes', *Communications of the ACM*, Vol. 37, No. 2, pp 76–86.

Machado, I., Martins, A., Paiva, A., (1999) 'One for all and all for one: a Learner Modeling Server in a Multi-agent Platform', Kay, J (Ed): *User Modeling: Proceedings of the Seventh International Conference*, Springer Wien, New York.

Ohene-Djan, J., Gorle, M., Bailey, C., P., Wills, G. B., Davis, H., C., (2003) 'Is it possible to devise a unifying model of adaptive hypermedia and is one necessary?', *AH2003*, pp 167-182.

Ohene-Djan, J., (2000) '*A Formal Approach to Personalisable, Adaptive Hyperlink-Based Interaction*', PhD thesis, Department of Computing, Goldsmiths College, University of London, University of London, available from http://homepages.gold.ac.uk/djan/jodthesis.pdf.

Opperman, R. (1994) 'Adaptively supported Adaptability', *International Journal of Human Computer Studies*, 40 (3), pp 455-472.

Orwant, J., (1995) 'Heterogeneous Learning in the Doppelgänger User Modeling System', *User Modeling and User-Adapted Interaction*, Vol. 4, no. 2, pp 59–106.

UML, (1999) 'Unified Modeling Language. Version 1.3', *The Object Management Group*. http://www.omg.org

Vrieze, de P., Bommel, van P., Weide, T., (2004) 'A Generic Adaptivity Model in Adaptive Hypermedia', *Adaptive Hypermedia and Adaptive Web-based Systems*, pp 344-347.

Vrieze, P., Bommel, P., Weide, Th., Klok, J., (2005) 'Adaptation in Multimedia Systems', *Multimedia Tools and Applications*, Vol. 25, No.3, Springer, pp. 333-343.

Warmer, J., and Kleppe, A. (1999) 'The Object Constraint Language: Precise Modeling with UML', *Object Technology Series*, Addison-Wesley.

Whalster, W. and Kobsa A., (1986) 'Dialogue-based User Models', *Proc. of IEEE*, Vol. 74, No.7, pp 948-960.

Wu, H., de Bra, P., and Houben, G. J., (1999) 'AHAM: A Dexter-based Reference Model for Adaptive Hypermedia', *Proc. of ACM Conference on Hypertext and Hypermedia*, pp 147-156.

Wu, H., (2002) '*A Reference Architecture for Adaptive Hypermedia Applications*', PhD thesis, Techniche Universiteit Eindhoven.

Wu, H., E. de Kort, and P. de Bra (2001) 'Design issues for general-purpose adaptive Hypermedia Systems', *Proc. Of ACM Conference on Hypertext and Hypermedia (Hypertext '01),* Aarhus, Denmark, pp 141-150.

In: Focus on Information, Intelligence and Knowledge
Editor: Rafiq Islam

ISBN 978-1-61209-979-8
2011 Nova Science Publishers, Inc.

Chapter 12

THE MULTI–CARRIER DS-CDMA SYSTEMIN A FADING RAYLEIGH CHANNEL WITH MULTIPLES ACCESS INTERFERENCE

I. Ikhlef*[1], *F. Soltani*[2], *A. Said*[*3] *and O. Chaalal

Faculty of Engineer's Sciences,
[1, 2]Department of electronics- University of Constantine,
25000-Constantine, Algeria
[3]UAE University

ABSTRACT

In this paper, we study the multi – carrier direct – sequence code division multiple – access system when the received signal is divided into predetermined spreading sequence, these sub - bands modulates M carriers. The receiver produces the correlation for each carrier. Because of the no – knowledge of the phase of the received signal in each sub - band, the receiver cannot coherently combine outputs of the correlators, we consider in this paper two non - coherent combining schemes; the equal gain combining and the selecting combining. The calculation of the probabilities of detection and false alarm for both multi - carrier and single carrier systems in a Fading Rayleigh channel demonstrate that the performance of the multi - carrier system is better than the performance of the single carrier in equal gain combining, while the performance of the multi - carrier system in the selecting combining has the same performance as the single carrier.

Keywords: CDMA, multi - carrier DS-CDMA, equal gain combining, selecting combining, non- coherent combing scheme.

---

* Corresponding author: Email: abdellatifs@uaeu.ac.ae

I. INTRODUCTION

The problem of allowing multiple users to simultaneously access a channel without causing an undue amount of degradation in the performance of any individual user is a classical one in communications systems.

The two most common techniques are, the frequency division multiple access (FDMA), and the time division multiple access (TDMA), attempt to solve the problem by spreading the signal in frequency and time respectively.

For example, in the FDMA access scheme the inter-modulation problem will be generated. However, In the TDMA access scheme the intermodulation problem does not exist, but an accurate synchronization of all the users becomes of paramount importance in the system.

Furthermore, if an interference or multipath is present (which is our case) degradation in system performance can result in both FDMA and TDMA systems [1]. For these reasons the code division multiple access (CDMA) has become a competitive multiple accessing scheme in certain situations [1].

In this paper, we propose a multi-carrier direct sequence (MC-DS-CDMA) system transmitting in a Fading Rayleigh channel. Because of anti-interference capability, and robustness against Fading and multiple access capability [2]. In a DS – CDMA system, the data sequence of each k user (k = 1… K) is multiplied by a specific and exclusive spreading code word and modulates disjoint multiple carriers. The receiver combines the outputs of the correlators for each carrier with maximal-ratio combiner (MRC) [2].

This type of acquisition system has the following advantages:

(1) A multi - carrier direct sequence system is robust to the multipath Fading effects.
(2) A multi-carrier system has a narrowband interference suppression effect
(3) A lower chip rate.

Anti - interference capability in a DS-CDMA system is achieved by correlating the received signal with predetermined spreading sequence, thus allowing the inherent processing gain of the system to attenuate the interference [4]-[5]. For this, the use of the chip waveform filter is necessary to reject the narrow band interference or self-interference. The bandwidth of the DS waveform exceeds the coherence bandwidth of the channel [2]-[3]-[6], and the DS bandwidth is divided into M equi–width frequency band (not necessarily continuous).

The serial acquisition performance of a MC-DS-CDMA system has been investigated in [11] where each sub-carrier signal was subject to frequency-selective Fading.

The performance of the parallel code acquisition of the MC-DS-CDMA system is studied in [2] when each sub-carrier signal was subject to frequency-nonselective and modeled as Fading or non-Fading (with or without partial band interference by computing the probability of error). However, in the MC-DS-CDMA system considered here, we study the performance of the MC-DS-CDMA and the single carrier SC-CDMA system in the presence of an additive white Gaussian noise (AWGN) and multiple access interference (MAI) and Fading Rayleigh channel for both non-coherent combining schemes , that is:

The equal gain combining (EGC) and selecting combining by computing the probability of detection and false alarm.

This is motivated by the simplicity of the derivations of the probabilities in comparison with the error probability. We aim to choose the best combining scheme for the MC-DS-CDMA system proposed.

This paper is organized as follows:

section II introduces the system model for the proposed multi - carrier DS-CDMA system. The statistics of the outputs of the non - coherent correlators and combiners for Fading Rayleigh channel is presented in section III as well as the calculation of the probability of detection and false alarm. In section IV and V, we compare the numerical results for both multi- carrier and single carrier systems for the two kinds of combination scheme proposed. Finally, we provide our conclusions in section VI.

II. System Model

A. System Overview

A multi-carrier system can be considered as one realization of such a wide – band DS system (as shown in Figure 1a) depicts a band – limited DS waveform of wide – band single carrier system in the frequency – domain, where the band wide W_{sc} is given by:

$$W_{sc}=(1+\alpha)\frac{1}{T_c} \quad (1)$$

where α is the roll off factor of the chip wave – shaping filter $0\leq\alpha\leq1$, T_c is the chip duration of the single carrier. In the case of a multi – carrier system proposed, we divide the total bandwidth W_{sc} into M sub-bands (equi – width frequency bands) as shown in (Figure 1b). The bandwidth of each frequency sub band W_{Mc} is given by:

$$W_{Mc}=(1+\alpha)\frac{1}{MT_c} \quad (2)$$

where M : The total number of the carriers and MT_c is the duration of the sub - band.

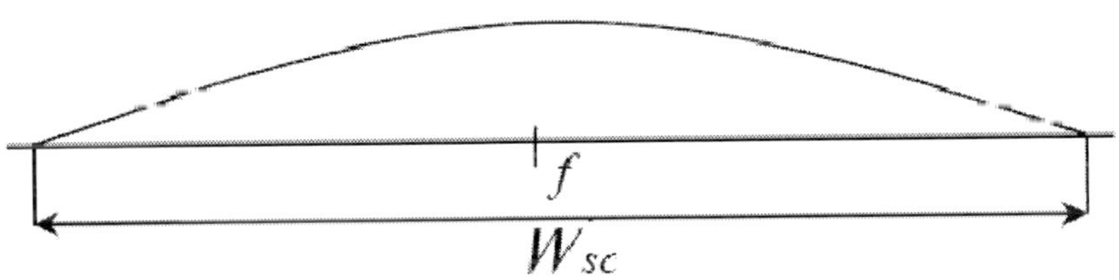

Figure 1a. Power spectrum density of single carrier DS-CDMA waveform.

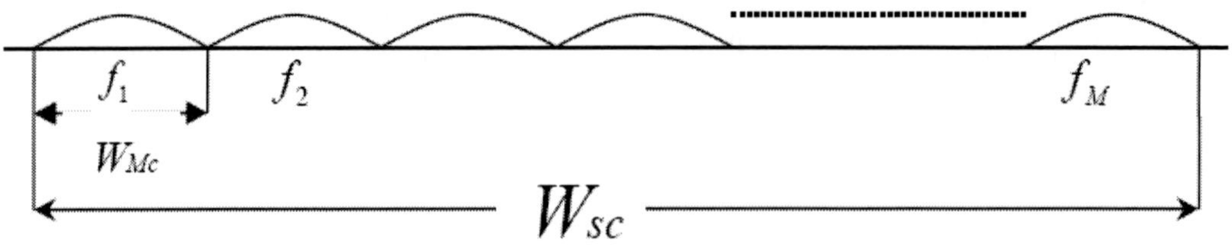

Figure 1b. Power spectrum density of multi - carrier DS-CDMA wave form.

B. The Transmitter

The transmitter for the kth users is shown in (Figure 2), where $d_{(n/N_D)}^{(k)}$ is a random binary sequence representing data, and $c_n^{(k)}$ is a pseudo – random spreading signature sequence. We assume that there are N_D chips per symbol; the $d_{(n/N_D)}^{(k)} c_n^{(k)}$ sequence modulates an impulse train, where the energy per chips is E_c.

The transmitted signal at the output of the chip wave – shaping filter modulates M carriers. The energy per-chip of the modulating signal over each carrier is $\frac{E_c}{M}$. The transmitted signal $S_k(t)$ is given by [2]-[3]-[7] and has the following expression:

$$S_k(t) = \sqrt{\frac{2E_c}{M}} \sum_{n=-\infty}^{+\infty} d_{(n/N_D)}^{(k)} c_n^{(k)} h(t - nMT_c) \sum_{m=1}^{M} \cos(\omega_m + \theta_{k,m}) \quad (3)$$

where $h(t)$ is the impulse response of the chip wave- shaping filter, and $\theta_{k,m}$ is a random phase uniformly distributed over (0, 2 π).

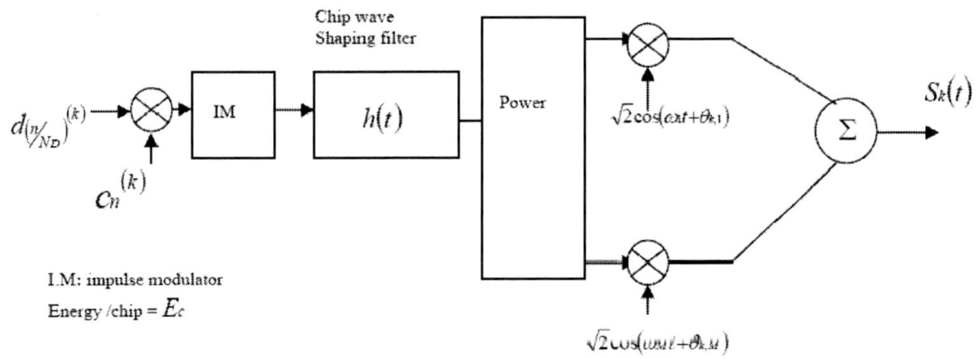

Figure 2. Block diagram of CDMA transmitter.

C. The Channel

The channel is assumed to be a slowly varying frequency selective Rayleigh channel with the delay spread of T_d [3]. The channel in DS-CDMA is characterized by the coherent bandwidth (Δf_c), which is given by [3]:

$$(\Delta f_c) = \frac{1}{T_d} \quad (4)$$

We choose M to satisfy the following conditions:

(i) Each sub band of the multi-carrier system has no selectivity i.e., $\frac{T_d}{MT_c} \leq 1$.

(ii) All the sub bands are independent and uncorrelated Fading i.e., $W_{Mc} \geq (\Delta f_c)$.

The tow conditions are satisfied if: $\frac{T_d}{T_c} \leq M \leq \frac{T_d}{T_c}(1+\alpha)$

The transfer function of the frequency band of the kth user is given by $\alpha_{k,m} \exp(j\beta_{k,m})$, where $\alpha_{k,m}$ and $\beta_{k,m}$ are independent and identically distributed (i.i.d) Rayleigh random variables with a unit second moment, and (i.i.d.) uniform random variables over [0,2 π], the received signal is given by:

$$r(t) = \sum_{k=1}^{K} \sqrt{\frac{2E_c}{M}} \left\{ \sum_{n=-\infty}^{+\infty} d_{(n+D_k)/N_D}^{(k)} c_{(n+D_k)}^{(k)} h(t - nMT_c - \zeta_k) \sum_{m=1}^{M} \alpha_{k,m} \cos(\omega_m + \theta'_{k,m}) \right\} + n_w(t) \quad (5)$$

We suppose in our case no-partial band interference (no-PBI). $n_w(t)$ is AWGN with a double side (PSD) density of $\eta_0/2$ and $\theta_{k,m} = \theta_{k,m} + \beta_{k,m}$ is uniformly distributed over [0, 2π], D_k is the phase of the code chip sequence, ζ_k is the unknown time given by $\zeta_k = D_k MT_c + \tau_k$, , here the value τ_k is the unknown chip delay assumed to be uniformly distributed over [0, MT_c].

D. The Receiver

The receiver side is shown in (Figure 3a, b). We assume the following characteristics for the chip wave-shaping filter: $X(f) = |H(f)|^2$ satisfied the Nyquist criterion, $\int_{-\infty}^{+\infty} X(f) = 1$, and $X(f)$ is band limited to W', where $W' \leq \frac{f_{m+1} - f_m}{2}$ and f_m is the mth carrier frequency. This implies that the DS waveforms does not overlap and therefore the adjacent channel interference may by ignored [2]. To ensure this assumption, $X(f)$ is a raised-cosine filter as shown in (6) [8].

$$X(f)=\begin{cases}\frac{1}{W'} & |f|\leq\frac{W'}{2}(1-\alpha)\\ \frac{1}{2W'}\left\{1-\sin\left[\frac{1}{2\alpha}\left(\frac{2\pi|f|}{W'}-\pi\right)\right]\right\} & \frac{W'}{2}(1-\alpha)\leq|f|\leq\frac{W'}{2}(1+\alpha)\\ 0 \quad elsewhere & \end{cases} \tag{6}$$

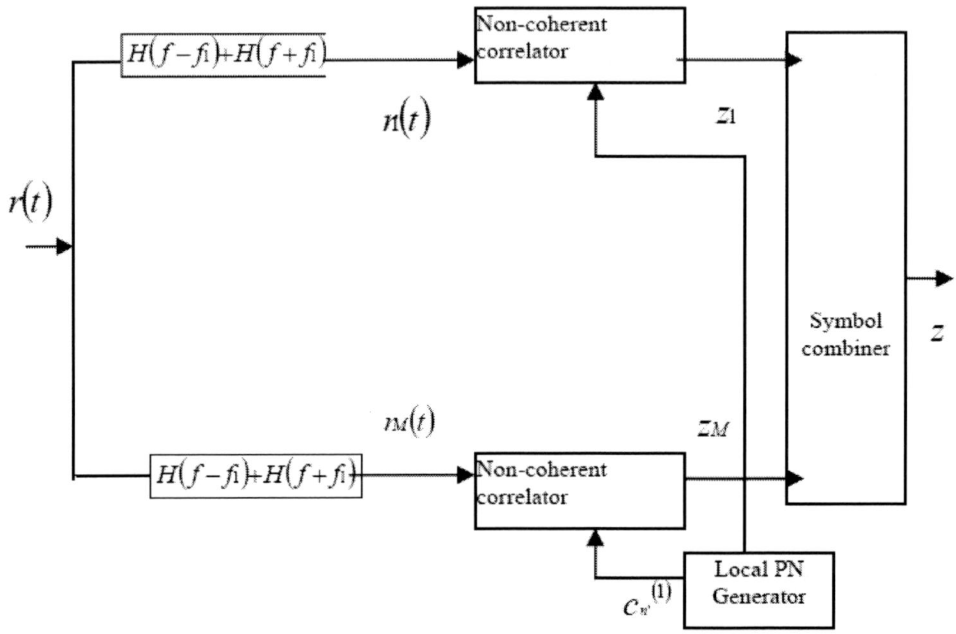

Figure 3a. Block diagram of a multi-carrier CDMA system receiver.

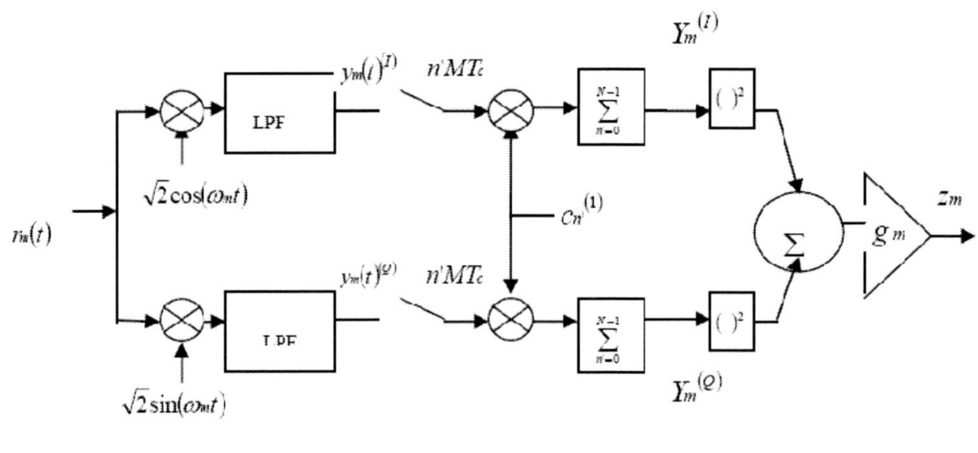

Figure 3b. Non-coherent (In phase, Quadratic) correlator for mth branch of the multi-carrier system.

The receiver of the kth user is shown in (Figure 3.a, b). There is an in-phase and quadratic correlator for each of the M carrier frequencies. Each in-phase (I) and quadratic (Q) non-coherent correlator is composed by a coherent demodulation and a low pass filter in such a way that double frequency terms can be ignored. The output of the low pass filter is sampled and despreaded by a local PN code replica. After correlation, the output is squared by a quadratic envelope detector and their outputs are summed in branch I and branch Q.

The result of summation passes through an adaptative gain amplifier for each of the M non-coherent correlators to produce the variables of decision combined in the symbol combiner. We suppose that U_1 is a total uncertainty region (M=1) and U : is the uncertainty region of each sub-band $\left(U_1/M\right)$.

The search technique for the proposed acquisition scheme is a sequential method (serial implementation of a maximum – likelihood algorithm [9]) where each correlator correlates serially the incoming signal with all the possible phases of the local PN code replica and stores the results in the symbol combiner. The outputs of the M correlators are with the same phase of the local replica [2].

In this paper, we consider two non-coherent combining schemes: equal gain combining and selecting combining. We assume that the receiver is chip-synchronization with the user to interest (for $k=1, \tau_1=0$). In initial acquisition the spacing between the two codes PN incoming and local is typically ($\frac{T_c}{2}$), to simplify the mathematical calculations, we choose the spacing (T_c). The correct phase (under hypothesis H_1) correspond to $\zeta_1=0$, and the others phases (under hypothesis H_0) correspond to $\zeta_1=D_1MT_c \; with\, D_1\neq 0$.

We consider the same time to collect the results for both the multi-carrier and single carrier systems. To have this, the period of correlation for the single carrier is M - times greater than the multi-carrier, since the chip rate of the later system is $1/M$ - times lower than the first one.

III. The Statistics of the Outputs of the Non-Coherent Correlator and Combiner

A. The Statistics of the Non-Coherent Correlator

We assume that during the acquisition process no data are sent i.e. $d^{(1)}_{(n+D_1)/N_D}=1$, the correlation obtained by integrating over N chips in the mth non-coherent correlator of the first user [2]. The in phase and quadratic components are:

$$Y_m^{(I)}=\alpha_{1,m}C(\zeta_1)\cos(\theta_{1,m})+I_{Y_m}^{(I)}+N_{Y_m}^{(I)} \tag{9a}$$

$$Y_m^{(Q)}=\alpha_{1,m}C(\zeta_1)\sin(\theta_{1,m})+I_{Y_m}^{(Q)}+N_{Y_m}^{(Q)} \tag{9b}$$

where:

$$C(\zeta_1)=\sqrt{E_c/M}\sum_{n'=0}^{N-1}\sum_{n=-\infty}^{+\infty}c_{n'}^{(1)}c_{n+D_1}x[(n'-n)MT_c-\zeta_1] \tag{10a}$$

$$I_{Y_m}^{(Q)}=\sum_{n'=0}^{N-1}c_{n'}^{(1)}\left\{\sum_{k=2}^{K}\sqrt{E_c/M}\alpha_{k,m}\sin(\theta_{k,m})\sum_{n=-\infty}^{+\infty}d_{\left[(n+D_k)/N_D\right]}^{(k)}c_{n+D_k}^{(k)}x[(n'-n)MT_c-\zeta_k]\right\}$$

$$N_{Y_m}^{(I)}=\sum_{n'=0}^{N-1}c_{n'}^{(1)}\left[Lp\{n'_{w,m}(t)\sqrt{2}\cos(\omega_m t)\}\right]_{t=n'MT_c} \tag{10b}$$

$$N_{Y_m}^{(Q)}=\sum_{n'=0}^{N-1}c_{n'}^{(1)}\left[Lp\{n'_{w,m}(t)\sqrt{2}\sin(\omega_m t)\}\right]_{t=n'MT_c}$$

In (10.a) and (10.b) the terms $n'_{w,m}(t)$ is $n_{w,m}(t)$ after passing through the mth wave shipping filter and $Lp\{\ \}$ represents a low pass filtering after which double frequency term can be ignored. Note that $N_{Y_m}^{(I)}$ and $N_{Y_m}^{(Q)}$ are Gaussian by assumption, and $I_{Y_m}^{(Q)}$ and $I_{Y_m}^{(I)}$ are asymptotically Gaussian [2], [3].We define new Gaussian random variables with zero-mean and variance σ_m^2, $\eta_m^{(I)}=I_{Y_m}^{(I)}+N_{Y_m}^{(I)}$ and $\eta_m^{(Q)}=I_{Y_m}^{(Q)}+N_{Y_m}^{(Q)}$, the two variables are not function of $\zeta_1,\theta_{1,m},\alpha_{1,m}$; the variance of the values $I_{Y_m}^{(Q)},I_{Y_m}^{(I)},N_{Y_m}^{(Q)}$, and $N_{Y_m}^{(I)}$ are derived from [2], [3] and equal to:

$$\sigma_{I_m}^2=\text{var}\left(I_{Y_m}^{(Q)}\right)=\text{var}\left(I_{Y_m}^{(I)}\right)=\frac{N(k-1)E_c}{2M}\left(1-\alpha/4\right) \tag{11}$$

$$\sigma_{N_m}=\text{var}\left(N_{Y_m}^{(Q)}\right)=\text{var}\left(N_{Y_m}^{(I)}\right)=\frac{N\eta_0}{2} \tag{12}$$

N : is the integration period of the correlation for the multi-carrier DS-CDMA system.

where: $\sigma_m^2=\sigma_{I_m}^2+\sigma_{N_m}^2$ (13)

We can rewrite (7.a) and (7.b) based on the new variables defined previously as:

$$Y_m^{(I)}=\alpha_{1,m}C(\zeta_1)\cos(\theta_{1,m})+\eta_m^{(I)} \tag{14a}$$

$$Y_m^{(Q)}=\alpha_{1,m}C(\zeta_1)\sin(\theta_{1,m})+\eta_m^{(Q)} \tag{14b}$$

In this paper, we study the statistics conditioned on $\alpha_{1,m}$ and the hypothesis H_i $(i=0,1)$ removing the condition on $\theta'_{1,m}$.Therefore equations (14.a) and (14.b) become:

$$Y_m^{(I)}=C(\zeta_1)\alpha_{1,m}+\eta_m^{(I)} \tag{15a}$$

$$Y_m^{(Q)}=C(\zeta_1)\alpha_{1,m}+\eta_m^{(Q)} \tag{15b}$$

We suppose here that the spreading sequence for (user 1) is deterministic and $d_{\left[(n+D_k)/N_D\right]}^{(k)} c_{(n+D_k)}^{(k)}$ for $k = 2, 3 \ldots K$ independent binary random sequences [2], [3].

$C(\zeta_1)$ is the correlation performed over N chips between the tow codes PN incoming and local, in a limit central theorem $C(\zeta_1)$ have a Gaussian distribution because $N \succ 1$.

The correlation statistics are given in the appendix A (mean and variance conditioned on hypothesis H_i and the offset normalized p_0).

The outputs of the non-coherent correlator of the mth branch are given by [2]:

$$z_m = g_m^{-1}\left\{\left(Y_m^{(I)}\right)^2 + \left(Y_m^{(Q)}\right)^2\right\} \tag{16}$$

where g_m^{-1} is an adaptive gain control (AGC) coefficient of the mth Branch assumed to be the perfect estimate of the average received power and is given by [2]:

$$g_m = \overline{y_m^{(I)}(n' MT_c)^2 + y_m^{(Q)}(n' MT_c)^2} \tag{17}$$

The terms $\overline{y_m^{(I)}(n' MT_c)^2}$ and $\overline{y_m^{(Q)}(n' MT_c)^2}$ can be obtained from (11), (12) and (20), g_m is given by:

$$g(C_m+I_m)=\frac{kE_c}{M}\left(1-\alpha/4\right)$$

$$g_{N_m}=\eta_0 \tag{18}$$

$$g_m=g_{N_m}+g(C_m+I_m)$$

In our case g_m is the same in all sub bands because there is no-PBI (partial band interference).

For a given $\alpha_{1,m}$ and $m_{m,i}$, $Y_m^{(I)}$ and $Y_m^{(Q)}$ are independent Gaussian random variables with the same variance $\sigma_{i,m}^2$ and the same mean $\sqrt{E_c/M}\,\alpha_{1,m}\, m_{m,i}$

$$\sigma_{m,i}^2=\sigma'^2_{m,i}+\sigma_{I_m}^2+\sigma_{N_m}^2 \tag{19}$$

The conditional probability density function (p.d.f) of z_m conditioned on $\alpha_{1,m}$ *and* $H_i(i=0,1)$, is a no central x^2 random variable with two degrees of freedom, we have:

$$p\left(z_m/\alpha_{1,m},H_i\right)=\frac{1}{2v_{m,i}}I_0\left(\frac{\sqrt{z_m}\,m'_{m,i}\alpha_{1,m}}{v_{m,i}}\right)\exp\left(-\frac{z_m+m'^2_{m,i}\alpha^2_{1,m}}{2v_{m,i}}\right) \tag{20}$$

$I_0(x)$: is the zeroth-order modified Bessel function of the first kinds

where $v_{m,i}=g_m^{-1}\sigma_{m,i}^2$ *and* $m'_{m,i}=\sqrt{\frac{E_c}{M}g_m^{-1}m_{m,i}^2}$.

For the Rayleigh Fading channel since $\alpha_{1,m}$ is a Rayleigh random variable with a unit second moment, the conditional (p.d.f) of z_m is given by [2]:

$$p\left(z_m/H_i\right)=\int_0^{+\infty}2\alpha_{1,m}\exp\left(-\alpha_{1,m}^2\right)p\left(z_m/\alpha_{1,m},H_i\right) \tag{21}$$

$$p\left(z_m/H_i\right)=\frac{1}{2v'_{m,i}}\exp\left(-\frac{z_m}{2v'_{m,i}}\right)$$

$$where\quad v'_{m,i}=v_{m,i}+\frac{m'^2_{m,i}}{2} \tag{22}$$

B. The Outputs of the Non-Coherent Combiner

In this paper, we consider two kinds of the non-coherent combiner schemes.

B.1. Equal Gain Combining

In this kind of the symbol combining, the output is given by [2]-[10]:

$$z=\sum_{m=1}^{M}z_m \tag{23}$$

The characteristic function conditioned on H_i of z is the product of all characteristics functions of z_m:

$$\Psi_{\left(z/H_1\right)}(s)=\prod_{m=1}^{M}\Psi_{\left(z_m/H_i\right)}(s) \tag{24}$$

where $\Psi_{\left(z_m/H_i\right)}(s)$ is the Fourier transform of the (p.d.f) of z_m witch is obtained from (22).

For a Rayleigh Fading channel, the characteristic function of z_m is given by [2]:

$$\Psi_{\left(z_m/H_i\right)}(s)=\frac{1}{1-2v'_{m,i}s} \tag{25}$$

The probability density function of z is the inverse Fourier transform of its characteristic function:

$$p\left(z/H_i\right)=TF^{-1}\left\{\prod_{m=1}^{M}\Psi_{\left(z_m/H_i\right)}(s)\right\} \tag{26}$$

From Cauchy residue theorem, we have:

$$p\left(z/H_i\right)=\frac{z^{M-1}\exp\left(-\frac{z}{2v'_{m,i}}\right)}{\left(2v'_{m,i}\right)^{M}Fact(M-1)} \tag{27}$$

B.2. Selecting Combining

We assume that all z_m are independent and identically distributed random variables. In the selecting combining, the phase corresponding to the largest test is assumed to be the correct phase (i.c. $z=\max(z_1,z_2.............z_M)$).

For this, we define the conditional joint distribution function as [10]:

$$F\left(z/H_i\right)=p\left\{\left(z_1\le z,z_2\le z.........z_M\le z/H_i\right)\right\} \tag{28a}$$

$$F\left(z/H_i\right)=1-\int_{z}^{+\infty}p\left(z_m/H_i\right)dz \tag{28b}$$

The (p.d.f) of z is given by:

$$p\left(z/_{H_i} \right) = \frac{\partial F(z_m \leq z)}{\partial z} \tag{29}$$

In the Rayleigh Fading channel, the conditional joint distribution function of z is obtained from (22) as [2]:

$$F\left(z/_{H_i} \right) = \left[1 - \exp\left(\frac{z}{2v'_{m,i}} \right) \right]^M \tag{30}$$

From (37), the conditioned (p.d.f) of z is given by:

$$p\left(z/_{H_i} \right) = \frac{M}{2v'_{m,i}} \exp\left(-\frac{z}{2v'_{m,i}} \right) \left[1 - \exp\left(\frac{z}{2v'_{m,i}} \right) \right]^{M-1} \tag{31}$$

IV. Calculation of the Probability of Detection and False Alarm for the Rayleigh Fading Channel

A. For the Equal Gain Combining

A.1. The Probability of Detection

The probability of detection p_D (under the hypothesis H_1) for the equal gain combining can be obtained from (28) as:

$$p_{Dfe} = \int_{\gamma}^{+\infty} \frac{z^{M-1} \exp\left(-\frac{z}{2v'_{m,1}} \right)}{(2v'_{m,1})^M \, Fact(M-1)} dz \tag{32}$$

where γ : the threshold of the decision process

$v'_{m,1} = v_{m,1} + \frac{m'^{\,2}_{1,m}}{2}$ and

$m'^{\,2}_{m,1} = \frac{E_c}{M} N g_m^{-1}$

N : *the integration period of correlation*

g_m^{-1} : *AGC coefficient of the mth branch*

$v_{m,1} = g_m^{-1} \sigma_{1,m}^2$

g_m *and* $\sigma_{m,1}^{\ 2}$ can be obtained from (26) and (25) respectively.

After calculation (see appendix C), we obtain:

$$p_{Dfe} = \sum_{m=0}^{M-1} \left(\frac{\gamma}{2v'_{m,1}} \right)^m \frac{\exp\left(-\frac{\gamma}{2v'_{m,1}} \right)}{Fact(m)} \tag{33a}$$

After normalizing the threshold over the correlation period N, we obtain: $\gamma^* = \frac{\gamma}{N}$

This implies that:

$$2v'_{m,1} = N(B+a)$$

where

$$B = \frac{\left(1+(K-1)v\left(1-\alpha/4\right)\right)}{\left(1+Kv\left(1-\alpha/4\right)\right)} \tag{33b}$$

$$and \quad a = \frac{Nv}{\left(1+Kv\left(1-\alpha/4\right)\right)}$$

$$v = \frac{E_{c1}}{\eta_0} \ with\ E_{c1} = \frac{E_c}{M}$$

This implies that p_{Dfe} is equal to:

$$p_{Dfe} - \sum_{m=0}^{M-1} \left(\frac{\gamma^*}{(B+a)} \right) \frac{\exp\left(\frac{\gamma *}{(B+a)} \right)}{Fact(m)} \tag{34}$$

A.2. The Probability of False Alarm

In a similar manner of calculation of p_{Dfe}, p_{Fafe} we can obtain:

$$p_{Fafe} = \sum_{m=0}^{M-1} \left(\gamma^*\right)^m \frac{\exp\left(\gamma^*\right)}{Fact(m)} \tag{35}$$

$$with\, v'_{m,0} = v_{m,0} + \frac{m_{m,0}^{\ \ 2}}{2} \ \sin ce\ m_{m,0} = 0$$

B. Selecting Combining

B.1. The Probability of Detection

From (38), we obtain the probability of detection (under hypothesis H_1) as:

$$p_{Dfs}=\int_{\gamma}^{+\infty}\frac{M}{2v'_{m,1}}\exp\left(-\frac{z}{2v'_{m,1}}\right)\left[1-\exp\left(-\frac{z}{2v'_{m,1}}\right)\right]^{M-1}dz \qquad (36)$$

$v_{m,1}'$ has been calculated in (40.b), as the same manner of the calculation of p_{Dfe}, we obtain:

$$p_{Dfs}=\int_{\gamma*}^{+\infty}\left(\frac{M}{(B+a)}\right)\exp\left(-\frac{z_*}{(B+a)}\right)\left[1-\exp\left(\frac{z_*}{(B+a)}\right)\right]^{M-1}dz_* \qquad (37)$$

with $z_*=\frac{z}{N}$, a *and* B have been defined in (40.b)

B.2. The Probability of False Alarm

Using the same technique of p_{Dfs}, p_{Fafs} is equal to:

$$p_{Fafs}=\int_{\gamma_*}^{+\infty}\frac{M}{2v_{m,0}'}\exp\left(-\frac{z}{2v'_{m,0}}\right)\left[1-\left(-\frac{z}{2v'_{m,0}}\right)\right]^{M-1}dz \qquad (38)$$

We, define a new variable z_* as:

$$z_*=\frac{z}{N}$$

$$p_{Fafs}=\int_{\gamma_*}^{+\infty}M\exp(-z_*)[1-\exp(-z_*)]^{M-1}dz \qquad (39)$$

V. Numerical Results

"the expressions of the probability of detection and false alarm contain too many parameters.In theory, calculations will be too long.That is why used the numerical aproach.

The obtained curves are resulting from a simulation using MatLab software, and making interpolation between points."

In this section, we make a comparison between the probability of detection and false alarm in Rayleigh Fading channel with no- partial band interference for both multi-carrier and single-carrier systems. The period of correlation of the single carrier has been chosen equal to $N_1=512$, but for the multi-carrier $N=\frac{N_1}{M}$, the roll-off factor α of the raised-cosine filter is chosen equal to 0.5 [2]-[8]. The probability of detection and false alarm for multi-carrier and single-carrier systems for both the equal gain combining and selecting combining are shown in Figures 4 and 5 respectively.

We can easily see that in the equal gain combining, the performance of the multi-carrier system is better than the single-carrier,

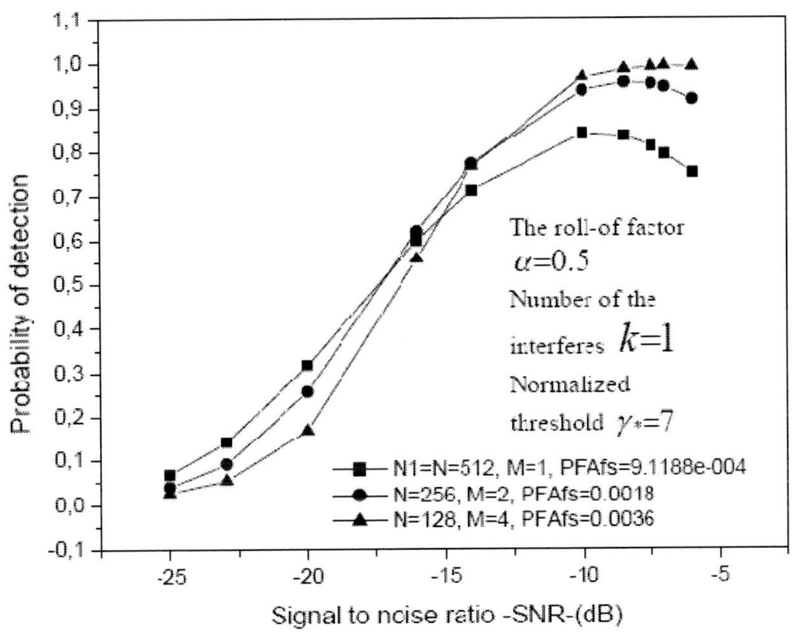

Figure 4. Probability of detection for Rayleigh Fading channel for both multi-carrier ($M\geq2$) and single carrier $(M=1)$ systems and selecting combining.

Since the number of sub bands (M) increases, the probability of detection increases. However, the performance of multi-carrier system is similar to the single carrier, because the probability of detection in single-carrier for low signal to noise ratio (SNR) [-30db—16db] is higher than the probability of detection for multi-carrier.

But for high signal to noise ratio [-16db—-10db], the opposite is true.

The effect of multiple access interference (MAI) on the probability of detection and false alarm, for both equal gain combining and selecting combining is shown in Figures 6 and 7. When the number of interference increases, the probability of detection decreases and the probability of false alarm have no dependence on (K).

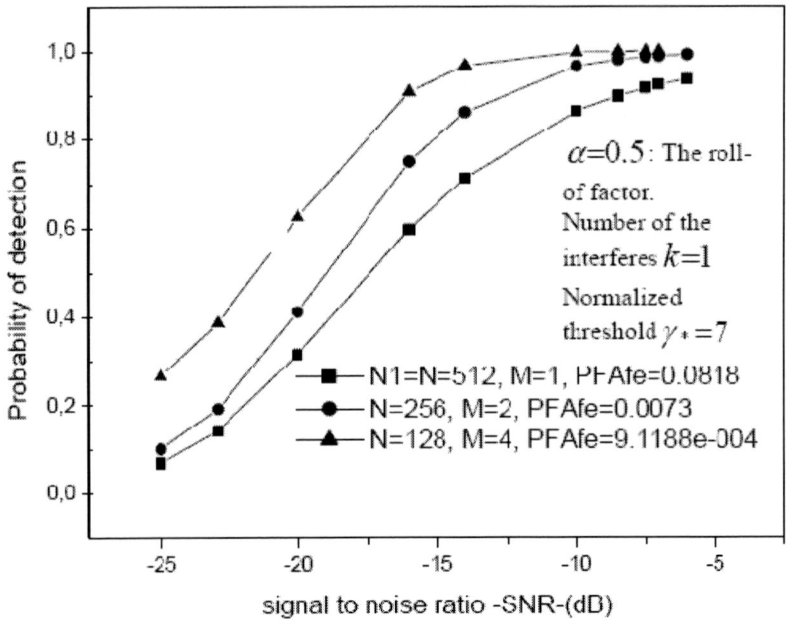

Figure 5. Probability of detection for Rayleigh Fading channel for both multi-carrier ($M \geq 2$) and single carrier $(M=1)$ systems and equal gain combining.

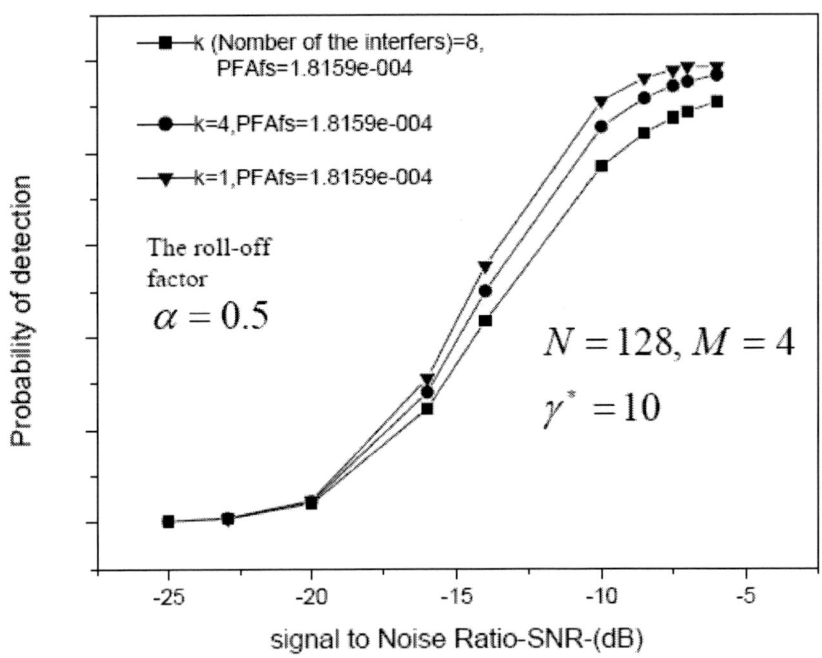

Figure 6. Probability of detection for Rayleigh Fading channel for both multi-carrier ($M \geq 2$) system and selecting combining for different value of interferes (k).

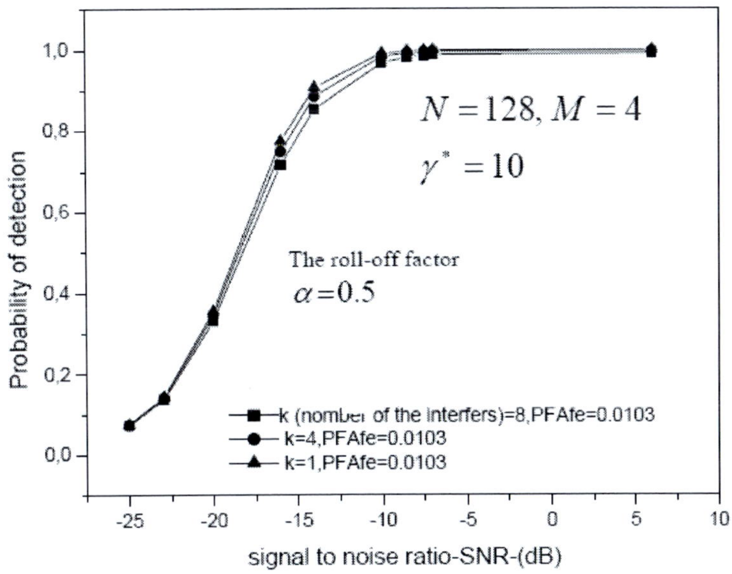

Figure 7. Probability of detection for Rayleigh Fading channel for both multi-carrier ($M \geq 2$) system and equal gain combining for different value of interferes (k).

The effect of the threshold on the probability of detection and false alarm for both equal gain combining and selecting combining is shown in Figures 8 and 9. When the probability of detection increases the threshold decreases, because when the threshold decreases the probability to cross the threshold is easy. The same thing is valid for the probability of false alarm.

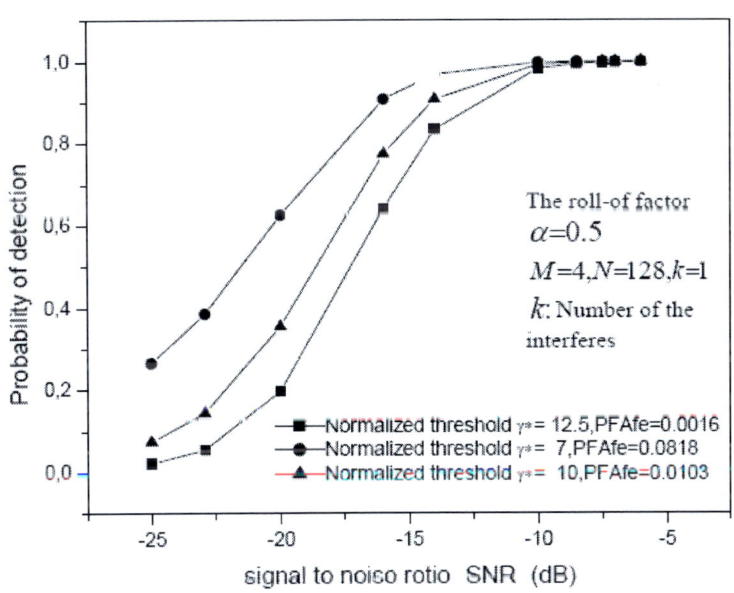

Figure 8. Probability of detection for Rayleigh Fading channel in multi-carrier ($M \geq 2$) system and equal gain combining for different value of threshold.

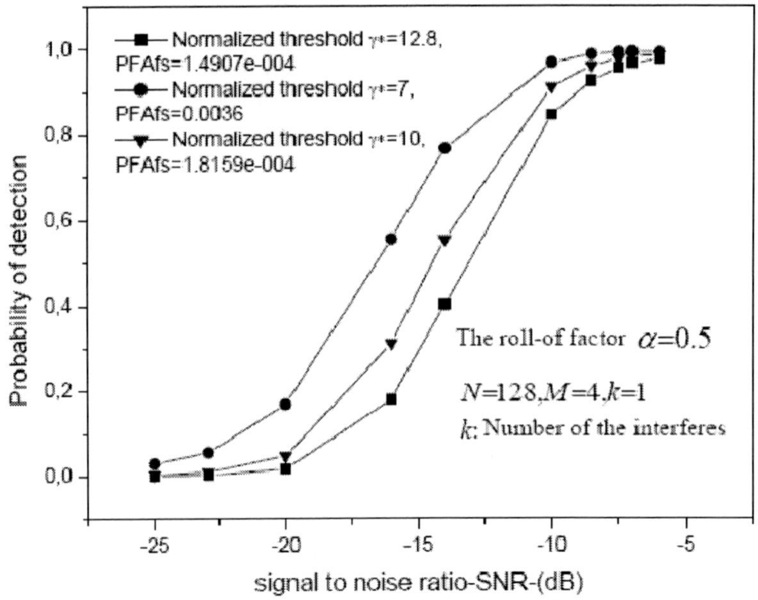

Figure 9. Probability of detection for Rayleigh Fading channel in multi-carrier ($M \geq 2$) system and selecting combining for different value of threshold.

CONCLUSION

In this paper, we have studied two types of combinations: the first is the equal gain and the second is the selecting gain.Both are used for multi-carrier and single carrier systems. In the Rayleigh Fading channel, the performance of the multi-carrier system is better than the single carrier system for the equal gain combining. However, the performance of the multi-carrier and single carrier systems is the same for the selecting combining, since in low SNR, the single carrier system has a greater probability of detection than in the multi-carrier for different values of M . However, for high SNR (SNR included in the DSSS system interval), the multi-carrier system has a greater probability of detection with respect to the single carrier.

Finally, we propose the equal gain combining for the multi-carrier system and the selecting combining for the single carrier in the Rayleigh Fading channel.

APPENDIX A

Calculation of the statistics of the correlation $C(\zeta_1)$.

The auto-correlation function of the PN code is [9]:

$$R_c(p_0) = \begin{cases} 1-|p_0| & H_1 \\ 0 & H_0 \end{cases} \tag{A1}$$

$$p_0 = \frac{\zeta_1}{N}$$

p_0: is the normalized offset between the tow codes PN incoming and local.

The mean of the correlation $C(\zeta_1)$ is given by:

$$\overline{C(\zeta_1)} = m_{m,i} = \sqrt{E_c/M}\, N R_c(p_0)\left(\frac{1-(-1)^i}{2}\right) \tag{A2}$$

Since $x[(n'-n)MT_c] = 0 \quad for\ n' \neq n$

The variance of the correlation conditioned on (hypothesis H_i and the offset p_0) is Similar to the calculation in [2] and [3] we have:

$$\sigma'^{\,2}_{i,m} = N\ R_{m,i}(0) + 2\sum_{l=0}^{N-1} R_{m,i}(lMT_c)\sum_{n'=l}^{N-1} c_{n'}^{(1)} c_{n'-l}^{(1)} \tag{A3}$$

$$for \quad l \neq 0 \quad R_{m,i}(lMT_c) = 0$$

$$\sigma'^{\,2}_{i,m} = N\ R_{m,i}(0)$$

$$R_{m,i}(0) = \int_{-\infty}^{+\infty} \Phi_{m,i}(f)\, df$$

where $\Phi_{m,i}(f)$ is the spectral density of the correlation $C(\zeta_1)$

We can rewrite (A_3) as:

$$R_{m,i} = \frac{E_c\ G(p_0)}{2\ M\ T_c}\int_{+\infty}^{-\infty} X^2(f)\, df \tag{A4}$$

$$\sigma'^{\,2}_{i,m} = \frac{E_c N}{2\ M} G(p_0)\left(1 - \alpha/4\right)$$

$$With \quad G(p_0) = \begin{cases} p_0^{\,2} & H_1 \\ 1 - 2|p_0| + 2(p_0)^2 & H_0 \end{cases}$$

We choose $p_0 = 1$ under the hypothesis H_0 and $p_0 = 0$ Under hypothesis H_1

APPENDIX B

For a Rayleigh Fading channel, $\alpha_{1,m}$ is a Rayleigh random variable with a unit second moment, the conditional (p.d.f) of z_m conditioned on hypothesis H_i is:

$$p\left(z_m/H_i\right)=\int_0^{+\infty} p(\alpha_{1,m})\,p\left(z_m/H_i,\alpha_{1,m}\right)d\alpha_{1,m} \tag{B1}$$

with : $p(\alpha_{1,m})=\alpha_{1,m}\exp\left(-\alpha_{1,m}^2\right)$ $for\ \alpha_{1,m}\succ 0$

$$p\left(z_m/H_i\right)=\int_0^{+\infty} 2\alpha_{1,m}\exp\left(-\alpha_{1,m}^2\right)\frac{1}{2v_{m,i}}I_0\left(\frac{\sqrt{z_m}\,m_{m,i}\alpha_{1,m}}{v_{m,i}}\right)\exp\left(\frac{-z_m-m_{m,i}^2\alpha_{1,m}^2}{2v_{m,i}}\right)d\alpha_{1,m}$$

$$=\int_0^{+\infty} 2\alpha_{1,m}\exp\left(-\alpha_{1,m}^2\left(1+\frac{m_{m,i}^2}{2v_{m,i}}\right)\right)\exp\left(-\frac{z_m}{2v_{m,i}}\right)I_0\left(\frac{\sqrt{z_m}\,m_{m,i}\alpha_{1,m}}{v_{m,i}}\right)d\alpha_{1,m} \tag{B2}$$

We divide and multiply the equation by $\left(1+\frac{m_{m,i}^2}{2v_{m,i}}\right)$, we choose a new variable and new variance $\alpha_{1,m}$ and $v'_{m,i}$ respectively as:

$$\alpha'_{1,m}=\alpha_{1,m}\left(1+\frac{m_{1,m}^2}{2v_{1,m}}\right)$$

$$v'_{m,i}=v_{m,i}\left(1+\frac{m_{m,i}^2}{2v_{m,i}}\right) \tag{B3}$$

(B.2) became:

$$p\left(z_m/H_i\right)=\frac{1}{\left(1+\frac{m_{m,i}^2}{2v_{m,i}}\right)}\int_0^{+\infty}\frac{2\alpha'_{1,m}}{2v'_{m,i}}\exp\left(\frac{-\alpha'^2_{1,m}}{\left(1+\frac{m_{m,i}^2}{2v_{m,i}}\right)}\right)\exp\left(-\frac{z_m}{2v'_{m,i}}\left(1+\frac{m_{m,i}^2}{2v_{m,i}}\right)\right)I_0\left(\frac{\sqrt{z_m}\,m_{m,i}\alpha'_{1,m}}{v'_{m,i}}\right)d\alpha'_{1,m} \tag{B4}$$

We define tow new variable $\alpha''_{1,m}$ and $\alpha'''_{1,m}$ respectively,

$$\alpha''_{1,m}=\frac{\alpha'_{1,m}}{\sqrt{1+\frac{m_{m,i}^2}{2v_{m,i}}}} \text{ and } \alpha'''_{1,m}=\sqrt{2}\,\alpha''_{1,m}$$

(B.4) is given by:

$$p\left(\frac{z_m}{H_i}\right)=\frac{1}{2v'_{m,i}}\exp\left(\frac{-z_m}{2v'_{m,i}}\right)\int_0^{+\infty}\alpha'''_{1,m}\exp\left(-\frac{\alpha'''^{2}_{1,m}+a^2}{2}\right)I_0\left(a\alpha'''_{1,m}\right)d\alpha'''_{1,m} \quad \text{(B5)}$$

with $a=\dfrac{\sqrt{z_m}\,m_{m,i}^{2}}{\sqrt{2}\,v_{m,i}\sqrt{1+\dfrac{m_{m,i}^{2}}{2v_{m,i}}}}$

Marcum's function is defined as:

$$Q(a,b)=\int_b^{+\infty}x\exp\left(-\frac{x^2+a^2}{2}\right)I_0(ax)dx \quad \text{(B6)}$$

For the special case when $b=0$, we have [10]:

$$Q(a,0)=1$$

Finally, we obtain the equation (22).

APPENDIX C

For Rayleigh Fading channel and equal gain combining, the probability of detection is given by:

$$p_{Dfe}=\int_{\gamma}^{+\infty}\frac{z^{(M-1)}\exp\left(\frac{-z}{2v'_{m,1}}\right)}{Fact(M-1)\left(2v'_{m,1}\right)^{M}}dz=I_1 \quad \text{(C1)}$$

By integrating by part, we put

$$L=\left(2v'_{m,1}\right)^{M}Fact(M-1) \text{ and } n=2v'_{m,i} \quad \text{(C2)}$$

$$u=y^{(M-1)}\Rightarrow u'=(M-1)y^{(M-2)}$$

$$v'=\exp\left(\frac{-y}{n}\right)\Rightarrow v=-n\exp\left(\frac{-y}{n}\right)\quad I1=\int vu-\int u'v \quad \text{(C3)}$$

$$I1=\exp\left(-\gamma/_n\right)+\frac{\gamma^{(M-2)}\exp\left(-\gamma/_n\right)}{(2v'_{m,1})^{(M-2)}Fact(M-2)}+\ldots\ldots+\frac{\gamma^{(M-1)}\exp\left(-\gamma/_n\right)}{(2v'_{m,1})^{(M-1)}Fact(M-1)} \quad (C4)$$

$$p_{Dfe}=\sum_{m=0}^{M-1}\left(\frac{\gamma}{2v'_{m,1}}\right)^{m}\frac{\exp\left(-\gamma/_n\right)}{Fact(m)} \quad (C5)$$

The same calculus for the probability of false alarm is done.

REFERENCES

[1] Charles E. Cook, Fred W. Ellersick, Laurence B. Milstein, Donald L. Schilling, “Spread-spectrum communications,” *IEEE press*, Inc., New york, copyright 1983.

[2] Dongwook. Lee, Laurence B. Milstein, Hun. Lee,”Analysis of multicarrier DS-CDMA code –acquisition system,” *IEEE Trans. Commun.*, vol.47, no.8, pp.1233-1243, Feb.1999.

[3] S. Kondo and L. B. Milstein, “On the performance of multicarrier DS-CDMA systems,” *IEEE Trans. Commun.*, vol.44,pp.238 - 246, Feb.1996.

[4] Douglas N. Rowitch, Laurence B. Milstein, “ Convolutional coded multicarrier DS-CDMA systems in – a multipath channel- Part I, *IEEE Trans. Commun.*, vol. 47, no.10, pp.1570-1582, Oct.1999.

[5] R.L. Pickholtz, D.L. Schilling, and L.B. Milstein,”Theory of spread spectrum communications-Atutorial,” *IEEE. Trans. Commun.*, vol. COM-30, no.5, May 1982.

[6] L.B. Milstein, “Interference rejection techniques in spread spectrum communications,“ *Proc. IEEE*, vol.76,Jun 1988.

[7] J. Bingham,” Multicarrier modulation for data transmission: An idea whorse time has come,” *IEEE. Commun. Mag.,* May 1990.

[8] Tan F. Wong, Tat M. Lok, James S. Lehnert,”Asynchronous multiple-access interference suppression and chip waveform selection with aperiodic random sequences, *IEEE, Trans. Commun*, vol. 47,no.1,pp.103-114, Jan 1999.

[9] Marvin K. Simon,Tun K. Omura, Robert A. Sholtz, Barvy K. Bevitt,” *Spread spectrum communications,*” volume III, Rockville, Maryland: computer science Press 1985.

[10] Frederic Coulon,”Theorie and traitment of signals, Edition Georgi. Paris: Dunod, 1984.

[11] Lie-Liang Yang, Lajos Hanzo,’’Serial acquisition performance of single-carrier and multicarrier DS-CDMA over Nakagami-m Fading channels’’ *IEEE. Trans. Commun.*vol.1, no.4, pp. 692-702, Oct 2002.

In: Focus on Information, Intelligence and Knowledge
Editor: Rafiq Islam

ISBN 978-1-61209-979-8
2011 Nova Science Publishers, Inc.

Chapter 13

RECONFIGURABLE SINGLE ELECTRON TRANSISTOR BASED MULTIPLE VALUED LOGIC MULTIPLIER

K. Rathnakannan* and P. Vanaja Ranjan
Department of Electrical and Electronics Engineering,
Anna University, Chennai, India-600025

ABSTRACT

The Single Electron Transistor is a nano electronic device; it has inherent multiple valued logic and re-configurable properties. It's a symmetrical device, when provide a back gate it acts as an n-SET, and if the back gate terminal is grounded the same device acts as a p-SET. This setup greatly reduces the number of component required to implement the same functionality. In this paper a Multiple Valued Logic Multiplier using reconfigurable-SET has been proposed. It has been found that p-SET and n-SET can be effectively combined into a single SET having dynamically reconfigured as n-SET or p-SET using an external voltage being applied to the back gate. This approach does not require any variance in the method of fabrication for different units or on any other parameters of the device. A thorough analysis on Eight Valued logic reconfigurable SET based Multiplier circuit has been performed and its performance in terms of area, delay and power consumption has been estimated.

1. INTRODUCTION

The single electron technology is a technology for manipulating electronic functions by controlling the transport of individual electrons. A Single Electron Transistor (SET) is a nano electronic device it changes its states to decrease its free energy; hence by nature it has multiple valued properties [1]. Exploiting this concept in an electronic circuit would lead to an increase in the density of integration, an increase in processing speed and a reduction in the number of the interconnections. The Multiple-valued logic (MVL) can be classified into

* Email: ckrkannan@yahoo.co.in

two approaches based on its application. The first approach uses multiple-valued logic to solve binary problems very efficiently [2]. The second approach is used for designing electronic circuits which employs more than two discrete levels of signals, such as multiple-valued memories, arithmetic circuits, Field Programmable Gate Arrays etc. Multiple-valued logic (MVL) circuits offer several potential opportunities for the improvement of integrated circuit designs [3-4]. The inverter is one of the basic logic gates, and it can be used to construct a threshold device that is useful for building a multiple-valued logic system. Three techniques for multiplications have been reported in the literature; one is binary multiplication using SET [5], which is a cumbersome approach because of the complexity in constructing the circuit, the second one is Electron counting (EC) Scheme[6] based high radix multiplication, this method is fast but requires very high gain analog amplifier at the intermediate stage and very tedious to realize and implement for real time applications. The third approach is based on complementary SET technology [7], which consumes more power and switches with less speed because of the circuit nature. The proposed reconfigurable SET based MVL multiplier is very compact, requires fewer elements and thus switches with high speed and consume ultra low power .In this chapter an eight valued reconfigurable SET and its design methodology for multiplier has been discussed. A thorough analysis on DAC and multiplier circuit has been done using simulation [8] studies.

2. Single Electron Technology

The SET is a quantum electro mechanical switching device, which can be used for controlling and measuring the movement of single electron [9]. The effects of single electron play an important role in devices with very small dimensions, which facilitate to miniaturize electronic circuits and to function with ultra low power. The characteristic feature of the SET technology is that transport of charge occurs in discrete quantities, i.e., single electron at a time, and this offers the ability to design circuits in which the transport of individual electrons can be accurately controlled .It consists of small Metallic Island coupled with source and drain electrodes through two small tunnel junctions and capacitively coupled with gates as shown in Figure.1.

The charge passes through the island is in quantized units. For an electron to tunnel onto the island, its energy should be equal to the coulomb energy .This additional energy suppresses electron transfer, unless it is overcome by either thermal excitation or by an externally applied by gate supply. This suppression is known as the Coulomb Blockade. When both the bias and gate voltages are zero, the energy level of electron is very low and the electron tunneling into the island does not take place. When the bias voltage between the source and drain is increased in such a way that the energy in the system reaches the Coulomb energy then an electron enters into the island, the critical voltage required to transport an electron onto the island is called as coulomb gap voltage.

When the gate voltage is increased, the initial energy of the system with no electrons on the island, steadily increases, while the system with one excess electron on the island gradually decreases. At the gate voltage corresponding to the point of maximum slope on the Coulomb staircase, both of these configurations equally qualify as the lowest energy states of the system. This lifts the Coulomb blockade, allowing electrons to tunnel into and out of island in discrete manner one after one.

If the Coulomb blockade is lifted then the gate capacitor is charged with half electron (negative charge). This means that the charge on it must be quantized in units of e, the charge on the gate capacitor merely represents a displacement of electrons relative to a background of positive ions. A further steady increase in the gate voltage, changes the charge of the gate capacitor into –e, the island again has only one stable configuration separated from the next lowest energy states by the Coulomb energy. The Coulomb blockade is set up again, but the island now contains a single excess electron.The tunneling is a discrete process, this favour the electric charge to flow through the tunnel junction in multiples of electrons, the charge of a single electron as $e = 1.602*10^{-19}$C. The transport of charge through a tunnel junction is referred to as tunneling and the transport of a single electron through a tunnel junction is referred to as a tunnel event [10].

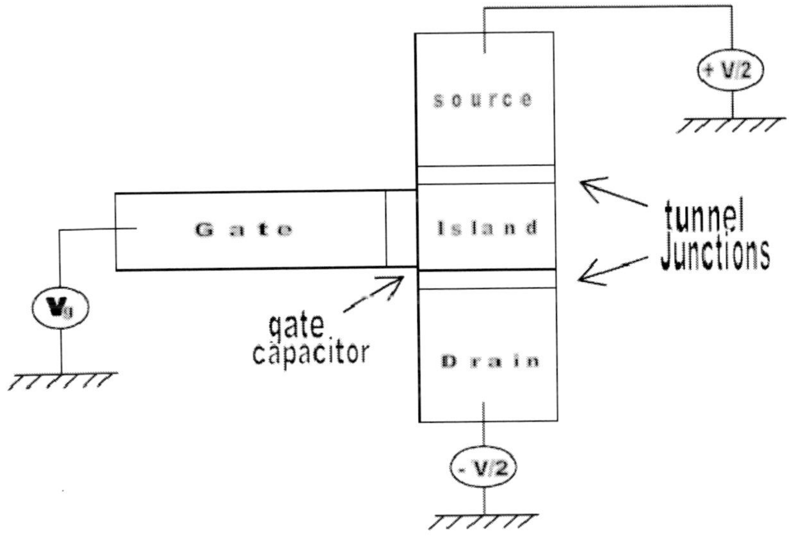

Figure 1. Single Electron Transistor.

The minimum voltage required for an electron to pass through the junction is called as critical voltage. The critical voltage V_{cr} across a tunnel junction is depends on the thickness of the tunneling junction. To determine the effective V_{cr} of a junction, the tunnel junction with a capacitance of C_{jn} and the equivalent capacitance C_{equ} of rest of the circuit have to be considered from the logic perspective. The calculation of critical voltage V_{cr} for the junction mathematically written as

$$V_{cr} = e/2(C_{equ} + C_{jn}) \quad (1)$$

If the magnitude of voltage across the junction V_{jn} is greater than the critical voltage then tunneling event will occur through the tunnel junction. The device is in stable state, if there is no occurrence of tunnel events in tunnel junctions. If it is initially in a stable state, then the V_{jn} is gradually increased or decreased to reach the point where its absolute value is larger than Vcr. This stable state determines a new output value from the distribution of charge in the circuit. If the input voltage V_{jn} is continuously increased, the electrons always tunnel away from the island, resulting in sudden increases of the island's charge.

3. Charging and Discharging Effect of SET

The inverter circuit is made into reconfigurable n/p-SET by removing the p-SET and connecting the drain of n-SET to Vdd as shown in figure 2.

A single SET can be act as a charging set i.e. p-SET when the voltage given to its back gate is "0" or it is grounded. In this condition the input voltage given to the Single Electron Transistor is stored as a charge. But only one disadvantage of this circuit is "It can't able to discharge the value already stored". To discharge the value stored in, it required the combination of n-SET as in the Tucker type inverter [11].

A single electron transistor can be act as a discharging set i.e. n-SET when the voltage given to back gate is "1" or back gate is connected to some voltage source. This type of connection is used to connect the circuit to ground. So the circuit can able to reset the available value to "0". To avoid the disadvantage of P-SET we can configure it as n-SET when reset is required.

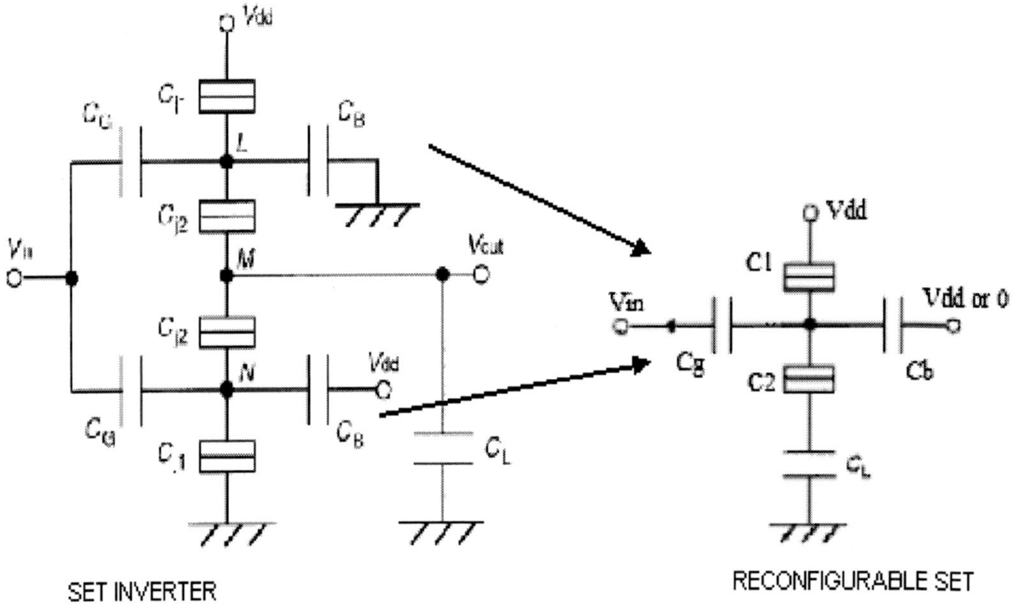

Figure 2. Reconfiguring the MVL Inverter.

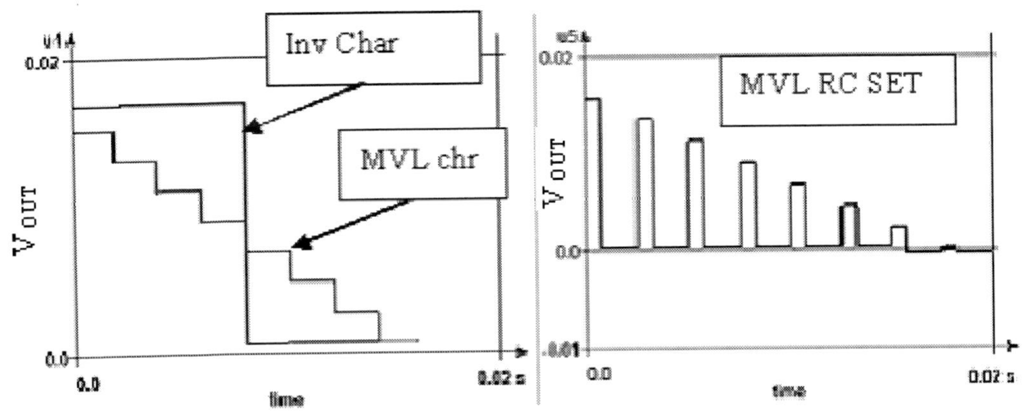

Figure 3. Simulation Result of n/p-SET MVL circuit.

As seen from the figure 3, the MVL inverter operation can be performed by a single reconfigurable n/p-SET. For proper functioning of the n/p-SET the charge accumulated must be drained properly.

4. Designing a Reconfigurable N/P-SET Circuit

In this section, the design methodology for an eight -valued reconfigurable SET inverter circuit has been discussed. A multiple-valued signal can be generated by adjusting the magnitude of the capacitances CG, CB, and CL, and fixing the values of Cj1 and Cj2. To obtain an n-Valued logic device, the value of CL should be designed to store n-1 holes to make the maximum output voltage.

An eight valued device can be designed as follows.

$$Q = C_L V \quad (2)$$

where

Q - No of holes

C_L -Load capacitance

V -Maximum Output voltage or Bias Voltage

By using the Equation (2) the Load capacitance is calculated. For eight states, the load capacitor should be capable to store 7 holes. Because '0' is also consider as one state. Then each increment of holes represents consecutive states. Here the maximum output voltage (or bias voltage) is taken as 16mv. Then the load capacitance value is designed as 70.875af by using the equation 1.Then the values of CG and CB should be adjusted so that the proper multiple-value could be generated. A useful tool for parameter adjustment is a stability diagram that shows Coulomb blockade condition on the bias voltage plane. This Plot clearly explains Coulomb blockade bias condition for several internal states. By plotting the stability diagram the values of the parameters *CL, CG* and *CB* can be finalized. For example, the stability diagram of an eight Valued CSET inverter operating at 16 mV is shown in figure 4.

This plot is drawn between Vin and Vdd (Vb1). The white region in the plot represents the stable region where the Coulomb blockade occurs. The black region in the plot represents an unstable region where tunneling frequently occurs.

The unstable region is started at 20 mv and ended with 16mv.The ending point of unstable region from up to down determine the operating voltage. Here, we have achieved the operating voltage as we have taken as 16mv. Near to this, the stable and unstable regions are overlapped, where we can get eight discrete triangle shaped white regions. The corresponding voltage available in Vin with respect to upper tip of the triangle is taken as one logic state. Like that we can get eight logic states. This logic states are more stable in between two black triangles. This Plot clearly explains Coulomb blockade bias condition for several internal states.

The internal state of the circuit is given by (l, m, n), where l, m, and n are the number of electrons that can exist at islands L, M , and N respectively, marked in figure.2. inverter section. In (l , m , n) 'm' mainly determines the output voltage, while l and n slightly modify the output voltage.

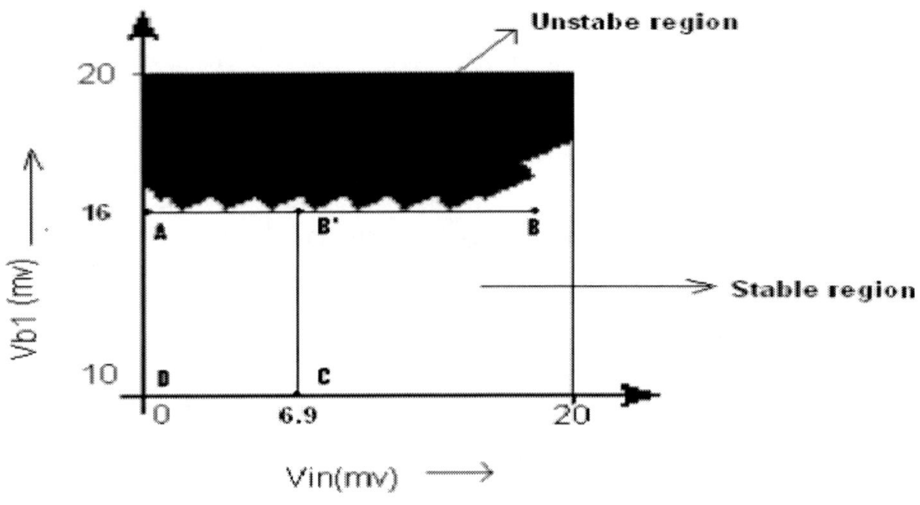

Figure 4. Stability diagram.

Since the 'm' node has major role in determining the output value, reconfigurable SET has been considered for computation, if we increase the 'm' value the output voltage decreases. The eight internal states available from left to right having number of electrons as (−7), (−6), (−5), (−4), (−3), (−2), (−1) and (0) respectively and are used to output multiple values "7,""6," "5," "4," "3," "2," "1," and "0," respectively for the corresponding input value (Vin), the response of the MVL inverter for all these input values is shown in figure 3.

The step by step approach for designing the reconfigurable n/p SET circuit is discussed below.

Step1:

Designing an n state MVL Inverter involves fixing the number of states. This is done by the important parameter CL in the circuit as they host the number of electrons which is directly proportional to the states.

Step2:

Design the n state MVL Inverter using the Stability Diagram. By plotting the stability diagram, fix the values of the parameters *CG* and *CB*, keeping the junction capacitances Cj1, Cj2 and *CL* constant. Once you finalize the CG and CB parameters, the CB remains untouched for the rest of the steps.

Step3:

CG is the input capacitance determine by the previous step to be threshold depending on the no. of inputs to the circuit. This can be done on the principle of Charge distribution on the Island. To design an n- input threshold unit,

$$\sum C_{Gi} V_{Gi} = C_G . V_G = \text{Charge at the Island for each state} \quad (3)$$

where,

C_G-Input capacitance initial
V_G- Gate voltage applied
C_{Gi}-Input capacitance after threshold
V_{Gi}- Fixed logic input after threshold
i- Number of inputs

In the above equation (3), the 1 and the 0 represent the voltage logic high and logic low which is 16mV and 0mV respectively. By solving these simultaneous equations you can arrive at the capacitance.

Step4:

After designing the MVL inverter just remove the p-set and connect the source terminal to Vdd. Now if we require n-SET operation the back gate voltage is applied. When n-SET operation is required the back gate voltage is made zero which acts as the virtual ground.

Thus a reconfigurable n/p-SET logic is designed by making use of Multigate and reusability properties of the SET device.

Once the Multiple-Valued inverter is designed, its application to digital to analog converter can be obtained by designing Multi-threshold circuitry. The threshold device receives more than one input signal($a_1..a_i..a_n$), weighs each input, sums up all the weighted-inputs ($w_1..w_i..w_n$), compares the sum with the threshold values *ti*, and then gives a multiple-valued signal Vout as output(f). This concept has been depicted in figure 5.

The value of *Vout* depends on the value of the sum of weighted input. For example, a quaternary threshold device changes its output as follows. Multigate can be easily realizable in SET Technology by connecting more than one capacitors to the Island. The design of multigate threshold system is very simple once the *CG* is determined. Each gate is distributed equivalent of binary weight represented by equation $\sum C_{Gi} V_{Gi}$ is equivalent to the state or threshold. Since VGi is fixed for logic high and logic low, the Capacitor design remains easy and illustrated in the next section.

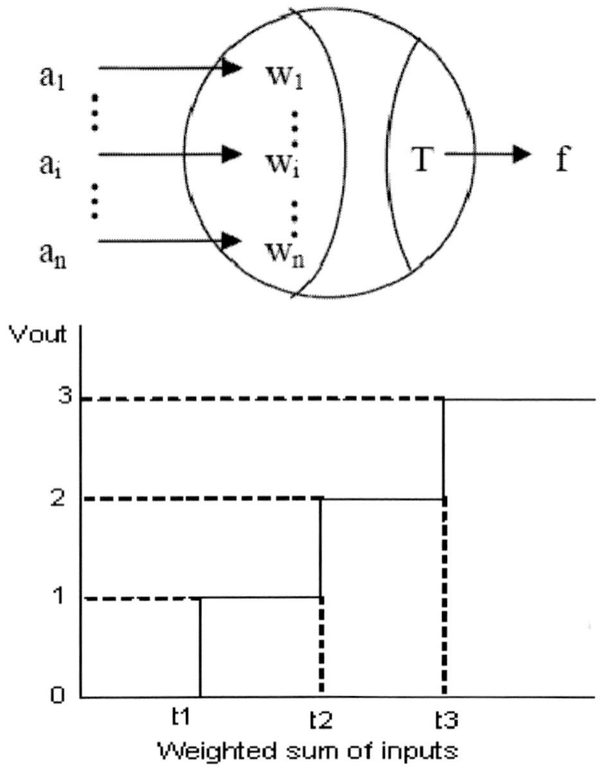

Figure 5. Schematic diagram of MVL Multi-threshold Unit.

5. Reconfigurable N/P-SET Based Digital to Analog Converter

The Digital to Analog Converter can be modeled and designed using reconfigurable n/p-SET as follows. In this chapter, a 3 bit DAC is taken into consideration. The DAC circuit has been designed using the same steps that are mentioned in the above section.

The 3-bit DAC has eight levels for all the three input combinations or min terms of three variables. Therefore the first step in designing a 3-bit DAC is designing an eight state MVL Inverter using step1 and step2 of above mentioned process. Then the process of thresholding it into 3 input signals has to be done. This process is illustrated by the equations (4-11)

$C_{G0}.\,(0) + C_{G1}.\,(0) + C_{G2}.\,(0) =$ charge at island at state 0 (4)

$C_{G0}.\,(0) + C_{G1}.\,(0) + C_{G2}.\,(1) =$ charge at island at state 1 (5)

$C_{G0}.\,(0) + C_{G1}.\,(1) + C_{G2}.\,(0) =$ charge at island at state 2 (6)

$C_{G0}.\,(0) + C_{G1}.\,(1) + C_{G2}.\,(1) =$ charge at island at state 3 (7)

$C_{G0}.\,(1) + C_{G1}.\,(0) + C_{G2}.\,(0)$ = charge at island at state 4 (8)

$C_{G0}.\,(1) + C_{G1}.\,(0) + C_{G2}.\,(1)$ = charge at island at state 5 (9)

$C_{G0}.\,(1) + C_{G1}.\,(1) + C_{G2}.\,(0)$ = charge at island at state 6 (10)

$C_{G0}.\,(1) + C_{G1}.\,(1) + C_{G2}.\,(1)$ = charge at island at state 7. (11)

Wherein charge at the island is shown in the table 1.

Table 1. State Island charge of eight-state MVL inverter

State	C_G	V_G	$Q=C_G.V_G$	Output
i7	2	16	32	i0
i6	2	13.4	26.8	i1
i5	2	11.2	22.4	i2
i4	2	9.1	18.2	i3
i3	2	6.9	13.8	i4
i2	2	4.7	9.4	i5
i1	2	2.6	5.2	i6
i0	2	0.8	1.6	i7

The value of C_1–3.250e-19, C_2–5.875e-19, C3–1.137e-18 are found by solving the above mentioned simultaneous equations. After summing C1, C2 and C3, we get only 2.045e-18 which is equal to the original gate capacitance. The circuit diagram of the reconfigurable n/p-SET based DAC is shown in figure 6.

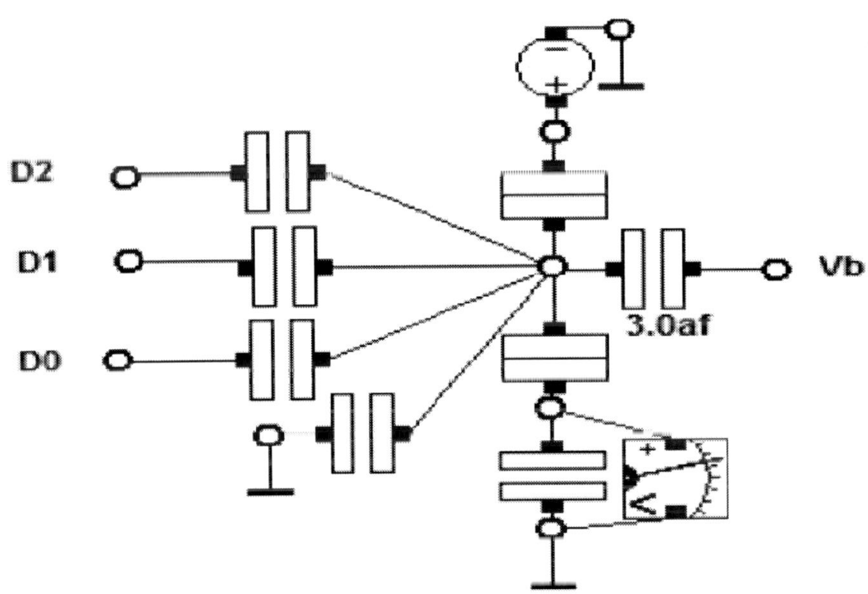

Figure 6. Circuit Diagram of 3-Bit DAC.

The 3 bit Multi-valued logic DAC can be designed just by understanding the relation between states and inputs. If there are n bits then MVL inverter must have 2^n states. For 3 bit DAC there need to be $2^3 = 8$ states. The eight state MVL inverter is designed from the C_L=7aF, C_G=2aF, C_B=3aF, C_{j1}=0.5aF, C_{j2}=2aF. Now in order to design the threshold input capacitance, the Formula $\sum C_{Gi}V_{Gi}$ = state charge, is being used. Therefore the values of the C0=0.325aF (LSB), C1=0.5875aF and C3= 1.135aF (MSB). The simulation result of the 3 bit DAC is shown in the figure 7.

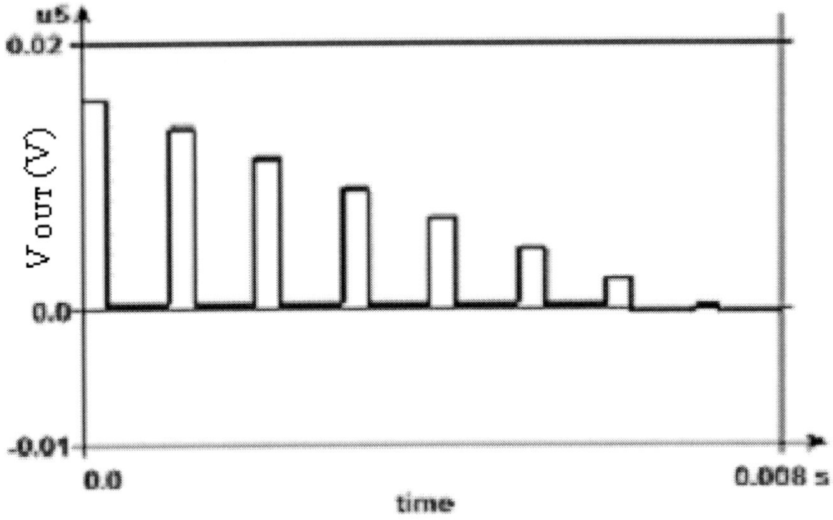

Figure 7. Simulation result of 3-Bit DAC.

The table 2 gives the exact quantum of the DAC in numerical values. The step size of the DAC is perfectly linear and also highly accurate.

Table 2. Results from Simulation of n/p-SET based 3-bit DAC

Binary	Decimal Equivalent	Analog value (V)	Step Size (V)
000	0	1.50E-04	-
001	1	2.37E-03	2.22E-03
010	2	4.59E-03	2.22E-03
011	3	6.81E-03	2.22E-03
100	4	9.04E-03	2.23E-03
101	5	1.13E-02	2.22E-03
110	6	1.35E-02	2.22E-03
111	7	1.57E-02	2.22E-03

6. Reconfigurable N/P-SET Based Single Electron Multiplier

The concept of reconfigurable n/p-SET multiplier has been also evolved from electron counting scheme, the DAC has been implicitly used to obtain the multiplier function. This is pretty much same as that of the MVL inverter based technique. This multiplier works on the invert logic i.e., highest logic state is 0 and lowest logic state is 16 mV and intermediates have their respective opposite state. The circuit diagram of the multiplier unit is shown in figure 8.

In this circuit, the first operand is given to the MVL inverter 1 which is designed for 4 state 16mV. The second operand is given to MVL inverter 2 through a switch which determines the final input. The MVL inverter 2 has 7state 16mV.

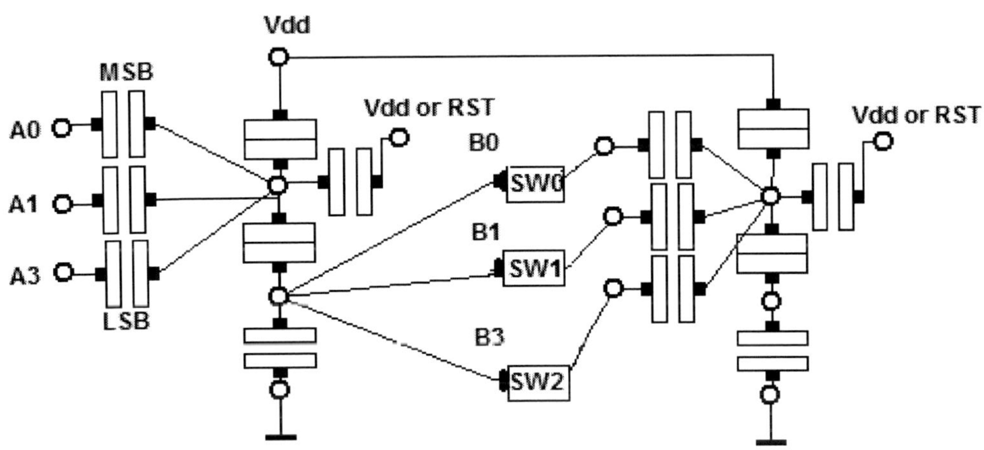

Figure 8. Circuit Diagram of 4X4 Multiplier.

The state of MVL inverter is chosen as 7 because the multiplier as diagonal elements that are unique while the non- diagonal elements both sides of the diagonal repeat itself which is been indicated in the table 3.

Table 3. Configuration of the n/p-SET Multiplier

X operand1->/operand2	0	1	2	3
0	i6 0	i6 0	i6 0	i6 0
1	i6 0	i5 1	i4 2	i3 3
2	i6 0	i4 2	i2 4	i1 6
3	i6 0	i3 3	i1 6	i0 9

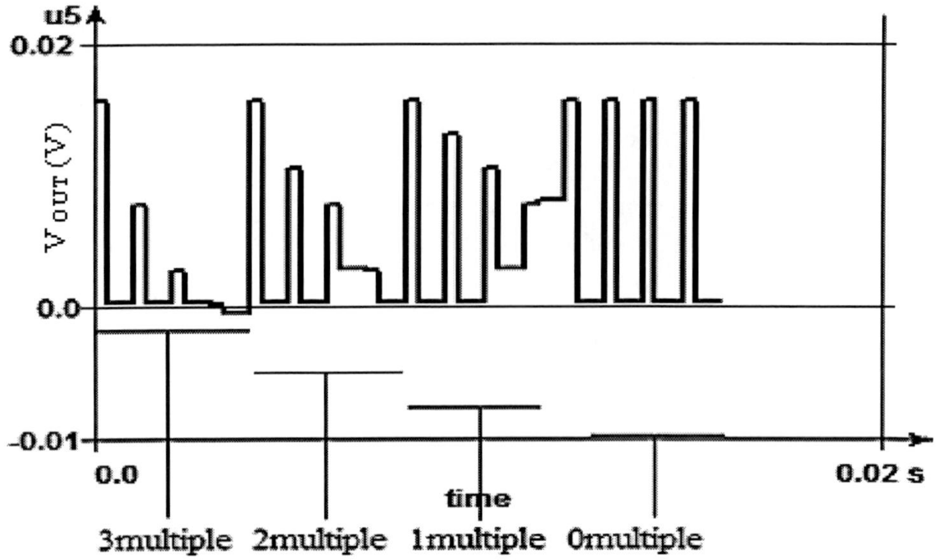

Figure 9. Simulation result of 2X2 Multiplier.

The simulation results of reconfigurable n/p SET multiflier is shown in figure.9 ,From the diagram we can conclude that the states of the multiplier are remaining the expected result. Since the multiplier works in the invert logic, the 3 multiples 0, 3, 6, 9 are represented by the states i6, i3, i1, i0. This can be noted from the table 3. Similarly the 2 multiples 0, 2, 4, 6 are represented by i6, i4, i2, i1. It can be noted from the diagram that there occurs an error in the 2x2=4, must produce a i2 but produces the i2. This is because of the capacitance shedding effect caused by neutralization three dissimilar simultaneous equations. But this error can be contented by intelligent sensing of input and providing it to the back gate. . The above result can be fed into a Periodic Symmetric Function (PSF) analog to digital converter [12-13] as in the Electron Counting method to obtain the Digital equivalent of the Analog voltage produced by the MVL Inverters. In this case, the electron trap unit of the PSF Analog to converter must be re-designed for the step size of the MVL inverter output.

7. Performance Analysis

The stability and accuracy in physical quantities are difficult to obtain in real circuits. They are affected by manufacturing tolerances, temperature, power supply voltage, cosmic rays and noise created by other circuits, among other thing.

Also, many mathematical and logical operations can be difficult or impossible to perform with analog quantities. Some of the important parameters are discussed below and the performance measures summarized in table 4.

Table 4. Performance Analysis

Various parameters	Reconfigurable MVL Multiplier
Area(no of elements)	SET- 2
Power	7.4 pW
Delay	2.5 ps
PDP	1.85E-23 J
Noise Margin	Low level = 0-0.00095v High level = 0.00095- 6.5mv
Sensitivity	0.0093 V
Charging Energy	6.417E-21J
Fan-in (aF)	2
Fanout (aF)	8

7.2. Propagation Delay

The minimal delay t_d is calculated using the resistance and capacitor of the junction terminals. The value of the junction capacitor is 100 Kohm and the capacitor value is 0.5 aF.

7.3. Fan-in

The amount of dynamic feedback that affects a buffered threshold gate in a network depends on two factors: fan-out and fan-in. Generally, fan-in of a gate is the number of inputs connected to the gate without any degradation in the voltage levels. Fan-in is measured in terms of the sum of a gate's input capacitors.

ΔV = fan-in $*V_1/10C$ (12)

The maximum deviation of the buffer's output voltage is 0.2 * V_1 and approximately 2*C*.

7.4. Fan-out

Generally, fan-out is the maximum number of similar logic gates that a gate can drive without any degradation in voltage levels. This is function of the output impedance of the driving gate and the input impedance of the driven gates. If their output impedance is low while their input impedance is high, then one gate can often drive many others.

Fan-out is measured as the total capacitive load attached to an output.

ΔV = fan-out $*V_1/10C$ (13)

The maximum deviation of the buffer's output voltage is 0.2 * V_1 and approximately 2*C*.

7.5. Speed-Power Product

The speed-power product refers to the product of the propagation delay and power consumption of a typical gate. The result is measured in joules (J). The joule is a unit of energy so the speed power product measures a sort of efficiency. It's obvious that lower the energy usage better the efficiency.

7.6. POWER CONSUMPTION

The maximum power consumption for the multiplier circuit can be estimated by the power consumption P_{max} by 7 electrons (8 valued MVL inverter) being charged and discharged in the load capacitance C_l by

$$P_{max} = ((7e^2 / C_L) * f \qquad (14)$$

where

e = charge of an electron

f = clock frequency, C_L -Load capacitor.

This equation gives a maximum power consumption of 2.7fw. The clock operation is 120ns for ON-OFF periods. This value is much smaller than typical values for CMOS MVL gates (Which usually ranges between μW and mW) and Electron counting scheme based multiplier circuits, though the power consumption of Electron Counting Scheme is relatively less than the MVL multiplier it requires an amplifier to magnify the output of multiplier by 100 times, which is practically impossible because amplifier part may consume a large amount of power. In MVL circuits the input is inversely proportional to the gain, since the operating voltage of reconfigurable SET MVL circuit is in the range of a few millivolts, the gain problem may not arise.

CONCLUSION

This paper mainly illustrates the re-configurability and multiple valued logic property of the single electron transistor. The Multiple Valued Logic operation is obtained by dynamically reconfiguring the SET as n-SET / p-SET. With this concept a DAC unit is built and a Multiplier is also designed. The simulation of these units also has been presented with neat description. The performance analysis of the reconfigurable SET based multiple valued logic circuit has been done and found its suitability for real time and higher end computations.

REFERENCES

[1] Masamichi Akazawa, , Kentarou Kanaami, Takashi Yamada, Yoshihito Amemiya, Multiple-Valued Inverter Using a Single-Electron-Tunneling Circuit, *IEICE trans. electronics,* Vol. 9, pg. e82–c, 1999.

[2] Inokawa, A. Fujiwara, and Y. Takahashi, A Multiple-valued logic and memory with combined single-electron and metal–oxide–semiconductor transistors, *IEEE Trans. Electron Devices*. Vol.50,No. 2, pg. 462–470, 2003.

[3] K. Degawa, T. Aoki, T. Higuchi, H. Inokawa, and Y. Takahashi, A single-electron-transistor logic gate family and its application — part I: Basic components for binary, multiple-valued and mixed-mode logic, in P*roc. to 34th IEEE Int. Symp.on Multiple-Valued Logic,* pg.262–268 2004.

[4] P. Glenn Gulak, A Review of Multiple-Valued Memory Technology, in. *Proc. to 28th IEEE international symposium on Multiple valued Logic*, 1998.

[5] C. R. Lageweg, S. D. Cotofana, S. Vassiliadis, Binary Multiplication Based on Single Electron Tunneling, in *Proc. to 15th International Conference on Application-specific Systems, Architectures and Processors*, (Galveston, USA) pg.152-166, 2004.

[6] C.H. Meenderinck, S. D. Cotofana, Electron Counting based High-Radix Multiplication in Single Electron Tunneling Technology, in *Proc.to IEEE International Symposium on Circuits and Systems* p.g . 4571-4574, 2006.

[7] Binary Coded Decimal Arithmetic Computation using Single Electron Transistor, Rathnakannan.K and Vanaja Ranjan.P, *International Journal of Nanotechnology and Applications*, Vol.1 ,No. 2 , pg. 61-72, 2007.

[8] C. Wasshuber, H. Kosina, and S. Selberherr, SIMON—A simulator for single-electron tunnel devices and circuits, *IEEE Trans. Com. Aided Des*, Vol.16,No. 9, pg. 937–944,1997.

[9] A.N.Korotkov, Single-Electron Logic and Memory Devices, *Int. J. Electr*, Vol. 86, No. 5, pg.511-547 ,1999.

[10] Konstantin K. Likharev, Single-Electron Devices and Their Applications, *IEEE Trans.*, Vol. 87, pg. 606-632, 1999.

[11] J.R. Tuker, Complementary Digital Logic based on the Coulomb Blockade, *J. App. Phy*, Vol.72, No. 9, pg. 4399-4413, 1992.

[12] C.H. Meenderinck, S. D. Cotofana, Computing Periodic Symmetric Functions in Single Electron Tunneling Technology, *Proceedings of International Semiconductor Conference* pg. 47-50, 2005.

[13] C. Hu, S. D. Cotofana, J. Jianfei, Single-Electron Tunneling Transistor Implementation of Periodic Symmetric Functions, *IEEE Tran. on Cir. and Sys*. II, Vol. 51, No. 11, pg. 593 – 597, 2004.

In: Focus on Information, Intelligence and Knowledge
Editor: Rafiq Islam

ISBN 978-1-61209-979-8
2011 Nova Science Publishers, Inc.

Chapter 14

STUDY OF TEMPERATURE INCREASE IN HUMAN HEAD DUE TO EM EXPOSURE

R. Seetharaman* and N. Kumaravel
Department of Electronics and Communication Engineering
College of Engineering, Anna University, Chennai -600 025 India

ABSTRACT

In this paper the thermal analysis has been performed considering the human head exposed to cellular phones and focusing the attention on the temperature increase and its distribution in the skin, gray matter, ear and skull. The temperature increase and its distribution are calculated at different power levels 0 mW, 600 mW and 1000 mW. It is well known that some significant thermal damage can occur in sensitive tissue under conditions of partial-body exposure to intense EM waves.

Keywords: Biological effects of electromagnetic radiation, Cell phone radiation, Temperature rise, Electromagnetic heating.

I. INTRODUCTION

Microwaves are used in scientific and industrial applications. But there are some adverse effects of high power microwave radiations. There is also the possibility that weak electric and magnetic fields from high power transmission lines may affect biological cellular processes at the cell nucleus. In fact, when a cellular phone is working, the transmitting antenna is placed very close to the user's head where a substantial part of the radiated power is absorbed.

Since biological substances such as blood, brain, bone, muscle and fat behave as conductive dielectrics, the microwave energy directed onto the body may be scattered, reflected and absorbed depending on the field strength, the frequency, the dimension of the

* E-mail : rseetharamanece@annauniv.edu

body and the electrical properties of the tissues. In fact, the tissue heating is strongly influenced not only by the power dissipated in the local tissue volume, but also by the way in which absorption is distributed in the surrounding area. A thermal analysis seems to be necessary, to assess the safety of an exposure. From the study of bio electromagnetic interaction, it is interesting to compare the obtained temperature increases.

The absorbed microwave energy produces molecular vibration and converts this energy into heat. If the organism cannot dissipate this heat energy as fast as heat is produced and hence the internal temperature of the body will rise. In this work, we have evaluated the temperature increases induced in human skin, gray matter, ear and skull when human head is exposed to the field emitted by a source operating at different power levels.

II. Calculation Methods

The thermal response as a function of time, until the steady state is reached, has been calculated. The head model was created by taking the superimposition of different tissues like skin, skull, bone, gray matter, white matter, fat, humor, cerebro spinal fluid (CSF). All the above tissues have different mass densities and dielectric properties at different frequencies and also they have different thermal properties. The head model had 500X341 pixels .According to range of pixel values; the different tissues of head are created. Sixteen different types of tissues and organs have been evidenced, and particular attention has been devoted to the modeling of the ear, skin, and skull, simulating its morphology during a phone call.

In some simulations, the hand grasping the phone has been considered. The hand tissue has been assigned electric properties equal to two-thirds of those used for the muscle. The hand is wrapped around the case on three sides, leaving free the side facing the head. In this paper particularly we concentrate on the temperature rise in human skin and gray matter during the time of excitation at different power levels.

If the lens of the eye is exposed to microwaves, its circulatory system would be unable to provide sufficient flow of blood for cooling and may cause cataract. Similarly, the stomach, intestines and bladder are especially sensitive to thermal damage from high power microwaves. Although the biological damage occurs mostly due to electric field coupling, low frequency magnetic field coupling can also produce damage when exposure time is large.

Since the conductivities of the muscle and blood are higher, microwaves dissipate more in these substances resulting in greater rate in heating. Sixteen different types of tissues and organs have been evidenced, and particular attention has been devoted to the skin, gray matter, ear and skull. Body tissues are typically classified according to their water content. High water content tissues such as muscle and skin can absorb more microwave energy than low water content tissues such as fat and bone or skull.

The thermal parameters used in the above equation have been extrapolated from existing literature [3]–[5] and are reported in Table1. The most interesting feature to be noted in Table1 is that the normal blood perfusion in the brain tissues is about ten times higher than in the other head tissues.This large blood perfusion compensates for the heat production caused by the high basal metabolic rate, maintaining a physiological temperature of about 37.5 C in the inner brain region.

Table 1. Thermal properties of the head tissues

Tissue	C [J/(kg.^{0}C)]	K [J/(S.m.^{0}C)]	A_0 [J/(S.m³)]	B [J/(S.m³.^{0}C)]
Skin	3500	0.42	1000	9100
Muscle	3600	0.50	690	2700
Bone	1300	0.40	0	1000
Blood	3900	0	0	0
Fat	2500	0.25	180	520
Cartilage	3400	0.45	1000	9100
Grey Matter	3700	0.57	10000	35000
White Matter	3600	0.50	10000	35000
Humor	4000	0.60	0	0
Lens	3000	0.40	0	0
Sclera/ Cornea	4200	0.58	0	0
Cerebellum	3700	0.57	10000	35000
Hypophysis	3700	0.57	10000	36000
CSF	4000	0.6	0	0
Parotid	3700	0.57	7000	25000
Tongue	3300	0.42	3700	13000

The temperature increase inside the head model has been obtained making use of the bio-heat equation

$$\nabla.(K\nabla T) + A_0 + Q_v - B(T-T_b) = C\rho\partial T/\partial t \qquad (1)$$

where K is the thermal conductivity, A_0 *and* Q_v are volumetric heat sources due to metabolic processes and microwave power deposition, respectively, B is a parameter proportional to the blood perfusion, T_b-T is the difference between blood and tissue temperature, and, finally, C and ρ are the tissue specific heat and density, respectively.

The boundary condition for (1) is given by

$$H(T_s - T_e) = -K\partial T/\partial n \qquad (2)$$

where H, T_s and T_e denote, respectively, the convection coefficient, the surface temperature and the fluid temperature.

III. Results and Conclusion

The temperature increase has been analysed in the human skin, gray matter, ear and skull by taking blood temperature as 36.8^{0}C Figure 1, Figure 2, Figure 3 and Figure 4 shows the temperature increase in the skin, gray matter, ear and skull respectively. It is observed that the steady state temperature at different power levels of 0 mW, 600 mW and 1000 mW for the above mentioned tissues is increased due to mobile phone radiation. By increasing the power from microwave sources the temperature rise in different head tissues is observed. When compared to all other tissues in the human head, the temperature rise in the ear (pinna) is more since it is nearer to the microwave source.

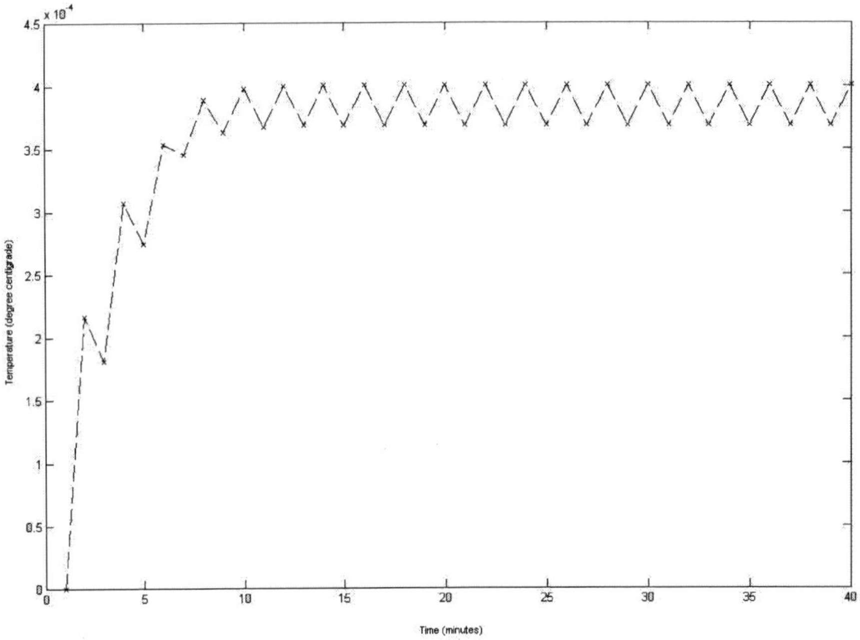

Figure 1(a). Temperature increase in the skin at 0 mW.

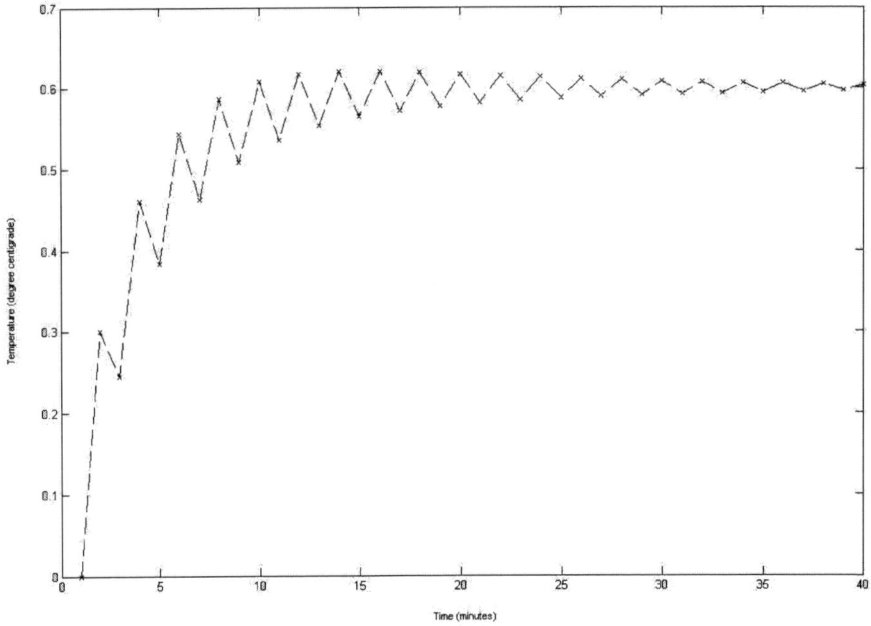

Figure 1(b). Temperature increase in the skin at 600 mW.

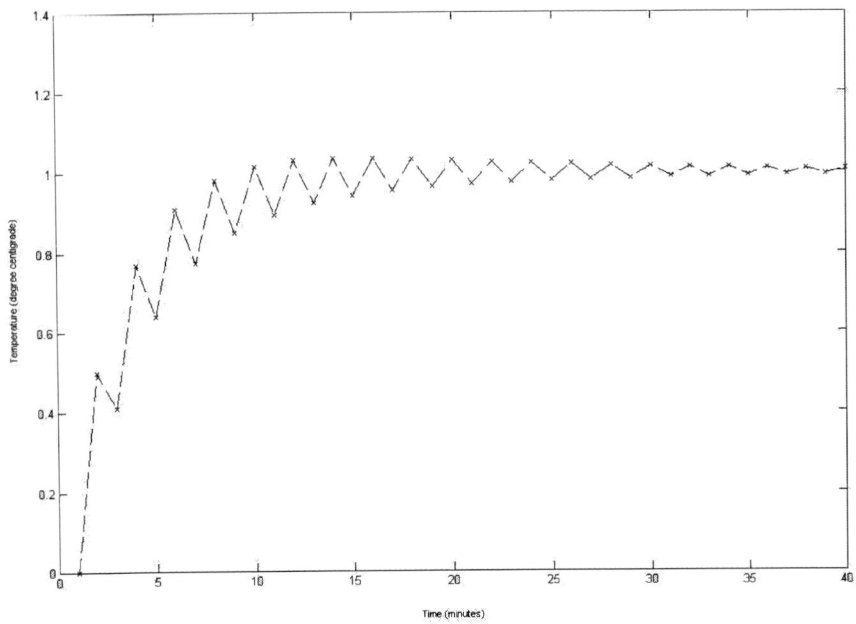

Figure 1(c). Temperature increase in the skin at 1000 mW.

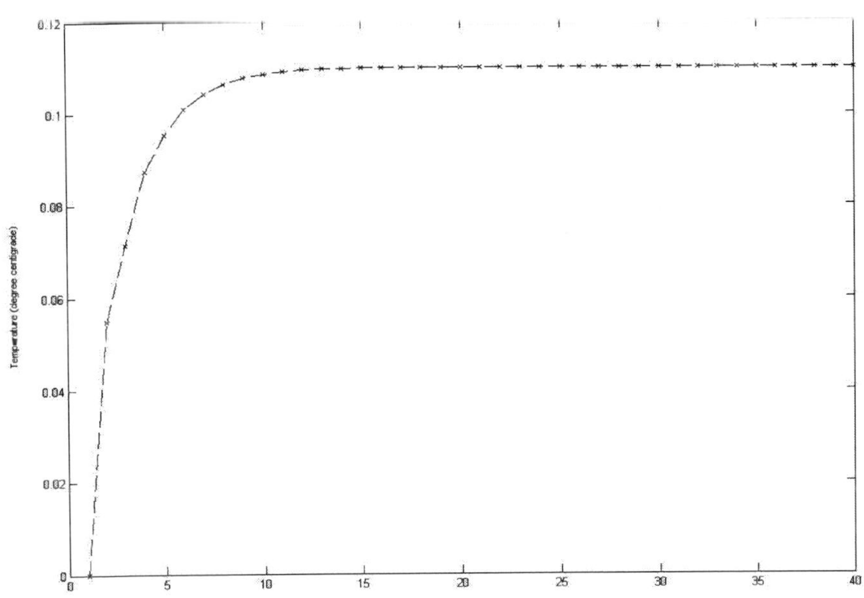

Figure 2(a). Temperature increase in the gray matter at 0 mW.

Figure 2. Cntinued on next page.

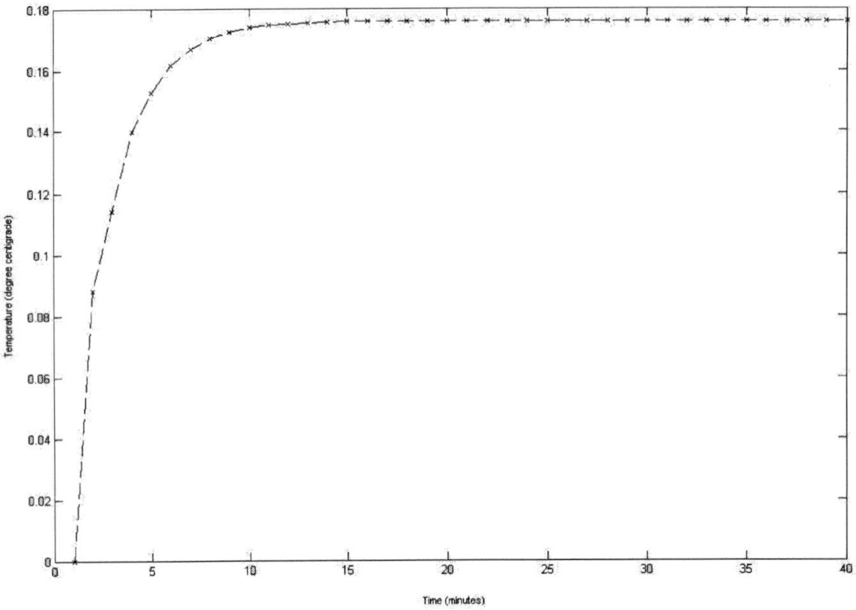

Figure 2(b). Temperature increase in the gray matter at 600 mW.

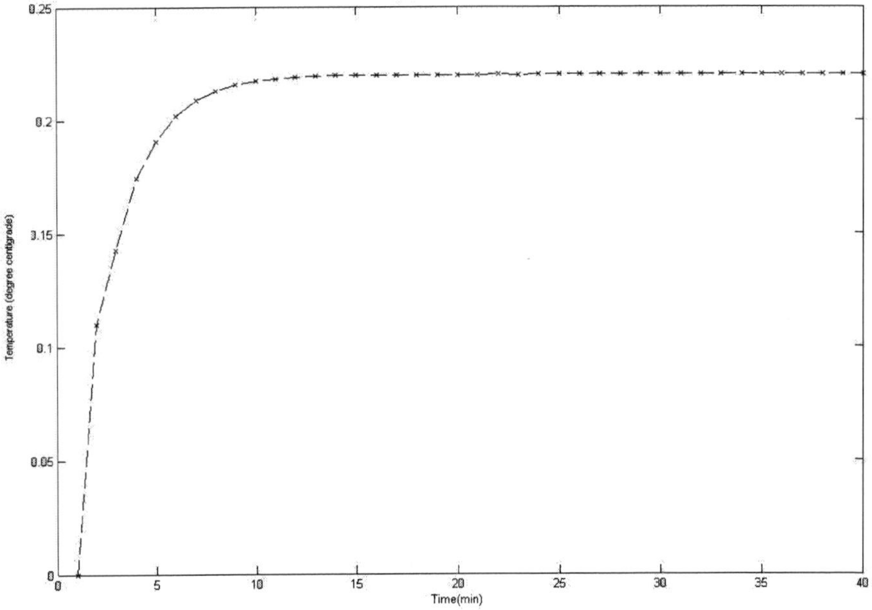

Figure 2(c). Temperature increase in the gray matter at 1000 mW.

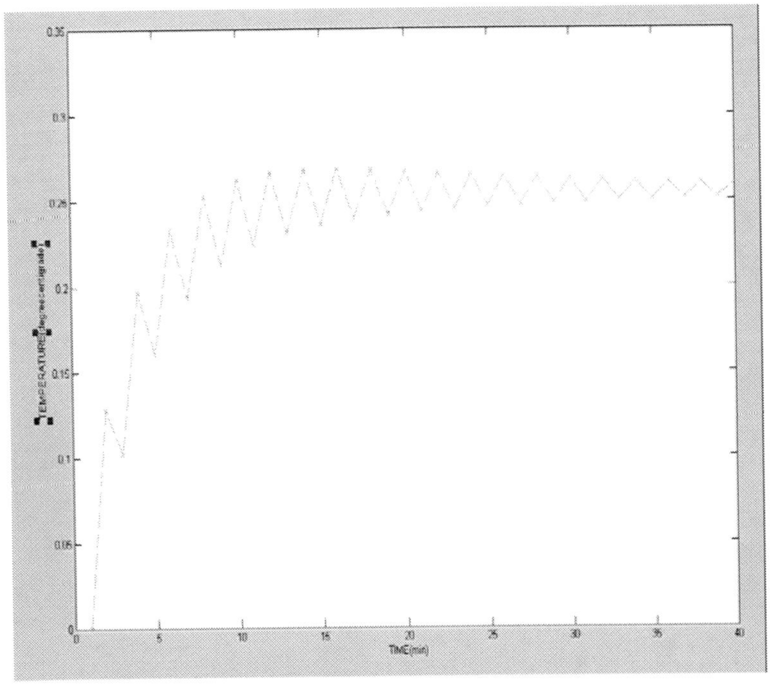

Figure 3(a). Temperature increase in the pinna (ear) at 0 mW.

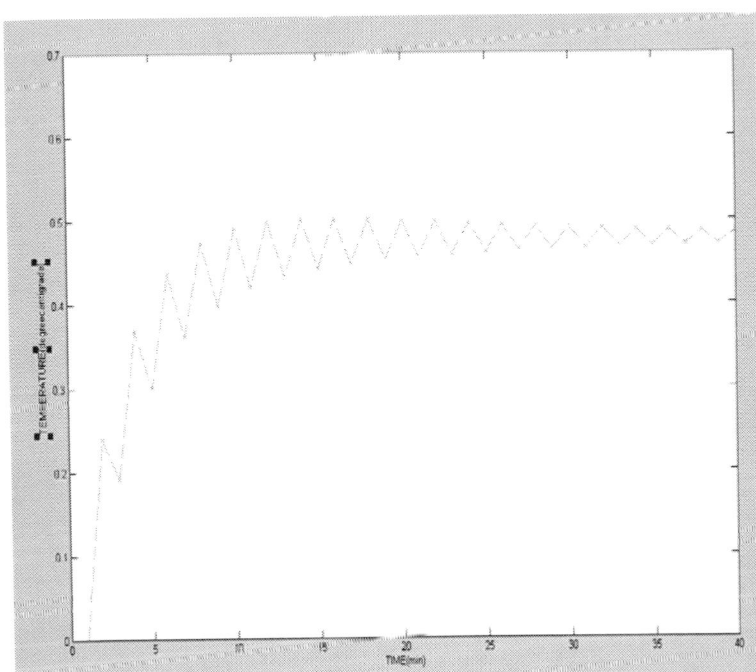

Figure 3(b). Temperature increase in the pinna (ear) at 600 mW.

Figure 3. Continued on next page.

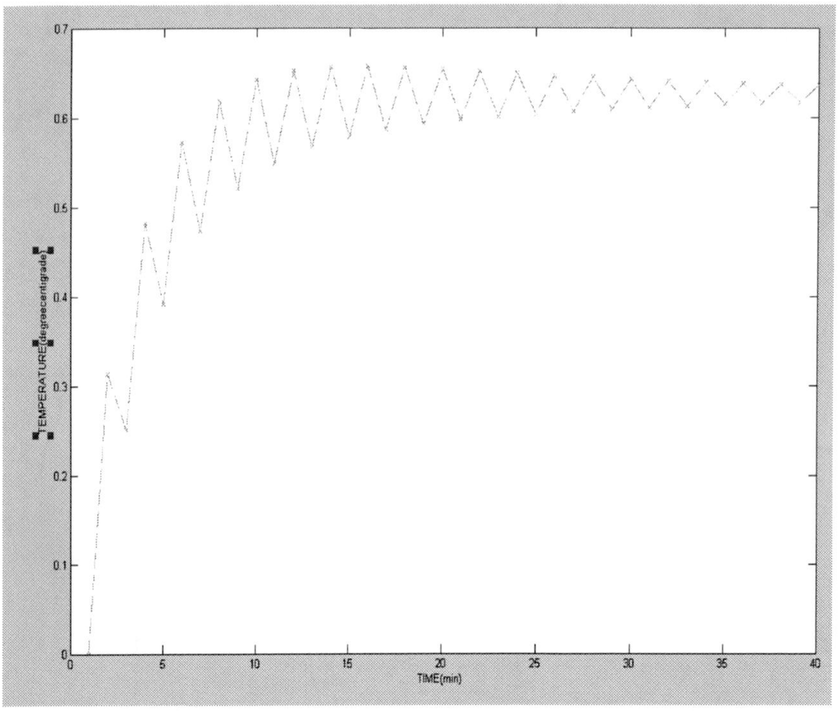

Figure 3(c). Temperature increase in the pinna (ear) at 1000 mW.

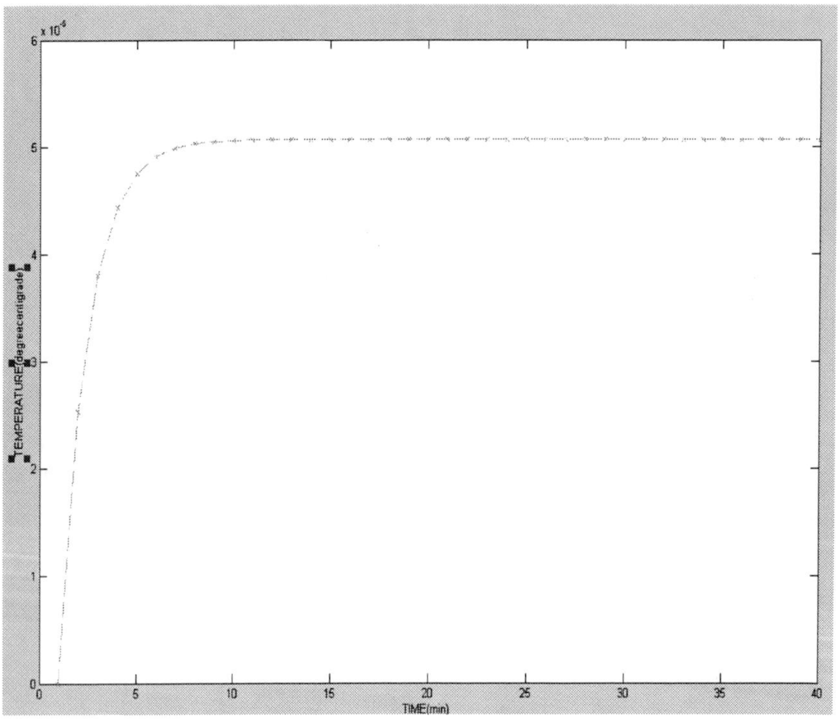

Figure 4(a). Temperature increase in the skull at 0 mW.

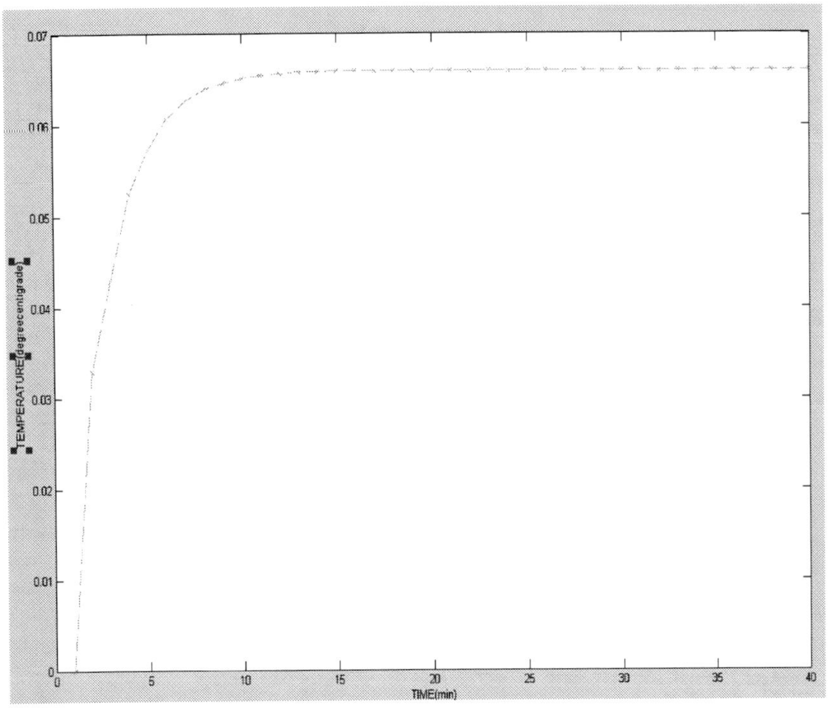

Figure 4(b). Temperature increase in the skull at 600 mW.

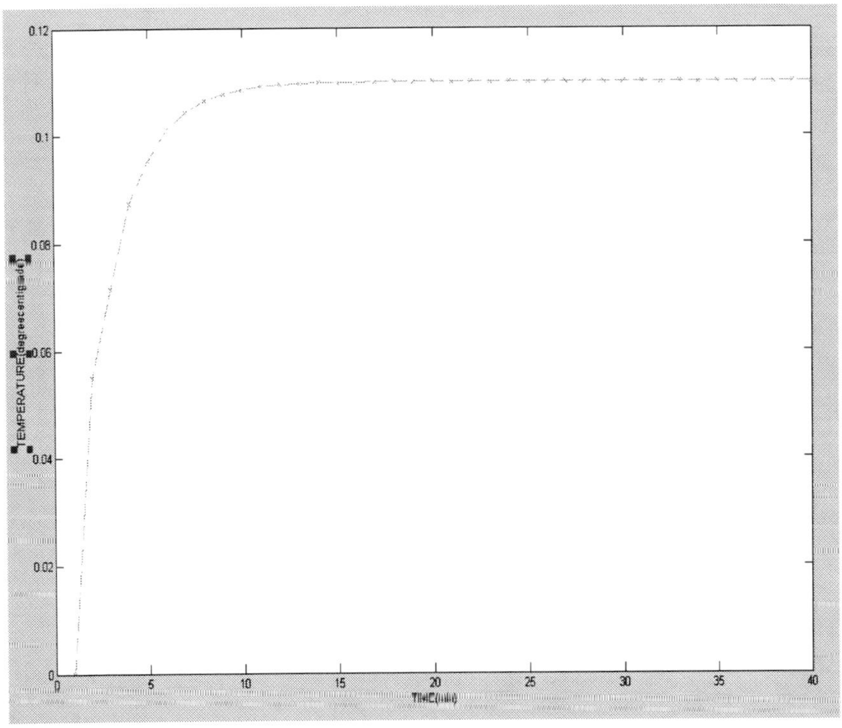

Figure 4(c). Temperature increase in the skull at 1000mW.

REFERENCES

[1] Hirata.A (2005),"Temperature Increase in Human Eyes Due to Near-Field and Far-Field Exposures at 900 MHz,1.5 GHz, and 1.9 GHz", *IEEE Trans.Electromagn. compat.*, Vol. EMC-47, no.1, pp.68-76.

[2] Wang J., and Fujiwara O(1999), "FDTD computation of Temperature rise in the human head for portable telephones," *IEEE Tran. Microwave Theory Tech.,* Vol. MTT-47, pp. 1528-1534.

[3] Guyton A. C. (1991), *Text book of Medical Physiology*.Philadelphia, PA: Saunders.

[4] Gordon R. G, Roemer R. B., and Horvath S. M(1976), "A Mathematical model of the human temperature regulatorySystem—Transient cold exposure response," *IEEE Trans.Biomed. Eng.*, Vol. BME-23, pp. 434–444.

[5] Dickinson R. J.(1984), "An ultrasound system for local hypothermia using scanned focused transducers," *IEEE Trans. Biomed. Eng.*, Vol. BME-31, pp. 120–125.

[6] Mur.G(1981), "Absorbing boundary conditions for the finer difference approximation of the time domain electromagnetic field equations," *IEEE Trans. Electromagn. Compat.,* Vol. EMC-23, pp. 377-382.

[7] Mittra.R and Pekel.U(1995), "A new look at the Perfectly Matched Layer (PML) concept for the reflection less absorption of electromagnetic waves," *IEEE Microwave and Guided Wave Lett.,*Vol.5,pp84-86.

In: Focus on Information, Intelligence and Knowledge
Editor: Rafiq Islam
ISBN 978-1-61209-979-8
2011 Nova Science Publishers, Inc.

Chapter 15

A SYSTEM OF SYSTEMS APPROACH TO TRANSPORTATION INFRASTRUCTURE SYSTEMS ASSESSMENT: A CONCEPTUAL SCHEMA

Kasthurirangan Gopalakrishnan* and Halil Ceylan
354 Town Engineering Building
Iowa State University,
Ames, IA 50011, USA

ABSTRACT

It is anticipated that recent advances in nanotechnology, distributed micro-electromechanical system (MEMS) applications, fiber optic sensors, and recent developments in smart structures and designer materials will play a key role in the near future in enhancing the functionality, serviceability, and increased lifespan of civil infrastructure systems. A multidisciplinary research front, represented by material scientists, physicists, chemists, biologists, and engineers from diverse disciplines, especially civil, electrical, computer and mechanical engineering, is required to integrate the multiple interactive component systems related to the design, construction and maintenance of highway infrastructure into a meta smart road system.

This paper reviews some of the latest developments related to the applications of smart devices for distributed monitoring of highway infrastructure systems. These include embedded wireless sensors to investigate the concrete mixing process leading to optimization of mix designs and mixers, measurement of subsequent compaction and shrinkage after placement of concrete, and detecting and identifying several deteriorations in concrete structures, including the presence of chloride ions, carbonation, and leaching of calcium hydroxide.

Potential futuristic applications which involve a combination of wireless sensors, advanced concrete pavement materials, fiber optic sensors, sophisticated data acquisition and processing systems, etc. are discussed in light of a recently released long-term plan for concrete pavement research and technology based on the concepts of pavement performance and systems integration.

* E-mail: rangan@iastate.edu

Keywords: MEMS, fiber-optic sensor, ANN, concrete pavement, GPR.

INTRODUCTION

Congested highways, overflowing sewers and corroding bridges are constant reminders of the looming crisis that jeopardizes the prosperity of United States. The American Society of Civil Engineers' *2005 Report Card for America's Infrastructure* was recently released. Over 27.1% of the nation's 590,750 bridges were rated structurally deficient or functionally obsolete. It will cost $9.4 billion a year for 20 years to eliminate all bridge deficiencies. Poor road conditions cost U.S. motorists $54 billion a year in repairs and operating costs—$275 per motorist. Americans spend 3.5 billion hours a year stuck in traffic, at a cost of $63.2 billion a year to the economy. Total spending of $59.4 billion annually is well below the $94 billion needed annually to improve transportation infrastructure conditions nationally.

While the security of United States' critical infrastructure has improved since Sept. 11, 2001, the information needed to accurately assess its status is not readily available to engineering professionals. This information is needed to better design, build and operate the nation's critical infrastructure in more secure ways. With the recent developments in nanotechnology based devices, it may now be possible to do real-time or quasi-real time monitoring and assessment of transportation infrastructure, one of the critical infrastructure assets of the nation. There is also a need to train a new generation of civil engineers and professionals in infrastructure deterioration science, assessment technologies, renewal engineering, and project management to provide feasible solutions to infrastructure problems.

Non-Destructive Test and Evaluation (NDT/NDE) methods are fast becoming choice alternatives for realistic characterization of materials and for assessing the integrity of structural systems like roads, bridges, and airport runways. The integrated nature of issues to be addressed in properly modeling, studying and evaluating such engineered systems requires incorporation of scientific knowledge and experience gained from a variety of disciplines into a coherent, unified framework using a system of systems approach.

Researchers from diverse disciplines especially electrical, computer and mechanical engineering, have been drawn into vigorous efforts to develop sensing technologies and nanotechnology in infrastructure condition assessment, crack detection and building monitoring. Recently, Micro-ElectroMechanical Systems (MEMS) and sensors have been proposed as a tool in infrastructure condition monitoring. MEMS are miniature sensing or actuating devices that can interact with other environments (provided no adverse reaction occurs) to either obtain information or alter it [1]. Based on results from other fields, MEMS have the potential to be embedded in transportation infrastructure members for real-time monitoring of transportation infrastructure health.

The in-service performance of the pavement depends on consistent, cost-effective and accurate monitoring of condition for early scheduling of repair and maintenance. It will be advantageous to have information on the state of health of infrastructure at all times in order to carry out effective on-demand maintenance. With the tremendous advancement in technology, it is possible to employ MEMS-based devices embedded in structural members for real-time monitoring of infrastructure health. The envisioned future for infrastructure monitoring uses a large number of inexpensive, spatially distributed, wirelessly powered, wirelessly networked, embedded sensing devices supporting frequent and on-demand acquisition of real-time information about the state of deterioration of structural material [2].

There are several advantages of using a MEMS-based monitoring system for highway infrastructure over other methods of condition assessment and monitoring. MEMS devices can be embedded, bringing their readings closer to the true *in situ* properties of the structure and not estimates based on external visual inspections or expensive nondestructive tests. MEMS technology has the potential to measure the properties of interest directly. Batch processes are used to fabricate MEMS devices, introducing high quantity and low cost and ensuring good coverage of a structure. Finally, MEMS devices are robust and bring improved ease of use. Their small, encapsulated nature makes them difficult to damage while wireless technologies make them easy to embed and interrogate [2].

This paper reviews some of the emerging applications of MEMS technology and other state-of-the-art developments in the construction and maintenance of highway infrastructure systems. It also highlights the possibility of integrating the various component systems related to pavement infrastructure to derive maximal benefit considering the interdependence and other critical relationships between the individual systems. A recently proposed long-term plan for concrete pavement technology partially provides the framework for developing a total performance-based concrete pavement system by making use of the latest sensing technologies and the concept of systems integration.

THE CONCRETE PAVEMENT ROAD MAP

The quality and performance of concrete plays a key role for most of infrastructure including commercial, industrial, residential and military structures, dams, power plants and transportation systems. Concrete is the single largest manufactured material in the world and accounts for more than 6 billion metric tons of materials annually. In the United States, federal, state, and local governments have nearly $1.5 trillion dollars in investment in the U.S. civil infrastructure. The worldwide use of concrete materials accounts for nearly 780 billion dollars in annual spending [3].

In the last 15 years, the concrete pavement industry has experienced more changes than those that occurred in the previous 80 years, and these changes are turning the process of building concrete pavements on end. In this environment, the old system for constructing concrete pavements does not work anymore [3].

A project to develop the Long-Term Plan for Concrete Pavement Technology (the Concrete Pavement, or CP Road Map) began in 2001 through an agreement between the Innovative Pavement Research Foundation and a team led by Iowa State University's National Concrete Pavement Technology Center (CP Tech Center) (previously known as Portland Cement Concrete Pavement Technology Center or PCC Center) [4]. In May 2003, Federal Highway Administration (FHWA) initiated a new agreement with the CP Tech Center to complete the work under a CP Technology Program contract (Task 15).

The overall vision of the CP Road Map is that "by 2015, the highway community will have a comprehensive, integrated, and fully functional system of concrete pavement technologies that provides innovative solutions for customer-driven performance requirements."

A schematic of the CP Road Map systems integration concept is displayed in Figure 1.

Figure 1. Concrete pavement road map systems integration concept [4].

An Iowa State University-led team facilitated development of the CP Road Map. They developed a database of existing research, and gathered input, face-to-face, from the highway community. The Iowa team identified gaps in research that became the basis for problem statements, which were organized into a cohesive, strategic research plan. To capture the cross-categories and the integrated nature of the problem statements, the problem statements were organized into 12, product-focused tracks of research within the database:

- Track 1: Performance-based concrete pavement mix design system
- Track 2: Performance-based design guide for new and rehabilitated concrete pavements
- Track 3: High-speed nondestructive testing and intelligent construction systems
- Track 4: Optimized surface characteristics for safe, quiet, and smooth concrete pavements
- Track 5: Concrete pavement equipment automation and advancements
- Track 6: Innovative concrete pavement joint design, materials, and construction
- Track 7: High-speed concrete pavement rehabilitation and construction
- Track 8: Long-life concrete pavements
- Track 9: Concrete pavement accelerated and long-term data collection
- Track 10: Concrete pavement performance
- Track 11: Concrete pavement business systems and economics
- Track 12: Advanced concrete pavement materials

It is noted that research in one track often affects or is affected by research in another track. In the CP Road Map, this interdependence and other critical relationships are outlined in the track and problem statement descriptions. The products developed through the first four tracks may be especially critical to helping the industry achieve the full potential of concrete pavements. Note that one subtrack in every phased track is devoted to developing innovative technology transfer, training tools, and methods to ensure that innovative research products are quickly and efficiently moved into practice.

The CP Road Map recognizes that new construction and intelligent rehabilitation of deteriorating concrete pavement systems include efficient and innovative use of high-performance designer materials, sensors, actuators, and mechanical and structural systems. With the new developments in nanotechnology and in materials sciences, increasing availability of advanced computing capabilities, both in hardware and software, researchers and engineers can now characterize processes, and design, model, optimize, construct and maintain smart pavement systems with desirable performance and properties, which is one of the main goals of the CP Road Map.

This could be seen as a research exploration to integrate advanced concepts, sophisticated modeling techniques, and emerging technologies from diverse scientific disciplines including non-destructive evaluation, numerical simulations, computational modeling, artificial intelligence, deterioration science, renewal engineering, and engineering mechanics into a comprehensive, unified framework for real-time assessment of transportation infrastructure systems. There are potential applications for MEMS-based devices, non-destructive testing and evaluation technologies, and so-called designer and smart materials under many of the research tracks listed in the CP Road Map which are discussed in the following sections.

MEMS Applications

The term micro-electromechanical systems (MEMS) refers to a collection of microsensors and actuators that sense their environment and have the ability to react to changes in that environment with the use of microcircuit control [5]. Typically no bigger than a grain of sand, these MEMS devices are complex machines that enable chips to become "intelligent". These devices act as the most direct links between digital electronics and the physical world, allowing the integration of electronics and mechanical systems on a single chipset [6].

MEMS merge the functions of sensing and actuating with computation and communication to locally control physical parameters at the micro scale. Numerous duplicate devices joined together can cause effects at much greater scales. Figure 2 shows the main components of MEMS [1]. In addition to reducing size and mass, other advantages of MEMS-based systems include low cost through mass production. Because MEMS are produced with the same processes as semiconductors, large-scale systems integration is possible on a single silicon chip.

It is important to note that while some MEMS devices have features that are measured in nanometers, MEMS and nanotechnology are fundamentally different. MEMS devices operate with dynamics that are determined by modeling matter as a continuum, while nanotechnology deals with matter's quantum mechanical behavior [1].

Distributed MEMS applications can be grouped into three classes [7]: (a) smart particles distributed in the environment, in which the location of the devices relative to one another varies over time; (b) active surfaces, in which the devices are permanently attached to a surface; and (c) smart structures, in which the MEMS element is fixed in place and interactions are coupled to one another through the dynamics of the material to which they are attached.

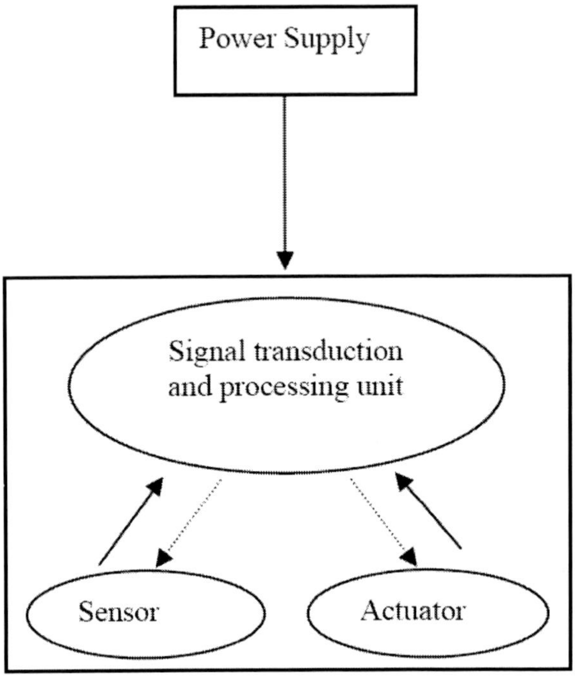

Figure 2. Main components of MEMS [1].

Smart materials and structures is a new field of study that is finding its way into many applications in transportation infrastructure systems, including structural control, condition or health monitoring, damage assessment, structural repair, and integrity assessment. The potential benefits here are improved system reliability, longevity and enhanced system performance, improved safety against natural hazards and vibrations, and a reduction in life cycle cost in operating and managing the infrastructure. The potential ability of MEMS to accelerate these benefits has been demonstrated in aerospace and automobile applications [8].

MEMS have been rigorously tested in harsh environments for defense and aerospace where they are used as navigational gyroscopes, sensors for border control and environmental monitoring, and munitions guidance. In medicine they are commonly used in disposable blood pressure transducers and weighing scales. In the transportation industry, MEMS and radio frequency identification tagging are used for railroad car management, traffic management, toll roads and fee collection, fare collection, equipment identification, solid waste hauling, fuel dispensing, and vehicle access [9].

Out of these practical experiences have grown a raft of other applications which are quickly approaching everyday use. For example, engineers are utilizing the ability of MEMS devices to collectively assemble information and applying it to "smart roads." It is envisioned that smart roads would be covered with millions of MEMS sensors. The sensors would act as a blanket of information, gathering and transmitting data about road conditions to maintenance engineers. Problems with roadways would be detected and repaired before they became serious. These smart roads could also send information to cars equipped with GPS devices, informing the on-board computer of road hazards, accidents, and traffic. Another safety application includes the detection of black ice on roadways and the development of windshields with automatic glare resistance [6].

There is no doubt that MEMS can lend itself to transportation infrastructure applications. However, there are some general drawbacks: there are no generic MEMS products – rather they are application specific – and the vast majority of applications require unique devices. Issues like survivability of sensor and materials for long-term stable operation have to be evaluated. Also, the impacts of embedding the sensor on the performance of a structure and vice versa have to be assessed [1].

Sackin et al. [2] performed an experiment on the potential application of embedded MEMS in concrete and termed the embedded microdevice "Smart Aggregate" (approximately 2.5 cm in diameter). This research project explored the possibility of using MEMS-based sensors and wireless communication technologies as a basis for distributed monitoring of concrete structures, primarily bridges and pavements. In several different experiments, these Smart Aggregates were embedded in concrete test specimens and their performances were observed as the concrete cured. It was argued that to move beyond the limitations, power supply techniques must provide in situ measurements of local material properties throughout a structure and with a minimum of effort. Figure 3 presents the overall Smart Aggregate system [2].

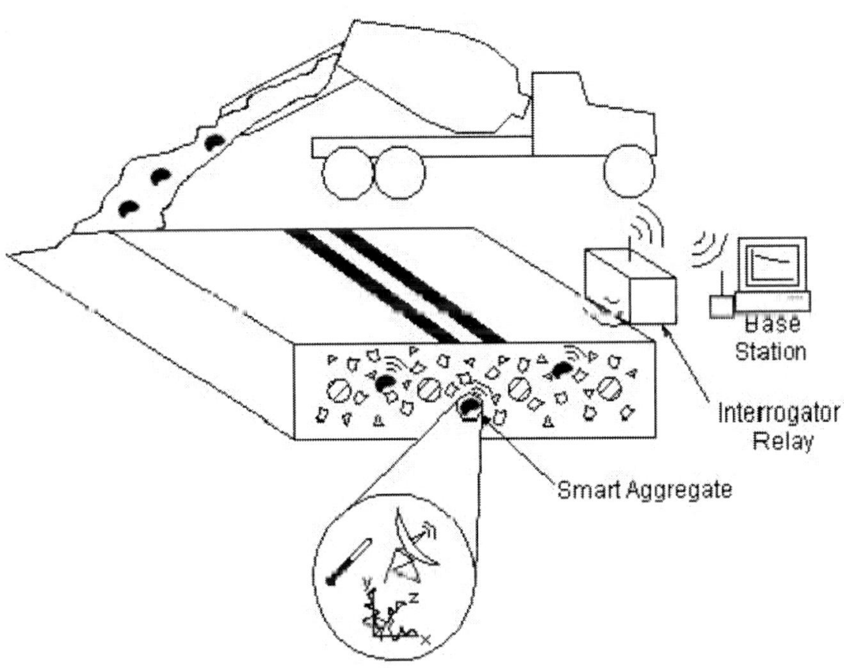

Figure 3. Smart aggregate system concept [2].

In collaboration with the California Department of Transportation, researchers at SRI International, an independent nonprofit RandD institute in Menlo Park, California, have developed "Smart Pebbles" – long-life wireless sensors that continually monitor the health of the bridge from deep inside the bridge deck [10].

Researchers call the sensor devices pebbles since they're roughly the size and weight of a typical piece of rock aggregate used in such structures. They're smart in that they each contain a chloride sensor and Radio Frequency IDentification, or RFID, chip that can be queried remotely to identify it and indicate chloride concentration levels in that part of the bridge. While driving over the bridge, the reader picks up information from the Smart Pebbles embedded in the bridge deck and sends them to a collection point. Using a global positioning system, the van could automatically update the health status of the bridge in the bridge's database. Smart Pebbles are in the early development stages. Laboratory tests have so far been very positive. Caltrans is formulating plans for a long-term evaluation of Smart Pebble prototypes in both the lab and in selected bridge decks [10].

Under the Federal Highway Administration's (FHWA's) Advanced Infrastructure Research Program, study is underway on a variety of nanotechnology applications to the highway and bridge industries. The feasibility of Cyberliths, or Smart Aggregates, as wireless sensors embedded in concrete or soil is being studied [11] under a project co-founded through the National Science Foundation (NSF).

Researchers at Johns Hopkins University's Applied Physics Laboratory have developed a robust Wireless Embedded Sensor Platform (WESP), suitable for long term field monitoring of corrosion in rebar, particularly in bridge decks [12]. These Smart Aggregates sensors can be embedded throughout a structure during concrete construction, added right to the mix before placement. The system is made up of the Smart Aggregates and a data reader that can be mounted on a car or truck. The reader powers the aggregates as it passes over them and collects the sensor data onto a PC.

Each Johns Hopkins Smart Aggregate contains wireless power receiver and data transmission coils, and is designed using ceramic hybrid integrated circuit technology to withstand mechanical stresses and the high pH environment of concrete. The aggregates are built to have a lifetime of over 50 years. The wireless power transmission and data collection approach eliminates the need for and potential problem with batteries, cables, and connectors. Prototype Smart Aggregates have been manufactured and are undergoing reliability measurements [11].

The NSF recently awarded a grant to Advanced Design Consulting (ADC) of Lansing, New York, to develop tiny passive sensors that can be mixed into concrete to collect data on the physical and chemical characteristics that affect its strength and performance. ADC used silicon-based MEMS to build the sensors. Each device consists of four or five MEMS sensors that measure information such as pH, moisture, temperature, and concentrations of chloride, sodium, and potassium ions within the concrete. It is paired with a radio frequency identification device (RFID) for ID and measurement purposes. The devices are encapsulated within a durable, non-corrosive material, similar to a pill, that would allow them to be mixed into concrete bridge decks, walls, or forms.

The University of California at Berkeley is experimenting with MEMS they call Motes. The Golden Gate Bridge now has an experimental sensor network of approximately 200 small Motes, each with an accelerometer that measures movement such as traffic, wind, or seismic loads. When all sensor readings are correlated, a three-dimensional picture is created which may portray structural abnormalities [11].

Attoh-Okine [13] discussed the potential application of MEMS in the management of transportation infrastructure assets and provided highlights of sensors and wireless monitoring applications. Mensah and Attoh-Okine [14] explored the possibility of adopting MEMS technology for microcrack monitoring in concrete, delving into both technical and technological issues that have to be surmounted in the process. They also proposed a protocol for the application of MEMS in microcrack monitoring. Mensah and Attoh-Okine [14] discussed both sensor communication and sensor fusion applications within MEMS applications in transportation infrastructure monitoring and emphasized the importance of wireless communication of data.

Saafi and Romine [15] studied the feasibility of embedding MEMS devices into concrete material for conditioning and health monitoring of civil infrastructure. A research program was conducted to evaluate the durability and sensing capability of different embedded MEMS devices. Saafi and Romine [15] presented a new technique that is currently under development for monitoring concrete setting and hardening. The new monitoring technique consists of embedding MEMS devices into concrete during construction to wirelessly monitor certain key parameters affecting the early age behavior of concrete structures such as in-place strength and shrinkage stresses. To test the feasibility of the proposed devices in monitoring temperature, moisture and early age shrinkage stresses as well as to evaluate their durability when embedded into concrete, experimental tests were performed.

NON-DESTRUCTIVE TESTING AND EVALUATION

Recent advances in hardware and software technology have significantly improved NDT equipment, data collection, and condition assessment methodologies. Using NDT data, it is possible to economically evaluate the load-bearing capacity of pavement systems, provide in-situ material properties for the design of infrastructure rehabilitation alternatives, and compare parts of a pavement system to each other to gain relative strength and/or condition within that section. Over the years, steady advances have been made in the modeling of pavement and geomechanical systems using finite element and discrete element techniques which account for pavement geomaterial non-linearity and anisotropy. Recent research studies have employed computational intelligent techniques as pavement structural analysis tools for the rapid and accurate prediction of critical structural responses and pavement layer stiffness based on NDT data.

Evaluating the structural condition of existing, in-service pavement infrastructure systems using *in-situ* NDT techniques has become an integral part of the pavement management process for selecting appropriate rehabilitation and reconstruction strategies by most transportation agencies. There are several advantages to using NDT, in lieu of, or supplement traditional destructive tests. Most important, is the capability to quickly gather data at several locations while keeping a highway or runway operational during these 2-minute to 3-minute tests with proper traffic control [16]. Nondestructive tests are economical to perform and data can be collected at up to 250 locations per day.

Nondestructive methods used for pavement evaluation are divided into two main categories: deflection and seismic methods. Among all the deflection based NDT methods, the Falling Weight Deflectometer (FWD) is the most widely used technique because of its ability to successfully simulate traffic loadings and capacity to produce larger amount of deflection data in unit time [17,18]. Recently, research studies have evaluated a new device known as Light Falling Weight Deflectometer (LFWD) for determining the mechanical properties of compacted unbound pavement layers. Other NDT equipments which have been studied by researchers for evaluating the structural integrity of pavements systems on a continuous basis include the Rolling Dynamic Deflectometer (RDD), Rolling Weight Deflectometer (RWD), the Rolling Wheel Deflectometer, the Swedish Rolling Deflection Meter, and the Danish High Speed Deflectograph.

The philosophy of the FWD methodology is that the structural integrity of the pavement system is inversely proportional with the amount of surface deflections utilized by applied load. During FWD testing (see Figure 4), a dynamic load is generated by a mass free falling onto a set of rubber springs and the device is set up to strike the pavement at a given force. Sensors placed around the plate and in a straight line radiating from the plate record the deflections in the pavement (analogous to ripples in a pond) induced by the falling weight. Thus, the FWD is an impulse-type testing device that imparts a transient load on the pavement surface, and the duration and magnitude of the force applied is representative of the load pulse induced by an aircraft or truck moving at moderate speeds. A schematic of the deflection basin obtained under FWD loading is shown in Figure 5.

Figure 4. Falling Weight Deflectometer (FWD) test equipment.

In recent years, several studies have focused on the on the development of Artificial Neural Networks (ANN) based pavement structural analysis tools. ANNs are valuable computational tools that are increasingly being used to solve resource-intensive complex problems as alternative to using more traditional techniques. The recent adoption and use of ANN modeling techniques in the recent Mechanistic-Empirical Pavement Design Guide [19] has especially placed the emphasis on the successful use of neural nets in geomechanical and pavement systems. Ceylan [20] employed ANNs in the analysis of concrete pavement systems and developed ANN-based design tools that incorporated the state-of-the-art finite element solutions into routine practical design at several orders of magnitude faster than those sophisticated finite element programs. Researchers at Iowa State University have focused on developing ANN based pavement structural analysis and backcalculation tools for flexible, rigid and composite pavements.

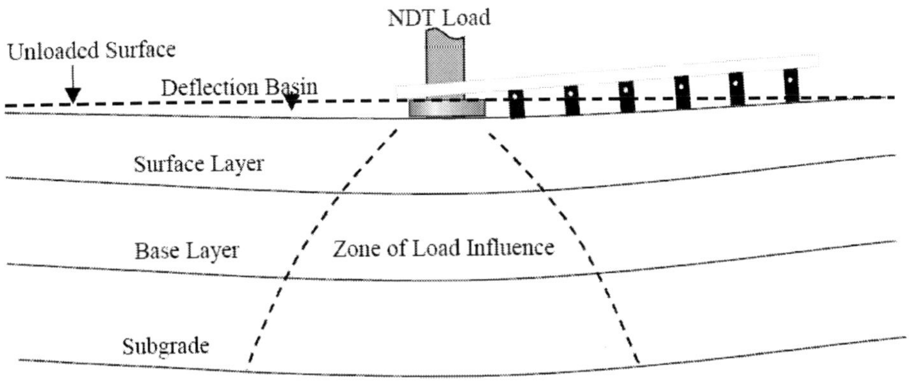

Figure 5. Schematic of FWD deflection basin [16].

The rapid backcalculation approach developed by the ISU research team is capable of analyzing 100,000 FWD deflection basins in a second in real-time. The models account for non-linear, stress-dependent behavior of unbound granular and subgrade layers and are based on validated finite element pavement structural models. They do not require seed moduli and offer many advantages over the conventional basin-matching programs. Further, the models are calibrated with field and laboratory data and have been validated and tested over numerous field datasets including the FWD data acquired at the National Airport Pavement Test Facility (NAPTF), LTPP sites, pavement sections in Iowa, etc.

Recently, all the NN based structural analysis tools develop at the ISU have been incorporated into a comprehensive user-friendly FWD analysis software. A screenshot of this rapid-approach FWD data analysis software are shown in Figure 6.

Such use of ANN models will enable pavement engineers to easily and quickly incorporate the needed sophistication in structural analysis, such as from finite element modeling with proper characterization of pavement layers, into routine practical mechanistic-based analysis and design and real-time pavement infrastructure systems condition assessment.

GEOPHYSICAL METHODS AND SENSING TECHNOLOGIES

Geophysical methods such as the Ground Penetrating Radar (GPR) allow for nondestructive investigation below the ground surface of a pavement, bridge deck, or other structure. To successfully backcalculate the in-situ moduli, accurate layer thickness information is essential apart from other considerations. Inaccurate layer thickness may lead to a significant error in the backcalculated layer moduli, and hence in the rehabilitation design. GPR is a noninvasive, nondestructive tool that has been used to map subsurface conditions in a wide variety of applications [21]. GPR has been successfully employed as a thickness evaluation method for pavements owing to their ability to survey pavements quickly, cost-effectively, and with minimal traffic disruption and safety risks. Research studies have concluded that the FWD and GPR are the best combination of technologies to effectively measure the AC stiffness and thickness considering the state-of-the-practice of all the NDT methodologies [22].

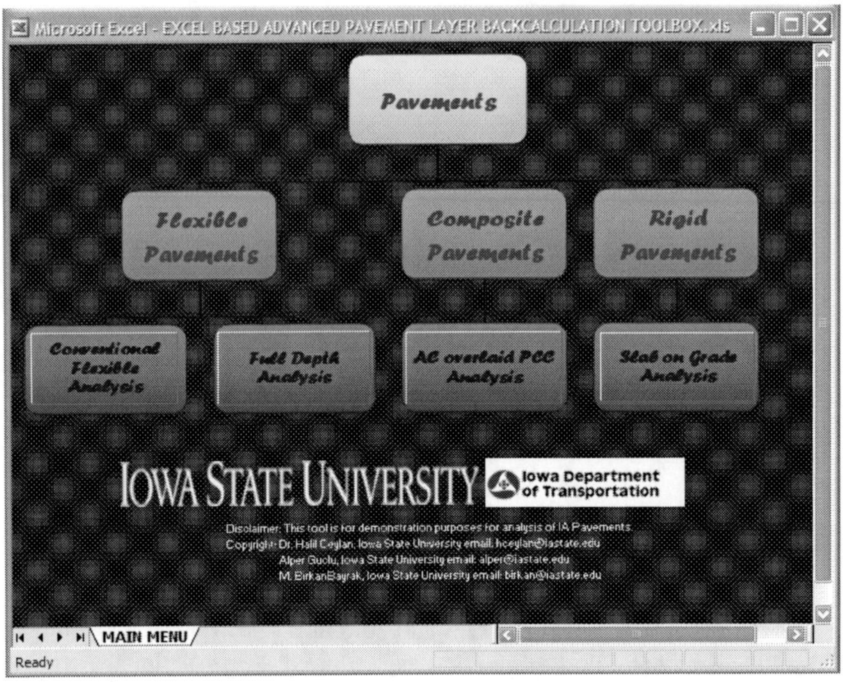

Figure 6. Screenshot of ANN-based FWD data analysis software.

In recent years there have been a number of research initiatives towards the development and deployment of fiber Bragg grating sensors (FBG) for sensing applications in civil and structural engineering [23]. Fiber-optic strain gauges based on Bragg gratings have proven to be effective replacements for conventional strain gauges for highway applications.

They create opportunities for monitoring on time and spatial scales that could not be achieved previously in civil engineering. The technology of fiber-optic strain gauges based on Bragg gratings permits as many as 100 gauges to be put on the same fiber, which can be calibrated to form a system that provides far more valuable information than individual gauges [24].

Researchers from Iowa State University's Bridge Engineering Center (BEC) recently demonstrated a novel approach to Structural Health Monitoring (SHM) data management, analysis and presentation based on fiber-optics sensor technology and FBG sensing applications [25].

TOWARDS A META SMART ROAD SYSTEM

It is conceived that a comprehensive framework could be developed for real-time non-destructive structural evaluation of transportation infrastructure systems by integrating advanced concepts, sophisticated modeling techniques, and emerging technologies described previously using a system-of-systems approach into a software workbench, referred to as Intelligent Transportation Infrastructure Assessment Integrated System (ITRAIS). Figure 7 shows a schematic of the modules which could constitute the proposed ITRAIS.

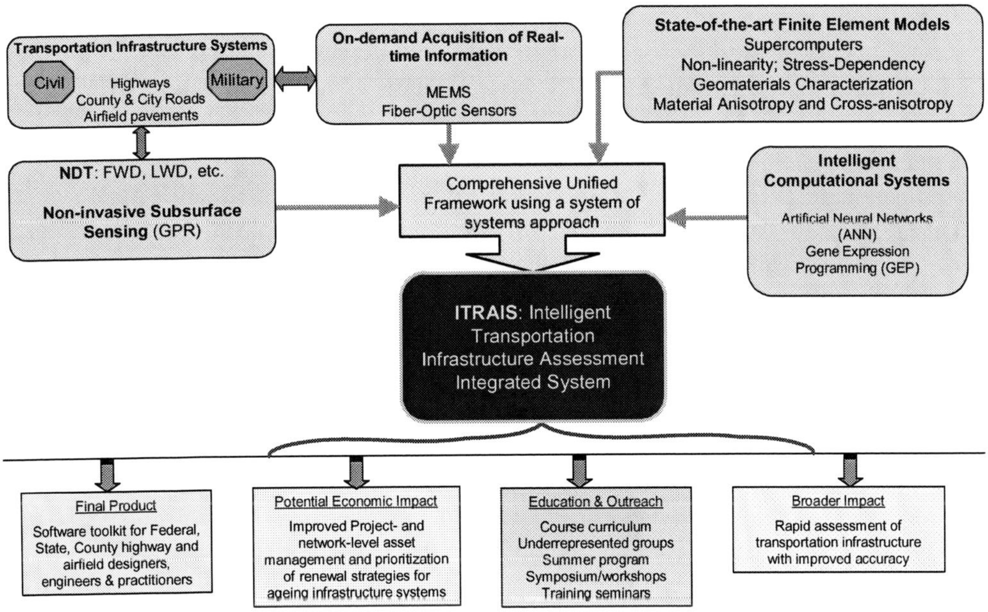

Figure 7. Schematic of proposed intelligent transportation infrastructure assessment integrated system (ITRAIS)

Such a scientific approach which seeks to comprehensively and mechanistically address the individual components of the problem at hand, but at the same time, integrates them into a user-friendly end product can provide highway and airfield pavement engineers and designers with the rapid ability for infrastructure systems condition assessment without the need for a high degree of expertise in the actual modeling of the problem. This will also address several of the research needs outlined in the CP Road Map.

REFERENCES

[1] N. O. Attoh-Okine, "The Future of MEMS in Transportation Infrastructure Systems." *Transportation Research Circular No. E-C056*, TRB, National Research Council, Washington, D.C.

[2] D. Sackin, J. Garret, K. Gabriel, and M. Patton. "MEMS for Civil Infrastructure: Smart Aggregate for Concrete." *Proc. Structures Congress 2000 – Advanced Technology in Structural Engineering*, ASCE, 2000, Philadelphia, PA.

[3] S. N. Vanikar "The Advances and Barriers in Application of New Concrete Technology." *Proc. International Workshop on Sustainable Development and Concrete Technology*, Beijing, China, May 2004.

[4] CP Road Map, *Long-Term Plan for Concrete Pavement Research and Technology: The CP Road Map*. Volumes I and II, HRT-05-047, Federal Highway Administration, U.S. Department of Transportation, 2005, Washington, D.C.

[5] V. K. Varadan and V. V. Varadan, "Microsensors, Microelectromechanical Systems (MEMS), and Electronics for Smart Structures and Systems." *Smart Mat. Structures*, Vol. 9, 2000, pp. 957–972.

[6] AllAboutMEMS, http://www.allaboutmems.com/, Accessed online Oct. 2006. Berlin and K. Gabriel, "Distributed MEMS: New Challenges for Computation." *IEEE Comp. Sci. and Engrg. Journal*, Vol. 4, No. 1, 1999, pp. 12–16.

[7] Oppenheim, I. J., "MEMS Devices and Applications in Aerospace and Civil Engineering." *ASCE Journal of Aerospace Engineering*, Editorial, 2003, pp. 45.

[8] Durfee and P. Goodrum, "RFID, MEMS, and Their Application in the Field of Construction" Unpublished White Paper, University of Kentucky, Lexington, KY, 2002.

[9] D. Watters, "Wireless Sensors will Monitor Bridge Decks." *Better Roads Magazine*, February 2003.

[10] T. Kuennen, "Road Science: Small Science will Bring Big Changes to Roads." *Better Roads Magazine*, July 2004.

[11] M. A. G. Darrin, B. G. Carkhuff, and T. S. Mehoke, "Future Trends in Miniaturization for Wireless Applications." *John Hopkins APL Technical Digest*, Vol. 25, No. 4, 2004.

[12] N.O. Attoh-Okine, "Potential Applications of Microelectromechanical Systems (MEMS) in the Management of Infrastructure Assets." *Proc. Fifth Int. Conf. on Managing Pavements*, 2001, Seattle, WA.

[13] S. Mensah and N.O. Attoh-Okine, "Civil Infrastructure Monitoring Using Microelectromechanical Systems (MEMS): A Case Study of Concrete Microcracking." *Proc. 82nd Annual Meeting of the Transportation Research Board*, 2003, Washington, D.C.

[14] M. Saafi and P. Romine, "Preliminary Evaluation of MEMS Devices for Early Age Concrete Property Monitoring." *Cement and Concrete Research*, 2005, Vol. 35, pp. 2158-2164.

[15] FAA, *Use of Nondestructive Testing in the Evaluation of Airport Pavements*. FAA Advisory Circular No. 150/5730-11A, Office of Airport Safety and Standards, Federal Aviation Administration, U.S. Department of Transportation, Washington, DC, 2004.

[16] M.S. Hoffman and M.R. Thompson, "Backcalculating nonlinear resilient moduli from deflection data." *Transportation Research Record 852*, TRB, National Research Council, 1982, Washington, D.C.

[17] Tholen, J. Sharma, and R. L. Terrel, "Comparison of FWD with other deflection testing devices." *Transportation Research Record 1007*, TRB, National Research Council, 1982, Washington, DC.

[18] NCHRP, *Guide for Mechanistic-Empirical Design of New and Rehabilitated Pavement Structures*. www.trb.org/mepdg. National Cooperative Highway Research Program, TRB, National Research Council, 2004, Washington, DC.

[19] H. Ceylan, "Analysis and design of concrete pavement systems using artificial neural networks." Ph.D. Dissertation, University of Illinois at Urbana-Champaign, Department of Civil Engineering, 2002, Urbana, IL.

[20] R. M. Morey, "Ground Penetrating Radar for Evaluating Subsurface Conditions for Transportation Facilities." NCHRP Synthesis of Highway Practice 255, National Cooperative Highway Research Program. TRB, National Research Council, Washington, DC.

[21] Saeed and J. Hall, "Determination of Insitu Material Properties of Asphalt Concrete Pavement Layers from Nondestructive Tests." *NCHRP Research Results Digest*, Number 271, TRB, National Research Council, 2002, Washington, D.C.

[22] D. Inaudi, N. Casanova, P. Kronenberg, S. Marazzi, and S. Vurpillot, "Embedded and surface-mounted fiber optic sensors for civil structural monitoring." *Proc. of the SPIE*, Vol. 3044, 1997, pp. 236-243.

[23] S.A. Wade, K.T.V. Grattan, B. McKinley, L.F. Boswell, and C. D'Mello. "Incorporation of fiber-optic sensors in concrete specimens: testing and evaluation." *IEEE Sensors Journal*, Vol. 4, No. 1, 2004.

[24] D. Hemphill, "Structural Health Monitoring System for the East 12th Street Bridge." *Proc. Transportation Scholars Conference*, Midwest Transportation Consortium, Iowa State University, Ames, IA, 2004.

In: Focus on Information, Intelligence and Knowledge
Editor: Rafiq Islam
ISBN 978-1-61209-979-8
2011 Nova Science Publishers, Inc.

Chapter 16

COMPUTER VISION AND ARTIFICIAL INTELLIGENCE IN FLIGHT CONTROL OF UNMANNED AERIAL VEHICLES

Vladimir Grishin*
Space Research Institute (IKI) of the Russian Academy of Sciences
117997, 84/32 Profsoyuznaya Str, Moscow, Russia

ABSTRACT

Applications of computer vision systems (CVS) in the flight control of unmanned aerial vehicles (UAV) are considered. In many projects, CVS are used for precision navigation, angular and linear UAV motion measurement, landing (in particular shipboard landing), homing guidance and others. All these tasks have been successfully solved separately in various projects. The development of perspective CVS can be divided into two stages. The first stage of perspective CVS development is the realization of all the above tasks in single full-scale universal CVS with acceptable size, weight and power consumption. Therefore, all UAV flight control tasks can be performed in automatic mode on the base of information that is delivered by CVS. All necessary technologies exist and the degree of its maturity is high. The second stage of CVS development is integration of CVS and control systems with artificial intelligence (AI). This integration will bring two great benefits. Firstly it will allow considerable improvement of CVS performance and reliability due to accumulation of additional information about the environment. Secondly, AI control system will obtain a high degree of awareness about the state of the environment. This allows the realization of a high degree of control effectiveness of the autonomous AI system in a fast changing and hostile environment.

* vgrishin@iki.rssi.ru

Introduction

The computer vision systems (CVS) revealed a great evolution during the last decades. This paper attempts to estimate the nearest perspective for its development. Further analysis will be dedicated to usage of CVS in mobile robot control systems, mainly in unmanned aerial vehicles (UAV). This problem is one of the most challenging tasks of control theory and practice. However, the main principles of this analysis are applicable to most of different kinds of mobile robots.

Present-Day Level of CVS Development

A large number of publications is devoted to the application of CVS to different tasks of UAV flight control. The enumeration of these publications may take many pages; so here we refer to a few arbitrarily chosen papers. Let's list the key tasks of such CVS.

- High precision navigation [1-6]. This task can be solved by means of recognition (matching) beforehand specified objects (landmarks) whose coordinates are known [1]. Another demand which is imposed on these landmarks is reliability of detection and recognition process. Reliability had to be guaranteed in conditions of possible imitation and masking. Since these landmarks are selected in advance and their reference patterns can be carefully prepared, the process of recognition (matching) can be reliable performed. Reference patterns are prepared with the account of different distances, perspective aspect angles of observation and observation conditions. Reliability of landmark recognition can be subsequently increased by joint usage of landmarks images and their 3D profiles. 3D profiles are widely used for navigation of missiles of different kinds (Tomahawk cruise missiles and others). The technologies for 3D profile reconstruction are well known. For instance, the complex of Israeli firm VisionMap [2] can be referred. The complex allows reconstructing of a 3D profile with precision about 0.2-0.3 m from the altitude of 3250 m. This complex is heavy enough and has considerable size. Some weakening of the precision requirement will allow significant decrease in weight and size. Further increasing of navigation reliability can be achieved by selection of a redundant number of landmarks in the whole area of observation. Information from CVS is usually integrated with information from inertial navigation system. Such integration allows serious decreasing of accumulated errors of the inertial navigation system [3]. The 3D profile of observed surface can be calculated simultaneously [4, 5]. In the presence of the on-board high precision inertial navigation system, it is possible to make a 3D reconstruction in monocular mode (with longitudinal stereo base) [6]. In this case it is possible to make a 3D reconstruction of very distant objects and surfaces.
- Flight stabilization and control of angular orientation [7-12]. This task is solved by the measurement of angular and linear UAV motion relative to the observed surface or objects [7, 8]. The set of features (points in frame with good localization) are selected and traced in subsequent frames. Observed shifts of a selected set of points are used for calculation of relative angular and linear motion. In this aspect, the star

trackers should be mentioned. These devices are the specialized CVS which are used in automatic spacecrafts for measurement of angular position with high precision and reliability [9, 10]. For flight control, it is important to estimate the UAV orientations with regard to local vertical (pitch and roll angles). These estimations are used for attitude stabilization and control. CVS sensors of local vertical use algorithms of horizon line detection, recognition and tracing [11-12]. Joint usage of CVS and inertial navigation systems allows significant improvement of precision and reliability of angular orientation [7, 8]. The pose estimation and 3D reconstruction are frequently realized in single algorithm, and this process is called SLAM (simultaneous localization and mapping) [4, 5, 7].

- Near-obstacle flight [13-17]. It is a very complicated flight control task. The control system had to guarantee high precision control with a short time delay. A good example of such control is the flight control of an airplane or helicopter between buildings in an urban environment in altitude about 10-15 m [15]. Another example is the ground hugging flight or on-the-deck flight. The CVS is able to provide all necessary information for solving such control tasks. In particular, the optical flow calculation allows to evaluate the risk of collision and to correct the direction of flight to avoid collision (obstacle avoidance). The distance and 3D profile of observed objects can be calculated by stereo pairs. The optical flow is used for estimation of flight altitude, too.
- Landing [18-26]. Landing is the most dangerous stage of flight. During this stage, UAV can crash. Moreover, some persons can be injured or property can be damaged. CVS can provide all necessary information for the control system which should realize this maneuver. This task includes the recognition of landing stripe or landing site and flight motion control relatively to the landing stripe. CVS is used also for landing site selection in condition of forced landing; see for example [19, 21 and 25]. In particular, 3D reconstruction of observed surface allows selection of the most proper landing area. In other words, CVS supports landing on unprepared (non-cooperative) sites. CVS are used for autonomous landing on surface of Solar system planets and small bodies [22, 24]. The most complicated task is the shipboard landing [26]. There are two main difficulties. The first – the landing stripe is small. The second – the ship deck is moving.
- Detection and tracking of selected moving targets [27-30]. A target can be a pedestrian, car, ship or other moving object. The selected target can make evolutions, attempts to hide from the observation and so on. In such a complicated condition, CVS should guarantee reliable tracing of specified target. In the case of automatic tracking collaps, the CVS should use effective search algorithms for target tracking restoration.
- Homing guidance to selected objects [31]. The homing task is similar to the task of navigation. It includes search, detection, recognition and tracking on the aim object. Significant problem is the multiple changing of distance during the homing process. During this process the observed size of the tracking object changes very significantly. In such case the correlation matching technologies are used. These technologies are used in smart weapons (precision-attack bombs). Another example is the Tomahawk cruise missile which is equipped with so-called DSMAC (Digital

Scene Matching Area Corellator) system. The other scene matching terminal guidance systems exist.

All these tasks have been successfully solved separately in different projects. The larger part of these projects belongs to the on-board real-time systems. The remaining part will be realized in the form of on-board real-time systems in near future.

During the last decades, great attention had been paid to the pattern recognition problem. From the flight control CVS view, the pattern recognition problem can be divided into two problems.

- Recognition of preliminary specified objects. This problem in fact is being solved in the high precision navigation task, tracking of selected objects and homing guidance.
- Recognition as a classification of observing objects [32-36]. This problem is being successfully solved during the development of automatic systems intended for the processing of visual information – space images of high resolution. The huge volume of such information makes the manual processing impossible. Such tasks should be realized in work stations as they require a great deal of calculation resources. But subsequent development of such systems and some simplification of task will allow realizing it in the form of on-board real-time systems. The possibility to recognize wide set of objects creates the necessary prerequisites for development artificial intelligence (AI) control systems.

FULL-SCALE UNIVERSAL CVS

The development of perspective CVS can be divided into two stages. The first stage of perspective CVS development is realization of all listed tasks in single full-scale system. In other words the task is to combine (to join) these technologies into the whole interrelated system. Mention should be made that all these technologies have many common methods and algorithms. All necessary technologies exist and the degree of its maturity is high. Many algorithms have been suggested for solution any of the listed tasks. One of the main aims is the selection of the most effective and reliable algorithms. Then it is necessary to develop appropriate hardware and software. The serious attention should be drawn to the development of CVS architecture [40] and special processors for most calculation consuming algorithms. It is highly probable that development and production of specialized chipsets for image processing in CVS will be required. Some attempts to move in that direction of the multifunctional CVS elaboration are currently appearing [37-42]. The cooperation of many groups of researchers, engineers and industry for realization this complicated CVS in acceptable size, weight and power consumption will be needed. These parameters should be acceptable for wide area of practical applications (robots of different kind). Another area of application is the safety systems which should be designed for preventing pilot errors in flight control or driver errors in car control. The accessibility of this task by means of present-day technologies is undoubted. Realization of the first stage will allow realizing complete automatic vision-based control of UAV and other mobile robots. Small size and cheap CVS will have very wide area of application which will be comparable with the area of GPS navigation receivers application.

INTEGRATION OF CVS AND AI CONTROL SYSTEM

One can say with confidence that the second stage of CVS development is the integration of CVS with AI control systems which realize the functions of tasks analysis, estimation of current situation, operation planning, realization, and correction of plans in real time. On the one hand such integration will allows the essential improvement of CVS performance and reliability due to accumulation of additional information about objects of environment, the possibility to undertake special action for investigation of the environment and the aggregation with the huge volume of non-visual information of AI control systems. On the other hand the AI control system will obtains high degree of awareness about the current state of environment. This allows to realize high degree of control effectiveness in uncertain, fast changing and hostile environment. The possibility of gaining and accumulation of AI system individual experience allows improving CVS reliability and effectiveness autonomously during the whole time of CVS and AI system operation. It stands to reason that CVS and AI control system have to be tightly integrated with the conventional flight control system.

It should be emphasized once more that CVS is the base for AI control systems development due to their high informativity and high awareness. From the other side, the requirements of AI system in solution of a task will stimulate the build-up of CVS functions and opportunities. On the certain stage of AI system development these processes can occur autonomously. In other words, the synergetic effect should takes place. In that way the close interaction between AI control system and CVS should be established.

There are other high informative sensors, such as radar sets and laser scanners. The modern radar set with electronically scanned array can provide the huge information flow. Laser scanner is capable to provide millions measurements per second. But its size, weight and power consumption are much larger then similar parameters of TV camera. The TV camera cost is much smaller then cost of a radar set or laser scanner. Moreover, a radar set and laser scanner produce electromagnetic wave emission what in some circumstances is highly undesirable. Thus the application area of CVS will be much wider then the application area of radar sets and laser scanners.

Some activities in the direction of integration some elements of AI and CVS are described in [43-49]. However, the realization of full-scale AI control system is still too complicated task. The effective methods for developing such combined CVS-AI systems are debatable and rather complicated.

The second stage is characterized by the considerable degree of uncertainty. But during the last 20-25 years the quite acceptable approaches to the development of AI control systems and its integration with CVS have been formulated. The most popular introduction to principles of artificial and natural intelligence structure and function is presented in [50]. This book is dedicated mainly to a brain functioning. A more complicated and advanced conception is presented in [51]. This book is dedicated to development of autonomous AI systems and their function. These AI conceptions are suitable for development of control system for mobile objects which are equipped with CVS. Mention should be made that these authors traditionally paid great attention to the artificial and natural neuron networks functioning. It seems that more effective special processors architectures should be used in developing of AI processors. Blind imitation of biological structures on the completely different technological base is ineffective. For instance, wheels allow such speed of motion, which is absolutely inaccessible for legs or its mechanical imitations. The present-day aircraft wings permit flying with the speed, which is absolutely inaccessible for birds with flapping wings or their mechanical imitations.

One can state that precedent thinking is the basis of natural and artificial intelligence. The development of the effective methods of permanent precedents accumulation and processing is required. These methods should include precedents data bases development, associative search, information and knowledge extraction methods from precedents data bases. Very significant precedent based methods of the situation evolution forecast should be developed. These methods should function in real time. Principles of precedent accumulation and processing are used by all living creature which have nervous system of any kind. Memory capacity and effectiveness of precedent processing determines the stage of the creature evolution. We describe here over-simplified constructive conception which is suitable for embedded autonomous AI system design. The concept "precedent thinking" is similar to the concept "case-based reasoning" [52-55]. Some distinctions between these concepts are unessential in this simplified description.

CVS can be considered as a basis for gradual *wide practical implementation* of AI. It will require the wide integration of different commands (groups) of researchers, engineers and industry. The second stage will require much more time, money and other resources then the first stage. But even during the development process it is possible to obtain practically useful and commercially applicable results. For instance vision system of such species as frog or crocodile are very primitive as well as their intelligence. However starting from survival task, admissible size, weight and power consumption such vision and intelligence systems are quite effective and useful. This fact is confirmed by wide spreading and quantity of frogs and crocodiles species. So even not very advanced and perfect CVS and AI systems can find their useful applications and be commercially successful.

Returning back to the UAV control, we see that the overwhelming majority of UAV operate under the remote control or preliminary prepared mission plan which is based on waypoints navigation. Possibilities of remote control are very limited. Remote control requires high reliability of information exchange with the UAV. Such information exchange is possible in limited range from the control station. Another drawback of remote control is the vulnerability to countermeasures. Remote control requires a highly skilled operator. Nevertheless the crash ratio of remote controlled UAV is relatively high. In the case of a complicated environment, the preparation of a flight plan for automatic UAV is a complicated and time consuming task. The initial information which is used for flight plan preparation can grow old rather fast. Any unmapped obstacle can cause the UAV to crash. Flight in a physically cluttered environment (such as streets of a city) on a preliminary prepared mission plan is impossible. Mention should be made about the high degree of vulnerability of such popular GPS navigation to countermeasures.

We have considered basic concept of artificial intelligence which is the precedent thinking. Accumulation of precedents allows forming the *individual experience* of AI UAV. Usage of analogy (associative search of similar precedents) allows using this individual experience in decision-making actions in real environment autonomously. Generalization of precedents permits to reveal the environment regularities and heuristics.

Precedent is the record of all information about external and internal environment and all control actions. Storage of this information requires considerable size of memory. For this reason the accumulated precedents should be periodically processed. During this operation recognized objects, events and interdependencies are replaced by references to their descriptions in data bases. Simultaneously these descriptions are refreshed. This process reminds the file compression, which is carried out by file archiver. In such a way the data bases of the artificial intelligence system are maintained in proper state. Additionally the multi-level generalization of precedents is fulfilled. The hypothesis exists that the night sleep of people and other living creatures is conditioned by the necessity of periodical precedents processing. The complicated system of the references, which is formed during this process, assists to search similar precedents during the precedent thinking.

Usage of CVS and precedents allows eliminating the "bottleneck" in information support of AI UAV. Presently the required information is prepared by a man. This is considerable keeping back the progress and applicability of AI UAV. We will not go into details of possible realization of on-board artificial intelligence control system architecture, but three significant aspects should be noted.

1. Hardware and software support of cooperative actions.
2. Hardware and software support of conflict actions.
3. Usage of reflexive control and recognition of hostile reflexive control.

1. Cooperative actions of AI UAVs are very important [56-60]. These actions may be a joint decision-making, joint information processing, joint coordinated flight and so on. Joint decision-making may include voting (taking into account different individual experience of different AI UAVs), or creation of hierarchical structure of AI UAVs for this process. Joint information processing is possible under condition of fast information exchange between AI UAVs. This exchange is possible by wireless local area network. Fast information exchange allows realizing cluster architecture for information processing. High productiveness of such systems and their high reliability allows significant increase the probability of the mission successful execution. Examples of such a mission are the search of definite objects and its tracing. The realization of a sensor network and a distributed detection system with high characteristic is possible in this way. Joint coordinated flights of AI UAVs with optimal special maneuvers contribute to the success of these tasks and other ones. AI UAVs are capable to realize the fast and highly coordinated group maneuvers and other actions. Such opportunities are absolutely inaccessible to aerial vehicles under control of pilots. So we can conclude that effeciency of practically all real missions can be significantly increased by cooperation between some AI UAVs. It is significant that it is not talking about the swarm of primitive UAVs but about advanced UAVs with artificial intelligence.
2. Since the conflict is the way of existence of complicated systems [61] it is impossible to omit this side of AI systems. During the mission AI UAV can meet with counteraction of different kinds. Skills and technical possibilities to resist to such counteraction will significantly increase the probability of the mission successful

execution. Conflicts spread in animate nature very wide. The struggle for existence represents conflicts in infinite variety of their forms and manifestation.

3. As far as AI UAV can self-dependently form personal data and knowledge bases from the precedents, the additional channels of external influence on AI UAV behavior appear. These channels should be used for influence on opponents' decisions and behavior in favorable for us direction [62]. Another task is detection of such influence from an opponent and prevention the negative consequences for us. Such reflexive control spreads in animate nature wide enough. For instance, birds pretend to be ill or injured to take away a beast from their baby birds. Harmless creatures use coloration of dangerous or poisonous ones.

Conclusion

One of the most difficult and attractive aims is the development of fully autonomous unmanned aerial vehicles and other robots. These vehicles should be capable effectively to solve the required tasks which should be carried-out in complicated unpredictable varying and hostile environment without remote control in any form. High awareness of computer vision systems is the necessary condition for the development of such advanced control systems.

References

[1] Xie, S.-R.; Luo, J.; Rao, J.-J.; Gong, Z.-B. Computer Vision-based Navigation and Predefined Track Following Control of a Small Robotic Airship. *Acta Automatica Sinica.* 2007, vol. 33, issue 3, pp. 286-291.

[2] Pechatnikov, M.; Shor, E.; Raizman, Y. Visionmap A3 - Super Wide Angle Mapping System Basic Principles and Workflow. *The International Archives of the Photogrammetry, Remote Sensing and Spatial Information Sciences.* 2008, vol. XXXVII, part B4.

[3] Veth, M. J.; Raquet, J. F. Two-Dimensional Stochastic Projections for Tight Integration of Optical and Inertial Sensors for Navigation. *Proceedings of the 2006 National Technical Meeting of the Institute of Navigation.* 2007, pp. 587–596.

[4] Karlsson, R.; Schon, T. B.; Tornqvist, D.; Conte, G.; Gustafsson F. Utilizing Model Structure for Efficient Simultaneous Localization and Mapping for a UAV Application. *Proceedings of 2008 IEEE Aerospace Conference.* 2008, pp. 1-10.

[5] Hygounenc, E.; Jung, I.-K.; Soueres, P.; Lacroix, S. The Autonomous Blimp Project of LAAS-CNRS: Achievements in Flight Control and Terrain Mapping. *The International Journal of Robotics Research.* 2004, vol. 23, issue 4-5, pp. 473-511.

[6] Borodovskiy, V. N.; Grishin, V. A. Estimation of Limitations in Precision for 3-D Surface Reconstruction. *Transactions on IEEE International Conference on Systems, Man and Cybernetics.* 1998, vol. 5, pp. 4387-4391.

[7] Baker, C.; Debrunner, C.; Gooding, S.; Hoff W.; Severson, W. Autonomous Vehicle Video Aided Navigation – Coupling INS and Video Approaches. *Advances in visual*

computing: Proceedings of the Second International Symposium on Visual Computing (ISVC 2006). 2006, vol. 4292, issue 2, pp. 534-543.

[8] Roumeliotis, S. I.; Johnson, A. E.; Montgomery, J. F. Augmenting Inertial Navigation with Image-Based Motion Estimation. *Proceedings of the IEEE International Conference on Robotics and Automation (ICRA '02).* 2002, vol. 4, pp. 4326–4333.

[9] Avanesov, G. A.; Forsh, A. A.; Bessonov, R. V.; Ziman, Ya. L.; Kudelin, M. I. BOKZ-M star tracker and its evolution. *Proceedings of the 14th Saint Petersburg International Conference on Integrated Navigation Systems.* 2007, pp. 219-224.

[10] SED 26 Star Tracker. (2006). http://www.sodern.fr/site/docs_wsw/fichiers_sodern/SPACE%20EQUIPMENT/FICHES%20DOCUMENTS/SED26.pdf

[11] Dusha, D.; Boles, W.; Walker, R. Fixed-Wing Attitude Estimation Using Computer Vision Based Horizon Detection. *Proceedings of the 12th Australian International Aerospace Congress.* 2007, pp. 1-19.

[12] Ettinger, S. M.; Nechyba, M. C.; Ifju, P. G.; Waszak, M. Vision-Guided Flight Stability and Control for Micro Air Vehicles. *Advanced Robotics*. 2003, vol. 17, issue 7, pp. 617-640.

[13] Oh, P. Y.; Green, W. E.; Barrows, G. Neural Nets and Optic Flow for Autonomous Micro-Air-Vehicle Navigation. *Proceedings of the ASME International Mechanical Engineering Congress and Exposition (IMECE 04).* 2004, vol. 2, pp. 1-7.

[14] Kellogg, J., Bovais, C.; Dahlburg, J.; Foch, R.; Gardner, J.; Gordon, D.; Hartley, R.; Kamgar-Parsi, B.; McFarlane, H.; Pipitone, F.; Ramamurti, R.; Sciambi, A.; Spears, W.; Srull, D.; Sullivan, C. The NRL MITE Air Vehicle. *Proceedings of the Bristol RPV/AUV Systems Conference.* 2001.

[15] Muratet, L.; Doncieux, S.; Briere, Y.; Meyer, J.-A. A Contribution to Vision-Based Autonomous Helicopter Flight in Urban Environments. *Robotics and Autonomous Systems.* 2005, vol. 50, issue 4, pp. 195-209.

[16] Ortiz, A. E.; Neogi, N. N. Object Detection and Avoidance Using Optical Techniques in Uninhabited Aerial Vehicles. *Proceedings of the AIAA Guidance, Navigation, and Control Conference and Exhibit.* 2007.

[17] Zufferey, J.-C.; Floreano, D. Toward 30-gram Autonomous Indoor Aircraft: Vision-based Obstacle Avoidance and Altitude Control. *Proceedings of the IEEE International Conference on Robotics and Automation (ICRA'2005).* 2005, pp. 2594-2599.

[18] Saripalli, S.; Montgomery, J. F.; Sukhatme, G. S. Visually-Guided Landing of an Unmanned Aerial Vehicle. *IEEE Transactions on Robotics and Automation.* 2003, vol. 19, issue 3, pp. 371-380.

[19] Fitzgerald, D.; Walker, R.; Campbell, D. A Vision Based Emergency Forced Landing System for an Autonomous UAV. *Proceedings of Australian International Aerospace Congress Conference.* 2005.

[20] Sharp, C. S.; Shakernia, O.; Sastry, S. S. A Vision System for Landing an Unmanned Aerial Vehicle. *Proceedings of the IEEE International Conference on Robotics and Automation. (ICRA'2001).* 2001, vol. 2, p. 1720-1727.

[21] Theodore, C., Rowley, D.; Ansar, A.; Matthies, L.; Goldberg, S.; Hubbard, D.; Whalley, M. Flight Trials of a Rotorcraft Unmanned Aerial Vehicle Landing Autonomously at Unprepared Sites. *Proceedings of the American Helicopter Society 62nd Annual Forum.* 2006, pp. 1250-1264.

[22] Zhukov, B.; Avanesov, G.; Grishin, V.; Krasnopevtseva, E. On-Board Real-Time Image Processing to Support Landing on Phobos. *Proceedings of the 7th International Symposium on Reducing Costs of Spacecraft Ground Systems and Operations (RCSGSO).* 2007, pp. 423-428.

[23] Grishin, V. A. A Cramer–Rao Bound for the Measurement Accuracy of Motion Parameters and the Accuracy of Reconstruction of a Surface Profile Observed by a Binocular Vision System. *Pattern Recognition and Image Analysis.* 2008, vol. 18, issue 3, pp. 507–513.

[24] Trawny, N.; Mourikis, A. I.; Roumeliotis, S. I.; Johnson, A. E.; Montgomery, J. F.; Ansar, A.; Matthies, L. H. Coupled Vision and Inertial Navigation for Pin-Point Landing. *Proceedings of the NASA Science and Technology Conference (NSTC'07).* 2007, paper B2P2.

[25] Fitzgerald, D. L.; Walker, R. A; Campbell, D. A Vision Based Forced Landing Site Selection System for an Autonomous UAV. *Proceedings of the 2005 International Conference on Intelligent Sensors, Sensor Networks and Information Processing (ISSNIP).* 2005, pp. 397- 402.

[26] Saripalli, S.; Sukhatme, G. S. Landing on a Moving Target using an Autonomous Helicopter. *Proceedings of the International Conference on Field and Service Robotics.* 2003, vol. 24, pp. 277-286.

[27] Helble, H.; Cameron, S. OATS: Oxford Aerial Tracking System. *Robotics and Autonomous Systems.* 2007, vol. 55, issue 9, pp. 661-666.

[28] Lenhart, D.; Hinz, S. Automatic Vehicle Tracking in Low Frame Rate Aerial Image Sequences. *International Archives of Photogrammetry, Remote Sensing, and Spatial Information Sciences.* 2006, vol. 36-3, pp. 203-208.

[29] Nguyen, T. T.; Grabner, H.; Bischof, H.; Gruber, B. On-line Boosting for Car Detection from Aerial Images. *Proceedings of the IEEE International Conference on Research Information and Vision for the Future (RIVF'07).* 2007, pp. 87-95.

[30] Zhao, T.; Nevatia, R. Car Detection in Low Resolution Aerial Image. ***Proceedings of the Eighth IEEE International Conference on*** *Computer Vision (ICCV 2001).* 2001, vol. 1, pp. 710-717.

[31] Min-Shou, T. (1996). The Application of Correlation Matching Technique in Image Guidance. http://handle.dtic.mil/100.2/ADA316752

[32] Rekik, A.; Zribi, M.; Hamida, A. B.; Benjelloun, M. An Optimal Unsupervised Satellite image Segmentation Approach Based on Pearson System and k-Means Clustering Algorithm Initialization. *International Journal of Signal Processing.* 2009, vol. 5, issue 1, pp. 38-45.

[33] Lari, Z.; Ebadi, H. Automatic Extraction of Building Features from High Resolution Satellite Images Using Artificial Neural Networks. *Proceeding of the Conference on Information Extraction from SAR and Optical data, with Emphasis on Developing Countries.* 2007.

[34] Kampouraki, M.; Wood, G. A.; Brewer, T. R. The Suitability of Object-Based Image Segmentation to Replace Manual Aerial Photo Interpretation for Mapping Impermeable Land Cover. *Proceedings of the 2007 Annual Conference of the Remote Sensing & Photogrammetry Society (RSPSoc2007).* 2007.

[35] Samadzadegan, F.; Abbaspour, R. A.; Hahn, M. Automatic Change Detection of Geospatial Databases Based on a Decision-Level Fusion Technique. *Proceedings of the*

XXth International Society for Photogrammetry and Remote Sensing Congress (ISPRS 2004). 2004, pp. 489-491.

[36] Li, Y.; Atmosukarto, I.; Kobashi, M.; Yuen, J.; Shapiro, L. G. Object and Event Recognition for Aerial Surveillance. *Proceedings of the SPIE Conference on Optics and Photonics in Global Homeland Security.* 2005, vol. 5781, pp. 139-149.

[37] Grishin, V. A; Aust, S. A. Architecture of Computer Vision Systems Intended for Control of Aircraft. *Proceedings of the Fourth International Conference Parallel Computations and Control Problems (PACO '2008).* 2008, pp. 1071-1086. (in Russian).

[38] Wagter, C. D.; Proctor, A. A.; Johnson, E. N. Vision-Only Aircraft Flight Control. *Proceedings of the 22nd AIAA Digital Avionics Systems Conference (DASC '03).* 2003, vol. 2, pp. 8.B.2-81-11.

[39] Montgomery, J. F.; Johnson, A. E.; Roumeliotis, S. I.; Matthies, L. H. The Jet Propulsion Laboratory Autonomous Helicopter Testbed: A Platform for Planetary Exploration Technology Research and Development. *Journal of Field Robotics.* 2006, vol. 23, issue 3-4, pp. 245-267.

[40] Wagter, C. D.; Bijnens, B.; Mulder J. A. Vision-Only Control of a Flapping MAV on Mars. *Proceedings of AIAA Guidance, Navigation and Control Conference and Exhibit.* 2007, pp. 1-7.

[41] Nordberg, K.; Doherty, P.; Farneback, G.; Forssen, P.-E.; Granlund, G.; Moe, A.; Wiklund, J. Vision for a UAV helicopter. *Proceedings of the IEEE/RSJ International Conference on Intelligent Robots and Systems (IROS'02). Workshop WS6 on aerial robotics.* 2002, pp. 29-34.

[42] Kurdila, A.; Nechyba, M.; Prazenica, R.; Dahmen, W.; Binev. P.; DeVore, R.; Sharpley, R. Vision-Based Control of Micro–Air–Vehicles: Progress and Problems in Estimation. *Proceedings of the 43rd IEEE Conference on Decision and Control.* 2004, vol. 2, pp. 1635– 1642.

[43] Kim, H. J.; Shim, D. H. A flight control system for aerial robots: algorithms and experiments. *Control engineering practice.* 2003, vol. 11, issue 12, pp. 1389-1400.

[44] Smith, N. W. (2006). Artificially Intelligent Autonomous Aircraft Navigation System Using a Distance Transform on an FPGA. www.altera.com/literature/dc/2006/a1.pdf

[45] Granlund, G.; Nordberg, K.; Wiklund, J.; Doherty, P.; Skarman, E.; Sandewall E. WITAS: An Intelligent Autonomous Aircraft Using Active Vision. *Proceedings of the UAV 2000 International Technical Conference and Exhibition.* 2000.

[46] Dickmanns, E. D. Vehicles Capable of Dynamic Vision. *Proceedings of 15th International Joint Conference on Artificial Intelligence (IJCAI-97).* 1997, vol. 2, pp. 1577-1592.

[47] Heintz, F.; Doherty, P. Managing Dynamic Object Structures using Hypothesis Generation and Validation. *Proceedings of the AAAI Workshop on Anchoring Symbols to Sensor Data.* 2004, pp. 54-62.

[48] Cliff, D.; Noble, J. Knowledge-based vision and simple visual machines. *Philosophical Transactions of the Royal Society. Biological Sciences.* 1997, vol. 352, issue 1358, pp. 1165-1175.

[49] Johnson, E. N.; Proctor, A. A.; Ha, J.; Tannenbaum, A. R. Development and Test of Highly Autonomous Unmanned Aerial Vehicles. *AIAA Journal of Aerospace Computing, Information, and Communication.* 2004, vol. 1, issue 12, pp. 485-501.

[50] Hawkins, J.; Blakeslee, S. *On Intelligence*; *Times Books; Henry Holt and Company: New York, NY, 2004.*

[51] Zhdanov, A. A. *Autonomous artificial intelligence*; Adaptive and Intellectual systems; Binom. Laboratory of knowledge: Moscow, Russia, 2008. (in Russian).

[52] Mata, A. A Survey of Distributed and Data Intensive CBR Systems. *Proceedings of International Symposium on Distributed Computing and Artificial Intelligence 2008 (DCAI'2008) (Advances in Soft Computing).* 2009; vol. 50, pp. 582-586.

[53] Nikitenko, A. Knowledge-based Robot Control. *International Journal "Information Theories & Applications".* 2008, vol.15, pp. 126-137.

[54] Aamodt, A.; Plaza, E. Case-Based Reasoning: Foundational Issues, Methodological Variations, and System Approaches. *AI Communications.* 1994, vol. 7, issue 1, pp. 39-59.

[55] Richter, M. M.; Aamodt, A. Case-based reasoning foundations. *The Knowledge Engineering Review.* 2005, vol. 20, issue 3, pp. 203–207.

[56] Bowen, D. G.; MacKenzie, S. C. (2003). Autonomous Collaborative Unmanned Vehicles: Technological Drivers and Constraints. http://pubs.drdc-rddc.gc.ca/BASIS/pcandid/www/engpub/DDW?W%3DKEYWORDS+INC+%27ugv%27+ORDER+BY+Repdate/Descend%26M%3D15%26K%3D519508%26U%3D1

[57] Pongpunwattana, A.; Rysdyk, R. Real-Time Planning for Multiple Autonomous Vehicles in Dynamic Uncertain Environments. *Journal of Aerospace Computing, Information, and Communication.* 2005, vol. 1, pp. 580-604.

[58] Ryan, A.; Zennaro, M.; Howell, A.; Sengupta, R.; Hedrick, J. K. An Overview of Emerging Results in Cooperative UAV Control. *Proceedings of the 43rd IEEE Conference on Decision and Control.* 2004, vol. 1, pp. 602-607.

[59] Iocchi, L.; Nardi, D.; Salerno, M. Reactivity and Deliberation: A survey on Multi-Robot Systems; In *Balancing Reactivity and Social Deliberation in Multi-Agent Systems: From RoboCup to Real-World Applications*; Hannebauer, M.; Wendler, J.; Pagello, E.; Eds.; Lecture Notes in Computer Science / Lecture Notes in Artificial Intelligence Series; Springer-Verlag: London, UK, 2001; Vol. 2103, pp. 9-34.

[60] Schmitt, T.; Hanek, R.; Beetz. M.; Buck. S.; Radig, B. Cooperative Probabilistic State Estimation for Vision-based Autonomous Mobile Robots. *IEEE Transactions on Robotics and Automation.* 2002, vol. 18, issue 5, pp. 670-684.

[61] Druzhinin, V. V.; Kontorov, D. S.; Kontorov, M. D. *Introduction in conflict theory*; Radio and Communications: Moscow, Russia, 1989. (in Russian).

[62] Thomas, T. L. Russia's Reflexive Control Theory and the Military. *Journal of Slavic Military Studies.* 2004, vol. 17, issue 2, pp. 237–256.

In: Focus on Information, Intelligence and Knowledge
Editor: Rafiq Islam

ISBN 978-1-61209-979-8
2011 Nova Science Publishers, Inc.

Chapter 17

EVERYDAY THEORIZING AND THE CONSTRUCTION OF KNOWLEDGE

Sherrie Bade*
Thompson Rivers University, Kamloops, BC, Canada

ABSTRACT

Diverse worldviews directly shape ontological and epistemological understandings related to what counts as valid theory and who is able to theorize. If valid knowledge development is believed to rest within the purview of professionals in the academy then the everyday theorizing capabilities of individuals within the community remain unrecognized or if recognized, discounted. However, if theory generation is viewed as a product of ordinary people's discovery of personal meaning within the experiential world then theory and knowledge generation become different in content but not in kind to that of the academic. Everyday theorizing results in practical knowledge generation that is no longer the singular domain of those who work in the academy but is instead a common and natural activity of daily life. Community Based Research (CBR) is one means by which those who engage in academic inquiry may partner with individuals and groups in the community to work together in a participatory manner. Participatory research values each person's knowing of the world and how this knowing may contribute to knowledge generation. Researching with people rather than on them is integral to the process of CBR. Inclusive grassroots research increases knowledge generation that can then be constructed into everyday theory. In this way the experiential knowledge of community members constructively critiques social inequities and may address social issues most effectively at the local community level.

INTRODUCTION

In modern societies non-academic citizens are rarely recognized for their ability to construct knowledge. Engagement with theory that contributes to knowledge development is

* Email: sbade@tru.ca

considered the mandate of professionals or those who are situated within academic circles. It is apparent to many individuals, including academics, that through day-to-day decision making the average person, although not associated with traditional research bodies, actively theorizes and constructs knowledge. A general acknowledgement of these abilities encourages collaboration among all citizens and prevents the unhealthy marginalization of a large segment of the population. As well, theory development and knowledge construction become demystified and result in communities of learning, collaboration and civility.

Construction of Knowledge

Diverse worldviews directly shape ontological and epistemological understandings related to what counts as valid theory and credible knowledge. The empiricist believes that true knowledge exists objectively outside of the person and is waiting to be discovered; those with a constructivist perspective believe that knowledge is relative, subjective and dependent on context (Crotty, 2007). The empirical world view values objectivity whereas more relative worldviews "reconnect what empiricism [has] separated–the dualism of the knowing subject and the object of knowing" thereby valuing subjectivity and lived experience (Smith & Hodkinson, 2005, p. 917). Kirby, Greaves and Reid (2006) discuss knowledge construction as a political process which has traditionally been conducted by universities and institutes of research. Those who drive the political process also have the power to control knowledge generation. A common yet little understood phrase is that "knowledge is power." One wonders if this is true. Are not professionals and academics who define what constitutes knowledge more powerful than knowledge itself? "Whoever is in a position of power is able to create knowledge supporting that power relationship. Whatever a society accepts as knowledge or truth inevitably ends up strengthening the power of some and limiting the power of others" (Brookfield, 2005, p. 136). To interrupt the power brokering of this political process, academic and non-academics alike can choose to work together to construct knowledge that retains power within the community by remaining responsive to identified community goals.

Partnerships between professionals, particularly those working within the system, and citizens that are developed in a respectful and thoughtful manner encourage mutual empowerment (Vanderplaat, 2002). Mutual empowerment is a relational endeavour that involves all who are involved in the community development process. "One can never be just an empowerer or a person in need of empowerment. In a mutual approach to empowerment, everyone involved, regardless of position of power and privilege, recognizes that she or he is both agent and subject … in a truly empowering process everyone changes" (Vanderplaat, 2002, pp. 93-94). Community Based Research (CBR) is one form of inquiry which includes both academic and experiential citizens working in collaboration to seek mutual empowerment. This occurs as local knowledge is valued as an alternative to mainstream research/knowledge production. Participatory research supports each person as they make meaning of their life experiences and supports them as these experiences are then expanded into useful knowledge. In this way the experiential knowledge of community members constructively critiques social inequities and may also address social issues more effectively at the local level.

The *Moving Women Beyond Bars: Building Community* initiative is one example of how those from academic circles are able to partner with vulnerable and experiential women in the community (Warnell, Bade & Carriere, 2008). The starting place for this initiative is bringing women having past conflict with the law and their mentors/allies into safe and communicative spaces where relationships may form and community develop. As trust builds it is anticipated that a participatory research team will evolve to identify the issues criminalized women experience as well as to generate useful grass roots knowledge. Foundations for this kind of research with women are currently being structured at the local level through other innovative initiatives some of which include *Linking Women to Health and Wellness* (Carriere, 2008) and *Voice: Challenging the Stigma of Addiction* (Paivinen & Bade, 2008). Initiatives such as *Moving Women Beyond Bars: Building Community* demonstrate that those who are experiential or have lived experiences outside of the mainstream may resist a system that silences them and respond by organizing into visible, knowledgeable and influential communities (Jurgens, 2005; Kerr, Small, Peeace, Douglas, Pierre & Wood, 2006; Rabinovitch & Lewis, 2001; Rans, 2005). Researching with people, rather than on them, produces healthy empowered communities and contributes to egalitarian relationships within a civil society.

Civil society is essentially comprised of "all those forms of collective human association not directly controlled by the state or corporations" (Brookfield, 2005, pp. 234-235). People come together and stand outside of the complex bureaucracies that control people's lives through the mechanisms of money and power (Habermas, 1984). Integral to civil society is the development of community spaces or what Habermas calls the public sphere which "is the civic space or 'commons' in which adults come together to debate and decide their response to shared issues and problems" (Brookfield, 2005, p. 230). Power is shared, not reluctantly but generously so that community is built and lives are open to common purposes. Civil society is a place where connections are made and relationships established. It does not require a formal gathering but occurs wherever people meet. Relationships that form in the public sphere are based on mutual respect and understanding as well as a valuing of the knowledge that comes from a range of lived experiences. Each brings what they can offer to mutually explore and theorize about what works and what does not work in everyday worlds. Grassroots knowledge is then constructed to determine avenues toward egalitarian communities in which civility and health are promoted at both the individual and community level.

Conclusion

If valid knowledge development is believed to rest within the purview of professionals in the academy then the everyday theorizing capabilities of individuals within the community remain unrecognized or if recognized, discounted. However, if theory generation is viewed as a product of ordinary people's discovery of personal meaning within the experiential world then theory and knowledge generation become different in content but not in kind to that of the professional or academic. With this understanding, partnerships between academic and non-academic citizens may be developed in a respectful and thoughtful manner to build a strong and healthy civil society.

ACKNOWLEDGMENT

The author gratefully acknowledges Star Mahara, a colleague at Thompson Rivers University, for her editorial support and conversations regarding this commentary.

REFERENCES

Brookfield, S. D. (2005). *The power of critical theory.* USA: Jossey-Bass.

Carriere, G. (2008). Linking women to health and wellness: Street ourtreach takes a population health approach. *International Journal of Drug Policy, 19*, 205-210.

Crotty, M. (2007). *The foundations of social research.* Thousand Oaks, CA: Sage Publications, Inc.

Habermas, J. (1984). *The theory of communicative action* (T. McCarthy, Trans. Vol. 2). Boston: Beacon Press.

Jurgens, R. (2005). *"Nothing about us without us" greater, meaningful involvement of people who use illegal drugs: A public health, ethical, and human rights imperative.* Toronto, Ont., Canada: Canadian HIV/AIDS Legal Network.

Kerr, T., Small, W., Peeace, W., Douglas, D., Pierre, A., & Wood, E. (2006). Harm reduction by a "user-run" organization: A case study of the Vancouver area network of drug users (VANDU). *The International Journal of Drug Policy, 17*(2), 61-69.

Kirby, S. L., Greaves, L., & Reid, C. (2006). *Experience research social change: Methods beyond the mainstream.* Peterborough, Ont., Canada: Broadview Press.

Paivinen, H., & Bade, S. (2008). Voice: Challenging the stigma of addiction; a nursing perspective. *International Journal of Drug Policy, 19*, 214-219.

Rabinovitch, J., & Lewis, M. (2001). *Impossible, eh? The story of peers.* Vancouver, BC: Save the Children Canada.

Rans, S. A. (2005). Hidden treasures: Building community connections by engaging the gifts of *people on welfare, people with disabilities, people with mental illness, older adults, young people. Evanston, IL: ABCD Institute.

Smith, J. K., & Hodkinson, P. (2005). Relativism, criteria, and politics. In N. K. Denzin & Y. S. Lincoln (Eds.), *The sage handbook of qualitative research* (3rd ed., pp. 915-932). Thousand Oaks, CA: Sage Publications, Inc.

Vanderplaat, M. (2002). Emancipatory politics and health promotion practice: The health professional as social activist. In E. Young & V. Hayes (Eds.), *Transforming health promotion practice* (pp. 87-98). Philadelphia, PA: F. A. Davis Company.

Warnell, R., Bade, S., & Carriere, G. (2008). *Moving women beyond bars: Building community.* Vancouver, BC: Women's Health Research Network: Team Community-based Research Funds.

INDEX

A

B

C

D

E

F

G

H

I

J

K

L

Q

R

S

T

U

V

W

X

Y